IET ENERGY ENGINEERING SERIES 237

High Voltage Power Transformers

Other volumes in this series:

Volume 1	**Power Circuit Breaker Theory and Design** C.H. Flurscheim (Editor)
Volume 4	**Industrial Microwave Heating** A.C. Metaxas and R.J. Meredith
Volume 7	**Insulators for High Voltages** J.S.T. Looms
Volume 8	**Variable Frequency AC Motor Drive Systems** D. Finney
Volume 10	**SF_6 Switchgear** H.M. Ryan and G.R. Jones
Volume 11	**Conduction and Induction Heating** E.J. Davies
Volume 13	**Statistical Techniques for High Voltage Engineering** W. Hauschild and W. Mosch
Volume 14	**Uninterruptible Power Supplies** J. Platts and J.D. St Aubyn (Editors)
Volume 15	**Digital Protection for Power Systems** A.T. Johns and S.K. Salman
Volume 16	**Electricity Economics and Planning** T.W. Berrie
Volume 18	**Vacuum Switchgear** A. Greenwood
Volume 19	**Electrical Safety: A guide to causes and prevention of hazards** J. Maxwell Adams
Volume 21	**Electricity Distribution Network Design, 2nd Edition** E. Lakervi and E.J. Holmes
Volume 22	**Artificial Intelligence Techniques in Power Systems** K. Warwick, A.O. Ekwue and R. Aggarwal (Editors)
Volume 24	**Power System Commissioning and Maintenance Practice** K. Harker
Volume 25	**Engineers' Handbook of Industrial Microwave Heating** R.J. Meredith
Volume 26	**Small Electric Motors** H. Moczala *et al.*
Volume 27	**AC-DC Power System Analysis** J. Arrillaga and B.C. Smith
Volume 29	**High Voltage Direct Current Transmission, 2nd Edition** J. Arrillaga
Volume 30	**Flexible AC Transmission Systems (FACTS)** Y.-H. Song (Editor)
Volume 31	**Embedded Generation** N. Jenkins *et al.*
Volume 32	**High Voltage Engineering and Testing, 2nd Edition** H.M. Ryan (Editor)
Volume 33	**Overvoltage Protection of Low-Voltage Systems, Revised Edition** P. Hasse
Volume 36	**Voltage Quality in Electrical Power Systems** J. Schlabbach *et al.*
Volume 37	**Electrical Steels for Rotating Machines** P. Beckley
Volume 38	**The Electric Car: Development and future of battery, hybrid and fuel-cell cars** M. Westbrook
Volume 39	**Power Systems Electromagnetic Transients Simulation** J. Arrillaga and N. Watson
Volume 40	**Advances in High Voltage Engineering** M. Haddad and D. Warne
Volume 41	**Electrical Operation of Electrostatic Precipitators** K. Parker
Volume 43	**Thermal Power Plant Simulation and Control** D. Flynn
Volume 44	**Economic Evaluation of Projects in the Electricity Supply Industry** H. Khatib
Volume 45	**Propulsion Systems for Hybrid Vehicles** J. Miller
Volume 46	**Distribution Switchgear** S. Stewart
Volume 47	**Protection of Electricity Distribution Networks, 2nd Edition** J. Gers and E. Holmes
Volume 48	**Wood Pole Overhead Lines** B. Wareing
Volume 49	**Electric Fuses, 3rd Edition** A. Wright and G. Newbery
Volume 50	**Wind Power Integration: Connection and system operational aspects** B. Fox *et al.*
Volume 51	**Short Circuit Currents** J. Schlabbach
Volume 52	**Nuclear Power** J. Wood
Volume 53	**Condition Assessment of High Voltage Insulation in Power System Equipment** R.E. James and Q. Su
Volume 55	**Local Energy: Distributed generation of heat and power** J. Wood
Volume 56	**Condition Monitoring of Rotating Electrical Machines** P. Tavner, L. Ran, J. Penman and H. Sedding
Volume 57	**The Control Techniques Drives and Controls Handbook, 2nd Edition** B. Drury
Volume 58	**Lightning Protection** V. Cooray (Editor)
Volume 59	**Ultracapacitor Applications** J.M. Miller
Volume 62	**Lightning Electromagnetics** V. Cooray

Volume 63 **Energy Storage for Power Systems, 2nd Edition** A. Ter-Gazarian
Volume 65 **Protection of Electricity Distribution Networks, 3rd Edition** J. Gers
Volume 66 **High Voltage Engineering Testing, 3rd Edition** H. Ryan (Editor)
Volume 67 **Multicore Simulation of Power System Transients** F.M. Uriate
Volume 68 **Distribution System Analysis and Automation** J. Gers
Volume 69 **The Lightening Flash, 2nd Edition** V. Cooray (Editor)
Volume 70 **Economic Evaluation of Projects in the Electricity Supply Industry, 3rd Edition** H. Khatib
Volume 72 **Control Circuits in Power Electronics: Practical issues in design and implementation** M. Castilla (Editor)
Volume 73 **Wide Area Monitoring, Protection and Control Systems: The enabler for smarter grids** A. Vaccaro and A. Zobaa (Editors)
Volume 74 **Power Electronic Converters and Systems: Frontiers and applications** A.M. Trzynadlowski (Editor)
Volume 75 **Power Distribution Automation** B. Das (Editor)
Volume 76 **Power System Stability: Modelling, analysis and control** A.A. Sallam and B. Om P. Malik
Volume 78 **Numerical Analysis of Power System Transients and Dynamics** A. Ametani (Editor)
Volume 79 **Vehicle-to-Grid: Linking electric vehicles to the smart grid** J. Lu and J. Hossain (Editors)
Volume 81 **Cyber-Physical-Social Systems and Constructs in Electric Power Engineering** S. Suryanarayanan, R. Roche and T.M. Hansen (Editors)
Volume 82 **Periodic Control of Power Electronic Converters** F. Blaabjerg, K. Zhou, D. Wang and Y. Yang
Volume 86 **Advances in Power System Modelling, Control and Stability Analysis** F. Milano (Editor)
Volume 87 **Cogeneration: Technologies, optimisation and implementation** C.A. Frangopoulos (Editor)
Volume 88 **Smarter Energy: From smart metering to the smart grid** H. Sun, N. Hatziargyriou, H.V. Poor, L. Carpanini and M.A. Sánchez Fornié (Editors)
Volume 89 **Hydrogen Production, Separation and Purification for Energy** A. Basile, F. Dalena, J. Tong and T.N. Veziroğlu (Editors)
Volume 90 **Clean Energy Microgrids** S. Obara and J. Morel (Editors)
Volume 91 **Fuzzy Logic Control in Energy Systems with Design Applications in Matlab/Simulink®** İ.H. Altaş
Volume 92 **Power Quality in Future Electrical Power Systems** A.F. Zobaa and S.H.E.A. Aleem (Editors)
Volume 93 **Cogeneration and District Energy Systems: Modelling, analysis and optimization** M.A. Rosen and S. Koohi-Fayegh
Volume 94 **Introduction to the Smart Grid: Concepts, technologies and evolution** S.K. Salman
Volume 95 **Communication, Control and Security Challenges for the Smart Grid** S.M. Muyeen and S. Rahman (Editors)
Volume 96 **Industrial Power Systems with Distributed and Embedded Generation** R. Belu
Volume 97 **Synchronized Phasor Measurements for Smart Grids** M.J.B. Reddy and D.K. Mohanta (Editors)
Volume 98 **Large Scale Grid Integration of Renewable Energy Sources** A. Moreno-Munoz (Editor)
Volume 100 **Modeling and Dynamic Behaviour of Hydropower Plants** N. Kishor and J. Fraile-Ardanuy (Editors)
Volume 101 **Methane and Hydrogen for Energy Storage** R. Carriveau and D.S-K. Ting
Volume 104 **Power Transformer Condition Monitoring and Diagnosis** A. Abu-Siada (Editor)
Volume 106 **Surface Passivation of Industrial Crystalline Silicon Solar Cells** J. John (Editor)
Volume 107 **Bifacial Photovoltaics: Technology, applications and economics** J. Libal and R. Kopecek (Editors)

Volume 108	**Fault Diagnosis of Induction Motors** J. Faiz, V. Ghorbanian and G. Joksimović
Volume 109	**Cooling of Rotating Electrical Machines: Fundamentals, modelling, testing and design** D. Staton, E. Chong, S. Pickering and A. Boglietti
Volume 110	**High Voltage Power Network Construction** K. Harker
Volume 111	**Energy Storage at Different Voltage Levels: Technology, integration, and market aspects** A.F. Zobaa, P.F. Ribeiro, S.H.A. Aleem and S.N. Afifi (Editors)
Volume 112	**Wireless Power Transfer: Theory, technology and application** N. Shinohara
Volume 114	**Lightning-Induced Effects in Electrical and Telecommunication Systems** Y. Baba and V.A. Rakov
Volume 115	**DC Distribution Systems and Microgrids** T. Dragičević, F. Blaabjerg and P. Wheeler
Volume 116	**Modelling and Simulation of HVDC Transmission** M. Han (Editor)
Volume 117	**Structural Control and Fault Detection of Wind Turbine Systems** H.R. Karimi
Volume 118	**Modelling and Simulation of Complex Power Systems** A. Monti and A. Benigni
Volume 119	**Thermal Power Plant Control and Instrumentation: The control of boilers and HRSGs, 2nd Edition** D. Lindsley, J. Grist and D. Parker
Volume 120	**Fault Diagnosis for Robust Inverter Power Drives** A. Ginart (Editor)
Volume 121	**Monitoring and Control using Synchrophasors in Power Systems with Renewables** I. Kamwa and C. Lu (Editors)
Volume 123	**Power Systems Electromagnetic Transients Simulation, 2nd Edition** N. Watson and J. Arrillaga
Volume 124	**Power Market Transformation** B. Murray
Volume 125	**Wind Energy Modeling and Simulation Volume 1: Atmosphere and plant** P. Veers (Editor)
Volume 126	**Diagnosis and Fault Tolerance of Electrical Machines, Power Electronics and Drives** A.J.M. Cardoso
Volume 127	**Lightning Electromagnetics, Volume 1: Return stroke modelling and electromagnetic radiation, 2nd Edition** V. Cooray, F. Rachidi, M. Rubinstein (Editors)
Volume 127	**Lightning Electromagnetics, Volume 2: Electrical processes and effects, 2nd Edition** V. Cooray, F. Rachidi, M. Rubinstein (Editors)
Volume 128	**Characterization of Wide Bandgap Power Semiconductor Devices** F. Wang, Z. Zhang and E.A. Jones
Volume 129	**Renewable Energy from the Oceans: From wave, tidal and gradient systems to offshore wind and solar** D. Coiro and T. Sant (Editors)
Volume 130	**Wind and Solar Based Energy Systems for Communities** R. Carriveau and D.S.-K. Ting (Editors)
Volume 131	**Metaheuristic Optimization in Power Engineering** J. Radosavljević
Volume 132	**Power Line Communication Systems for Smart Grids** I.R.S. Casella and A. Anpalagan
Volume 134	**Hydrogen Passivation and Laser Doping for Silicon Solar Cells** B. Hallam and C. Chan (Editors)
Volume 139	**Variability, Scalability and Stability of Microgrids** S.M. Muyeen, S.M. Islam and F. Blaabjerg (Editors)
Volume 142	**Wind Turbine System Design, Volume 1: Nacelles, drive trains and verification** J. Wenske (Editor)
Volume 143	**Medium Voltage DC System Architectures** B. Grainger and R.D. Doncker (Editors)
Volume 145	**Condition Monitoring of Rotating Electrical Machines** P. Tavner, L. Ran, C. Crabtree
Volume 146	**Energy Storage for Power Systems, 3rd Edition** A.G. Ter-Gazarian
Volume 147	**Distribution Systems Analysis and Automation, 2nd Edition** J. Gers
Volume 151	**SiC Power Module Design: Performance, robustness and reliability** A. Castellazzi and A. Irace (Editors)
Volume 152	**Power Electronic Devices: Applications, failure mechanisms and reliability** F. Iannuzzo (Editor)

Volume 153 **Signal Processing for Fault Detection and Diagnosis in Electric Machines and Systems** M. Benbouzid (Editor)
Volume 155 **Energy Generation and Efficiency Technologies for Green Residential Buildings** D. Ting and R. Carriveau (Editors)
Volume 156 **Lithium-Ion Batteries Enabled by Silicon Anodes** C. Ban and K. Xu (Editors)
Volume 157 **Electrical Steels, 2 Volumes** A. Moses, K. Jenkins, Philip Anderson and H. Stanbury
Volume 158 **Advanced Dielectric Materials for Electrostatic Capacitors** Q. Li (Editor)
Volume 159 **Transforming the Grid Towards Fully Renewable Energy** O. Probst, S. Castellanos and R. Palacios (Editors)
Volume 160 **Microgrids for Rural Areas: Research and case studies** R.K. Chauhan, K. Chauhan and S.N. Singh (Editors)
Volume 161 **Artificial Intelligence for Smarter Power Systems: Fuzzy logic and neural networks** M.G. Simoes
Volume 165 **Digital Protection for Power Systems, 2nd Edition** Salman K Salman
Volume 166 **Advanced Characterization of Thin Film Solar Cells** N. Haegel and M. Al-Jassim (Editors)
Volume 167 **Power Grids with Renewable Energy Storage, Integration and Digitalization** A.A. Sallam and B.O.M.P. Malik
Volume 169 **Small Wind and Hydrokinetic Turbines** P. Clausen, J. Whale and D. Wood (Editors)
Volume 170 **Reliability of Power Electronics Converters for Solar Photovoltaic Applications** F. Blaabjerg, A.I. Haque, H. Wang, Z. Abdin Jaffery and Y. Yang (Editors)
Volume 171 **Utility-Scale Wind Turbines and Wind Farms** A. Vasel-Be-Hagh and D.S.-K. Ting
Volume 172 **Lighting Interaction with Power Systems, 2 volumes** A. Piantini (Editor)
Volume 174 **Silicon Solar Cell Metallization and Module Technology** T. Dullweber (Editor)
Volume 175 **n-Type Crystalline Silicon Photovoltaics: Technology, applications and economics** D. Munoz and R. Kopecek (Editors)
Volume 178 **Integrated Motor Drives** Xu Deng and B. Mecrow (Editors)
Volume 180 **Protection of Electricity Distribution Networks, 4th Edition** J. Gers and E. Holmes
Volume 182 **Surge Protection for Low Voltage Systems** A. Rousseau (Editor)
Volume 184 **Compressed Air Energy Storage: Types, systems and applications** D. Ting and J. Stagner
Volume 186 **Synchronous Reluctance Machines: Analysis, optimization and applications** N. Bianchi, C. Babetto and G. Bacco
Volume 190 **Synchrophasor Technology: Real-time operation of power networks** N. Kishor and S.R. Mohanty (Editors)
Volume 191 **Electric Fuses: Fundamentals and new applications, 4th Edition** N. Nurse, A. Wright and P.G. Newbery
Volume 193 **Overhead Electric Power Lines: Theory and practice** S. Chattopadhyay and A. Das
Volume 194 **Offshore Wind Power Reliability, Availability and Maintenance, 2nd Edition** P. Tavner
Volume 196 **Cyber Security for Microgrids** S. Sahoo, F. Blaajberg and T. Dragicevic
Volume 198 **Battery Management Systems and Inductive Balancing** A. Van den Bossche and A. Farzan Moghaddam
Volume 199 **Model Predictive Control for Microgrids: From power electronic converters to energy management** J. Hu, J.M. Guerrero and S. Islam
Volume 201 **Design, Control and Monitoring of Tidal Stream Turbine Systems** M. Benbouzid (Editor)
Volume 204 **Electromagnetic Transients in Large HV Cable Networks: Modeling and calculations** Ametani, Xue, Ohno and Khalilnezhad
Volume 208 **Nanogrids and Picogrids and their Integration with Electric Vehicles** S. Chattopadhyay
Volume 210 **Superconducting Magnetic Energy Storage in Power Grids** M.H. Ali

Volume 211	**Blockchain Technology for Smart Grids: Implementation, management and security** H.L. Gururaj, K.V. Ravi, F. Flammini, H. Lin, B. Goutham, K.B.R. Sunil and C. Sivapragash
Volume 212	**Battery State Estimation: Methods and models** S. Wang
Volume 215	**Industrial Demand Response: Methods, best practices, case studies, and applications** H.H. Alhelou, A. Moreno-Muñoz and P. Siano (Editors)
Volume 213	**Wide Area Monitoring of Interconnected Power Systems, 2nd Edition** A.R. Messina
Volume 214	**Distributed Energy Storage in Urban Smart Grids** P.F. Ribeiro and R.S. Salles (Editors)
Volume 217	**Advances in Power System Modelling, Control and Stability Analysis, 2nd Edition** F. Milano (Editor)
Volume 225	**Fusion-Fission Hybrid Nuclear Reactors: For enhanced nuclear fuel utilization and radioactive waste reduction** W.M. Stacey
Volume 227	**Processing and Manufacturing of Electrodes for Lithium-Ion Batteries** J. Li and C. Jin (Editors)
Volume 228	**Digital Technologies for Solar Photovoltaic Systems: From general to rural and remote installations** S. Motahhir (Editor)
Volume 236	**Green HV Switching Technologies for Modern Power Networks,** K. Niayesh (Editor)
Volume 238	**AI for Status Monitoring of Utility Scale Batteries** S. Wang, K. Liu, Y. Wang, D. Stroe, C. Fernandez and J.M. Guerrero
Volume 239	**Intelligent Control of Medium and High Power Converters** M. Bendaoud, Y. Maleh and S. Padmanaban (Editors)
Volume 905	**Power System Protection, 4 Volumes**

High Voltage Power Transformers

State of the art and technological innovations

Stefano Zunino, Marino Sforna and Fabrizio Ferrari

The Institution of Engineering and Technology

Published by The Institution of Engineering and Technology, London, United Kingdom

The Institution of Engineering and Technology is registered as a Charity in England & Wales (no. 211014) and Scotland (no. SC038698).

First published 2023

The Institution of Engineering and Technology
Futures Place
Kings Way, Stevenage
Hertfordshire, SG1 2UA, United Kingdom

www.theiet.org

British Library Cataloguing in Publication Data
A catalogue record for this product is available from the British Library

ISBN 978-1-83953-811-7 (hardback)
ISBN 978-1-83953-812-4 (PDF)

Typeset in India by MPS Limited

Contents

Preface

This book aims to describe the technical characteristics of the design, construction, and operational choices of power transformers (TRs) used in high-voltage electrical grids; such topics that are not included generally in the teaching material of TRs' textbooks. The electrical devices covered in this book are large power TRs and autotransformers (ATRs), mainly with power ratings over 100 MVA.

The various chapters do not propose the theory of TRs, which is already covered extensively in numerous specialized texts, but direct the reader's attention to some significant details about the available design solutions, the choice of components and materials, and the management of ATRs, which are useful topics for the power utility's expert as well as academically, in the stages of drafting the technical specifications for the purchase of power TRs, assuring the quality of their design and manufacturing, and allowing an efficient operational management once installed in the grid. A greater knowledge of these topics makes it possible both to prevent possible risks resulting from a suboptimal choice of the characteristics of these electrical devices at the procurement stage and to take advantage of the opportunities offered by technological developments in the TR industry.

The chapters are mainly independent and self-contained and therefore can be read in any order; however, cross references among them have been provided where useful; the main topics covered in the book are briefly described in the following paragraphs.

Influence of magnetic material on the energy efficiency of power transformers

Chapter 2 proposes an overview of the technological solutions available for the choice of magnetic core's material of power TRs that allow to achieve higher energy efficiency, in order to comply with the possible regulatory "Ecodesign" prescriptions, such as the European Regulation No. 548/2014.

In fact, Ecodesign requirements limit the designer in the choice of the allowable values for no-load and load losses of TRs, which was a matter traditionally regulated by the power utilities in the purchase contract, who checked the quality of TRs through the measurement of losses during the factory acceptance tests.

The reduction of the design values of losses (especially no-load losses), necessary to meet the "*peak efficiency index*" criterion or similar requirements, can be a major challenge in the procurement, design, and construction of the TRs, whose dimensions and masses can pose a problem during transport and installation in electrical substations or power plants. Consequently, it is necessary to increase the performance requirements of the materials used for the magnetic core already at tender stage, in order to ensure uniform product quality and place all manufacturers in the same competitive conditions.

Conductors for power transformer windings

Chapter 3 describes the current state of the art of technology of conductors (wires) adopted for TR's windings and particularly copper wires, mainly used in large TRs.

A detailed insight into the characteristics of the different wires is useful to achieve an accurate specification of conductors so to ensure their correct usage and the best performance of power TRs, as well as a high quality of the final product.

Insulating materials and dielectric withstand of transformers

The dielectric design of TRs is of paramount importance in order to guarantee their operational reliability.

The optimal design includes the correct choice of insulating materials as well as the design calculation for their best use, which are both necessary conditions for meeting the dielectric withstand requirements under all testing and operational conditions of the TR.

Chapter 4 offers a panorama of the available insulating materials and indicates the current state-of-the-art criteria in the dielectric design of TRs.

Vegetable oils as dielectric insulators in transformers

The active part of TRs (windings and ferromagnetic core) is contained in a tank filled with an insulating fluid, normally hydrocarbon-based mineral oil, obtained by refining petroleum. The oil allows the insulation and cooling of the active part.

To ensure the insulating properties of oil, it is necessary to maintain its dielectric withstand, which can be affected by moisture, high temperature, and oxygen that degrade it and cause the formation of acid compounds and solid precipitates.

In addition, the considerable thermal capacity of the oil allows the heat produced in the active part due to losses to be extracted by oil's circulation through convection. However, in order to have an effective heat dissipation, it is necessary for the oil to have low viscosity at all operating temperatures of the TR.

Mineral oil possesses the abovementioned characteristics, but it is flammable and hardly biodegradable. There are commercially available alternative liquids with higher flash points, among them silicone oils and natural and synthetic esters. In particular, the latter have greater biodegradability and are increasingly preferred by power utilities when fire protection and sustainability are the main drivers of choice.

Chapter 5 describes the characteristics and conditions of use of the different insulating liquids.

Shell-type transformers

Shell-type technology has significant differences from the more widely used core-type technology of TR's design, with interesting advantages for operational capacity, short circuit withstand, and the flexibility to change the turn ratio.

Chapter 6 aims to describe the peculiar features of this technology, highlighting both its merits and shortcomings, and the possible more widespread use of this type of TR in the transmission grid.

Thermal aspects of power transformers

The design and operation of TRs are affected by the heating of the active part that occurs during their operation. There are numerous characteristics and variables that influence the temperature of a TR. Consequently, the design must include an effective and efficient cooling system.

Chapter 7 describes the main problems in the thermal management of TRs and provides technical reasons for the specification of the most suited cooling system, according to the specific application of the TR.

Calculation and measurement of the zero sequence parameters of transformers

Chapter 8 is devoted to the analysis of the electrical circuit at the zero sequence of TRs, which is significantly different from the circuits at the positive and negative sequence, which are mostly employed in the modeling of the electrical grid.

The analysis is relevant because the zero sequence is indispensable for the accurate calculation of ground faults and the related setting of protection relays, but measurement of TR's impedances at the zero sequence can be challenging, and sometimes approximate values are used instead that do not fit the need for sufficient accuracy.

Short circuit tests for transformers

TRs are subjected to severe stresses during their operation, in particular the transient feeding of a fault in the grid.

Beginning in the 1970s, the demonstration of short circuit withstand capability was introduced in the technical standards of power TRs, first for TRs in the distribution grids and, since the 1980s, for ATRs in the transmission grid. At the same time, dedicated test laboratories were built, with the purpose of subjecting TRs to actual short circuit tests, either by drawing power from an electrical node of a connected grid or by using suitable synchronous generators.

The short circuit withstand of a TR under test allows to verify the quality of both its design and construction. Therefore, it is important to document the associated knowledge in the choice of the most appropriate tests and the evaluation of the test's results.

Chapter 9 describes the state of the art on the subject, the test requirements, and the prospects for the evolution of testing techniques and related standards.

Experiments to reduce noise emitted by power transformers

Electric substations of the transmission grid are normally located outside the urban perimeter; however, the increasing demand for land urbanization implies both the proximity of homes to existing substations and the increased difficulty in finding areas without human presence to build new substations.

All electrical devices (ATRs, TRs, and reactors) are a source of noise that propagates beyond the perimeter of electrical substations.

This issue causes a problem for power utilities who must comply with legislation about acoustic emissions. In addition, power companies have adopted new environmental goals, and the public acceptance of electrical installations is important due to social sensitivity to this matter. Research and development departments of power utilities are working to find technological solutions to address these issues, without penalizing the operational capability of the electric system.

Chapter 10 describes the various possible solutions for mitigating noise emitted by electric substations, currently employed or experimented by power utilities. The results of the experiments allow some evaluations to identify the most promising solutions and future study activities.

DC core saturation of converter transformers

Renewable energy plants and High Voltage Direct Current (HVDC) links are connected to the High Voltage Alternating Current (HVAC) transmission grid with AC/DC conversion systems, which feed the grid with harmonic and DC components; the DC component is harmful to power transformers because it can cause core saturation and possible localized heating inside the active part.

Chapter 11 describes the problem of possible saturation of converter transformers in HVDC substations which, at present, are the transformers mostly affected by this problem.

Shunt reactors

The shunt reactors are large inductors used in electric grids to store reactive power. Reactors can be used also for other purposes, such as limiting fault currents, earthing, and filtering, although those applications are less common in HV grids.

Chapter 12 describes the design principles and the constructive type of shunt reactors, considering the main parameters as reactance, linearity, losses, noise, and the others that influence reactor characteristics and the manufacturing typology.

Industrial transformers

In Chapter 13, we will briefly discuss some basic concepts useful in defining industrial transformers, mainly represented by two broad categories: furnace transformers and rectifier transformers.

Furnace transformers are used to feed electric arc furnaces that use the heating of the electric arc to melt the materials contained in the furnace.

Rectifier transformers are used to feed rectifiers, composed of electronic valves, so to produce the direct current needed in several industrial processes (e.g., electrolytic processes).

Industrial transformers have peculiar working conditions, specific to their intended application. This peculiarity requires the employ of detailed simulation tools and specific design solutions, different from the grid's interconnection transformers, to guarantee the transformer's reliability and durability under all the working conditions of the industrial plant.

About the authors

Marino Sforna is presently head of the power system risk management unit at TSO TERNA, Italy, one of the largest transmission system operators in the world. He joined the former national power company ENEL in 1982; from 1987 to 1997, he was with the R&D Dept., and from 1997 to 2001, deputy manager of an Italian area control center and later, responsible for grid defense, restoration, protection systems, and control.

Stefano Zunino is a senior expert on power system components and electric devices and risk manager at TSO TERNA, Italy. Previously, he worked on power system protection at Siemens. He has been a member of CIGRE since 2008 and of CEI since 2009. He is an expert in several IEC technical committees in charge of standardization of electric components, including TC 14 about power transformers, and he is Secretary of TC 33 about power capacitors.

Fabrizio Ferrari is a senior consultant working with organizations such as CESI, major utilities such as EGP and A2A, and for transformer manufacturers including TAMINI Trasformatori S.r.l. He has experience in small, medium, and large power transformers and their main HV, MV, and LV components. He serves as a member of several specialist groups such as CIGRE A.2 and T&D Europe, and he is the chairman of the "Transformers Group" in ANIE Energia.

Acknowledgements

Writing a technical book has been an enlightening journey, guiding authors through the process of shaping an empty canvas into a repository of valuable and intriguing knowledge. This endeavor gains added significance in an era where aesthetics often competes with substance.

It is an acknowledged truth that the creation of something valuable is seldom the work of a solitary individual; rather, it typically emerges from the collective efforts of a select few, often with contributions from a multitude of individuals. Our book is no exception.

The content within these pages also stands as a testament to the experiences generously shared by colleagues within our organization and external experts, gathered over years of collaborative endeavors. We extend our gratitude to all of them for their generosity.

We wish to express our sincere thanks to the many individuals, publishers, scholars, industrial experts, and representatives who graciously provided valuable additional material and granted us permission to reproduce their original pictures and photographs, enriching this book.

Special acknowledgment is due to the team at the Institution of Engineering and Technology (IET) Publishing Group. Sarah Lynch and Olivia Wilkins have been constant guides throughout the preparation of this manuscript. We want to highlight the invaluable support provided by Christoph von Friedeburg, the Consulting Senior Commissioning Editor at IET Publishing, whose remarkable skill and persuasion played a pivotal role in convincing us to undertake this demanding endeavor. We thank him for believing in our project. Gratitude is also extended to N. Srinivasan from MPS Limited, whose efforts have contributed to the appealing presentation of this book.

Our debt of gratitude extends to our respective companies, Terna and Tamini, whose support allowed us to delve deeper into the knowledge entwined with our daily work activities, expanding our professional capabilities.

This book is intended to be a valuable resource for those interested in the multifaceted aspects and industrial practices associated with the manufacturing and operation of large transformers and reactors. We are confident that both the broad perspectives and the finer details contained within it will address the needs encountered in dealing with these electrical devices.

Dedicated to our friends and colleagues who have been a constant source of encouragement throughout this journey, and especially to our families, whose patience and unwavering support sustained us even when it meant sacrificing time with them.

Chapter 1
Introduction

The power transformer is an indispensable equipment[1] in the system of production, transmission, and distribution of alternating current electric power. In fact, the ability to change the amplitude of alternating voltage and current, while maintaining the same amount of transmitted power, results in significant savings in terms of material and power dissipation.

The invention and development of the transformer enabled the wide dissemination of electricity. It was a long and complex process conducted by many scientists. The first studies on magnetic induction, i.e., the physical phenomenon upon which the design principle of the transformer is based, were performed by M. Faraday in 1831 and refined the following year by J. Henry. These discoveries, through further studies, led to the conception of both the *electric generator* (the induction motor) and the *secondary generator*, as the antecedent of the transformer was called by L. Gaulard in 1882.

The first prototype of a real transformer was patented in 1885 by three engineers from the Hungarian firm Ganz: Ottó Bláthy, Károly Zipernowsky, Miksa Déri. These pioneers developed both the design that would later be called the *core-type* design, in which the primary and secondary circuit conductors are wound on a core of ferromagnetic material, and the *shell-type* design, in which the core surrounds the wire conductors (Figure 1.1). In both cases, the core is intended to maximize power transfer between primary and secondary windings by means of the magnetic flux.

Subsequently, the efforts of various firms that specialized in the design and construction of power transformers, including Westinghouse in the United States, Ferranti in Great Britain, Siemens in Germany, Schneider in France, and ASEA in Sweden, enabled the establishment of transformer design principles which reached around 1920 the level of evolution that is the current benchmark for this equipment (Figure 1.2).

[1]Although transformers are often referred to as "machines" or "machinery," according to the International Electrotechnical Vocabulary (IEV) published by the IEC these terms refer only to apparatuses that convert energy between different forms (e.g., kinetic energy to electrical energy). Transformers, on the other hand, fall under the classification of "electric energy converter" (IEV 151-13-36), or the more generic classifications of "equipment" (IEV 151-11-25) and "device" (IEV 151-11-20).

Figure 1.1 Specimens of the prototype transformers of core-type (on the left, conductors are wound on the toroidal core) and shell-type (on the right, the core consisting of iron wires surrounds the inner conductors); both manufactured by Ganz, 1885

(a) (b)

Figure 1.2 (a) Tank and active part of a three-phase transformer with a power rating of 36 MVA and a transformation ratio of 237 kV/11 kV, built in 1925 at the TIBB factory in Milan, Italy and (b) step-up transformer with power rating 670 MVA and transformation ratio 22 kV/400 kV, built in 2023 at the Tamini factory in Legnano, Italy

Currently, core-type transformers are significantly more common than shell-type transformers, mainly because of less use of materials and less manufacturing complexity, which translate into a lower cost.

The principles underlying the design of power transformers have not substantially changed since 1920; however, technology has evolved mainly in the following areas [1]:

- improved understanding of the physical phenomena of transformer operation such as the distribution of electric and magnetic fields, mechanical stresses, heating and thermal dissipation, and externally generated voltage surges;
- improved design with increasingly rigorous and accurate calculations and automated calculators, which made it possible to reduce design's overestimates and thus decrease material's use and transformer's size[2];
- the introduction of the *autotransformer* concept, which provides significant manufacturing savings when the rated voltages of the two circuits are not very different and share the same state of the neutral point;
- optimization of the manufacturing process, with specialization in the construction of the transformer's component parts: conductors, insulators, cores, tap changers, insulators, etc., and standardization and partial automation of the production stages: winding, core construction, active part's assembly, etc.;
- the use of components with increasingly high-performance, reliable characteristics, and wider limits of use, even under special operating conditions, resulting from continuous research and development in the field of materials.

The European Union Regulation No. 548/2014 [2], a consequence of the "Ecodesign" directive, gave a strong impetus to research into materials and design solutions to increase the energy efficiency of transformers, which already have efficiencies above 99%, as well as research into methods of measuring losses with minimal uncertainty [3].

In addition, increasing attention to environmental and safety issues has led to the development of components and materials with low fire risk and reduced impact on the ecosystem. Such is the case with *vegetable oils*, which appear as a possible alternative to the use of traditional mineral oil as the transformer's insulation fluid.

The evolution of nonprogrammable renewable energy sources (e.g., solar and wind power), the spread of very small generation plants controlled by power electronics, and the massive use of air-conditioning systems have introduced new and increased demands on the performance and reliability of transformers, which must cope with higher power peaks, more demanding duty cycles, grid voltages with non-negligible harmonic content, and increasingly high short circuit powers on the grid.

Therefore, modern design techniques employing computer simulations using finite element methods cannot avoid to consider all possible operating conditions of the transformer, especially abnormal conditions (overvoltages, overloads, short circuit events), which can threaten the reliability of the transformer.

Several following chapters describe the main topics of the state-of-the-art and technological challenges concerning materials and thermal, dielectric, and mechanical design of the transformers. Indeed, the correct choice of materials and

[2]The dimensions of the active part in Figure 1.2 are now associated with power ratings four to five times those of the transformer shown.

consequent design, together with a judicious specification of performance under normal and abnormal service conditions of the transformer, is critical to ensure reliability over time.

Each choice related to performance: power losses, impedance, acoustic limits, withstand voltages, permissible overloads and overtemperatures, cooling system, etc., affects the design and construction choices and constraints across the board. Therefore, it is important to know what impact the requirements have and the limits within which the desired characteristics can be achieved.

Although today's computational models and programs allow detailed analyses by coupling even different physical phenomena (electromagnetic, mechanical, thermal) and sometimes even allow automatic finding of the optimal value of some design parameters given an objective function (such as loss minimization), these simulations need to be validated through experimental tests.

Transformers, especially those with high power ratings, are still manufactured with a manual process, especially in the construction of the windings and the related connections. Thus, attention to manufacturing tolerances and strict quality control are required to prevent manufacturing defects (especially if latent) from compromising the achievement of the transformer's required performance and durability [4].

Indeed, although the transformer is an almost entirely static piece of equipment (excluding the on-load tap changer and cooling system), which can be attributed to an estimated useful life of about 40 years under normal operating conditions, recent statistical failure's analyses [5] indicate that young equipment, up to 10 years old, has failure rates comparable to equipment with a longer in-service period, between 20 and 25 years (Figure 1.3).

Excluding the initial spike of "infant mortality" due to construction defects, the graph shows that the reliability of recently built transformers has the same level as those with more seniority, whereas it would be reasonable to expect higher initial reliability. This observation leads to think that the construction margins of modern transformers are very small, due to the ever-increasing optimization of design, so much so that they have eroded the safety margin that guaranteed reliability even under abnormal and unexpected service conditions, although the assessment of this conclusion would require further data to be confirmed.

A requirement of prolonged reliability influences, when specifying and procuring materials, the adoption of special solutions. Therefore, the manufacturer must carefully balance the optimization of design and material's choices with the adoption of special solutions for which enough manufacturing and operational experience exists.

In any case, the ever-increasing complexity and width of available choices require knowledge of the risks and opportunities of the several technological alternatives, thus enabling informed choices to be made at all stages of the transformer life cycle: specification, design review, operation, maintenance, and end-of-life investigation.

It is hoped, therefore, that the topics discussed further on will provide a contribution to the knowledge of the risks and technological opportunities related to power transformers.

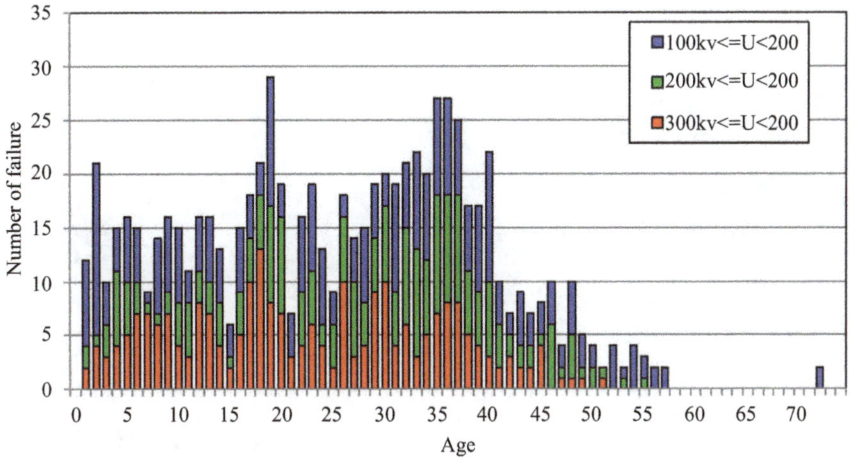

Figure 1.3 Failure rate of HV transformers as a function of age

References

[1] M. Hrkac, A. Baggini, F. Mauri, and R. Zannol: *Beyond of ecodesign and future of transformers*, 3rd International Colloquium Transformer Research and Asset Management, Split (Croatia), 15–17 October 2014.

[2] Regulation (EU) No. 548/2014: *Detailed rules for the implementation of Directive2009/125/EC of the European Parliament and of the Council with regard to small, medium and large power transformers*, OJ L 152, 22.5.2014, as amended.

[3] European metrology project 17NRM01: *Loss Measurements on Power Transformers and Reactors (TrafoLoss)*, EURAMET.

[4] I. Dasgupta: *Power Transformers Quality Assurance*, New Age Publishers, 2009.

[5] Technical Brochure 642: *Transformer Reliability Survey*, Cigré, 2015.

[6] R.M. Del Vecchio, B. Poulin, P.T. Feghali, D.M. Shah and R. Ahuja: *Transform Design Principles*, CRC Press, 2001.

[7] S.V. Kulkarni, and S.A. Khaparde: *Transformer Engineering*, Marcel Dekker, 2004.

[8] M.J. Heathcote: *The J & P Transformer Book*, Newnes, 12th ed. 1998.

[9] G. Bertagnolli: *Short Circuit Duty of Power Transformers*, ABB, 3rd ed., 2006.

[10] Alstom: *Power Transformers – Vol. 1 Fundamentals, Vol. 2 Expertise*, 2012.

Chapter 2

Magnetic materials for power transformers

2.1 Generalities

This chapter describes some technological solutions available for the choice of magnetic core's material for large power transformers in order to improve energy efficiency and possibly meet the requirements of new Regulations, e.g., EU Regulation No. 548/2014 [1] (referred to as Ecodesign).

In fact, Ecodesign requirements constrain the designer the range of the allowable values for no-load and load losses of power transformers, in addition to the purchaser's demands.

Simply lowering the design values of losses (especially no-load losses), necessary to meet, e.g., the *peak efficiency index* criterion mandated by the EU Regulation, can cause changes in transformer design and construction to the point of imposing dimensions and masses that are not compatible with the installation sites. Consequently, it is necessary to increase the performance requirements of the ferromagnetic materials employed in the core in order to avoid unmanageable increases of dimensions of the active part.

2.2 Introduction

2.2.1 Magnetic core of transformers

All power transformers have a core of material with high magnetic permeability onto which the windings are wound, or which surrounds them (Figure 2.1).

The magnetic core mainly performs the functions to:

- confine the magnetic flux created by the currents circulating in the windings to a space concatenated with them, reducing the leakage inductance;
- limit the magnetization current required to achieve the rated no-load magnetic induction;
- ensure the mechanical support of the windings.

Conventionally, energy losses due to the magnetic core are measured in the transformer no-load test, during which the only current circulating is the magnetization current in the energized winding. This current is small compared to the rated current of the winding wires, so the total losses are only those due to magnetic flux

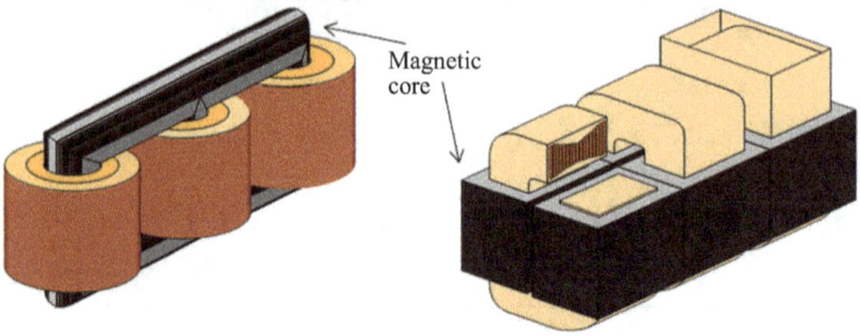

Figure 2.1 Magnetic cores of three-phase core-type (left) and shell-type (right) transformers

in the core. This correlation is so close that no-load losses are also called iron losses.

2.2.2 Magnetization characteristic

The magnetic field is unequivocally described by three physical quantities:

- *magnetic induction*, or *magnetic flux density*, denoted by the vector symbol $\boldsymbol{B}$ and measured in tesla (T). It represents the effects caused by the passage of electric current in a conductor, in the case of interest a winding, immersed in a space having certain magnetic properties;
- *magnetizing force*, or *magnetic field strength*, denoted by the vector symbol $\boldsymbol{H}$ and measured in amperes per meter (A/m). It represents the magnetic effects due to the passage of electric current in a conductor.
- *magnetization* denoted by the vector symbol $\boldsymbol{M}$ and measured in A/m. It represents the magnetic effects of materials with magnetic properties.

The relationship among the three previous magnetic quantities is as follows:

$$\boldsymbol{B} = \mu_0(\boldsymbol{H} + \boldsymbol{M}), \tag{2.1}$$

where $\mu_0 \simeq 4\pi \cdot 10^{-7}\mathrm{H/m}$ is the magnetic permeability of the vacuum[1]. In isotropic bodies, magnetization is proportional to the magnetizing force:

$$\boldsymbol{M} = \chi \boldsymbol{H}, \tag{2.2}$$

where the dimensionless scalar quantity χ, called *magnetic susceptibility*, can be dependent, with a nonlinear relationship, or independent of the $\boldsymbol{H}$ field.

Defining *relative magnetic permeability* as

$$\mu_r = \chi + 1, \tag{2.3}$$

[1]Following the 26th General Conference on Weights and Measures, the numerical value of the magnetic permeability of the vacuum is no longer fixed a priori but is subject to a measurement uncertainty, albeit a very small one.

this can be written for isotropic bodies:

$$\boldsymbol{B} = \mu_0 \mu_r \boldsymbol{H} = \mu \boldsymbol{H}. \tag{2.4}$$

In the case of a nonlinear relationship, we distinguish the *apparent* permeability given by (2.4), from the *differential* permeability given by

$$\mu = \frac{dB}{dH}. \tag{2.5}$$

Apparent permeability represents the slope of the intercept of the working point in the H–B plane with the origin. In contrast, differential permeability represents the slope of the tangent to the magnetization curve at the considered working point. The latter variable is more significant than the former and will be considered in the following unless otherwise specified.

The magnetic core must possess high magnetic permeability and is therefore made of ferromagnetic steel, which also provides mechanical strength. Ferromagnetic materials possess a nonlinear magnetization characteristic (Figure 2.2), in which three traits are identified as a function of the magnetization force value H:

- up to ~5 A/m, the magnetic permeability is high ($\mu_r \simeq 1000$) and weakly increasing. This is the initial trait where the magnetizing force leads the material to assume uniform magnetization;

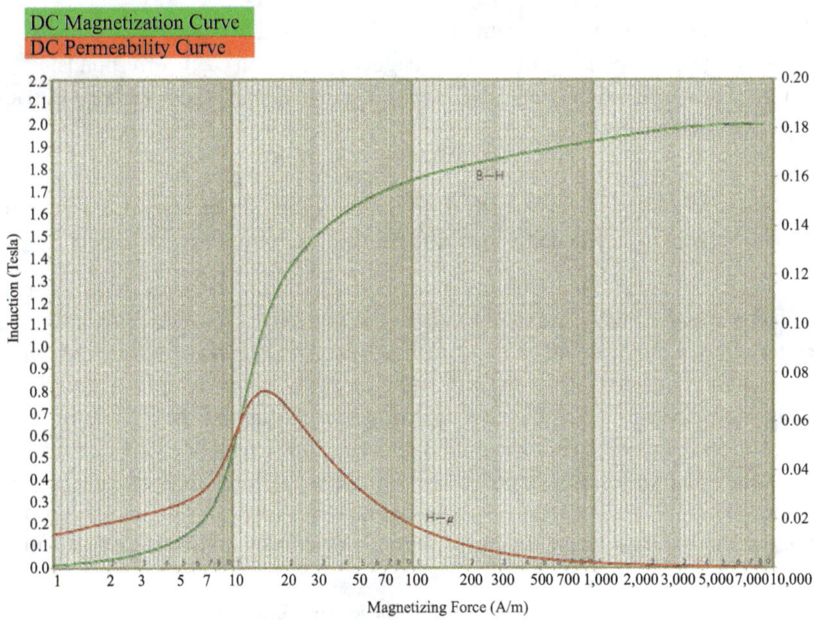

Figure 2.2 Magnetization characteristic and magnetic permeability of a ferromagnetic steel

- in the range of 5–100 A/m, the magnetic permeability is very high ($\mu_r > 10{,}000$). This is the trait whose magnetic induction reaches high values without an appreciable increase in magnetizing force;
- above ~100 A/m, the magnetic permeability rapidly decreases to very low values ($\mu_r \rightarrow 1$). This is the *saturation* trait where the magnetic relationship becomes linear.

The magnetic characteristic of Figure 2.2 is called *first magnetization*, i.e., that resulting from a material that has never been magnetized or otherwise lacks residual magnetization.

The voltage U induced in a winding immersed in a magnetic field depends on the magnetic induction according to a relationship called *Faraday's law*:

$$U = -N\frac{d}{dt}\Phi = -N\frac{d}{dt}\int_{\Sigma} \boldsymbol{B} \cdot d\boldsymbol{S},$$

where Σ is the area bounded by a coil, $d\boldsymbol{S}$ is the differential area element oriented in the right-hand direction[2] of the coil, and N is the number of coils in the winding. For the reason stated above, to achieve high magnetic induction of the area bounded by the coils, the ferromagnetic core should not be subjected to saturation.

Differently from the first magnetization characteristic of Figure 2.2, the magnetization curve that results from repeated cycles of applied alternating magnetizing force is shown in Figure 2.3 and is called the *hysteresis cycle*. The relationship between H and B depends on:

- the value of maximum magnetizing force (H_m) achieved in the previous half-cycle, to which the corresponding maximum magnetic induction (B_m) corresponds;
- the derivative of the magnetizing force, whether it is positive (increasing) or negative (decreasing).

Therefore, the complete hysteresis cycle encompasses a closed curve in the H–B plane (instead of a single line), which does not pass through the origin but is symmetric with respect to it. In such a curve, the following values are identified:

- the value of magnetic induction corresponding to the zero-magnetization force, called *remanence* (B_r). It corresponds to the magnetization that remains in the material after the removal of the external field, i.e., *remanence magnetization*. In permanent magnets, the magnetization does not change over time unless the body is heated above the *Curie temperature*, at which the magnetization cancels out. In magnetic steels, residual magnetization slowly decreases over time.
- the value of magnetizing force at which magnetic induction cancels, called *coercivity* (H_c). It corresponds to the magnetizing force of the opposite direction to the residual magnetization required to cancel it. Materials with high

[2]That is, if U is defined as the difference between the voltage at the final terminal and the voltage at the initial terminal of the loop (conventionally chosen), the path of the loop from the initial terminal to the final terminal occurs counterclockwise when viewed from the direction toward which the normal to the surface bounded by the loop points.

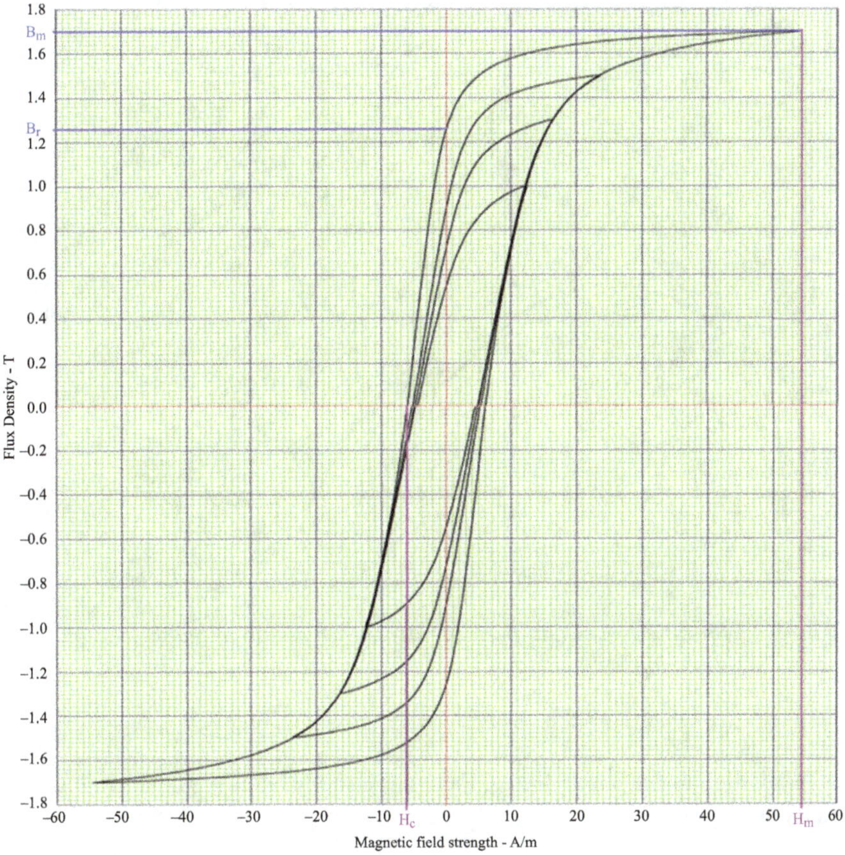

Figure 2.3 Hysteresis cycle (noteworthy are the induction values at the knee of the curve and at saturation)

coercivity are called *magnetically hard materials*, and this property is useful for permanent magnets. Materials with low coercivity are called *magnetically soft materials*, and this property is desirable for magnetic steels.

2.2.3 Losses in iron

It can be shown that the work done by the magnetizing force is given by [2] (31, 2)

$$\mathcal{W} = \int_{\mathcal{V}} \int_{\mathcal{C}} \boldsymbol{H} \cdot d\boldsymbol{B} dV \tag{2.6}$$

where $\mathcal{C}$ is the hysteresis loop and $\mathcal{V}$ the space in which the magnetic field is present. The inner integral is equivalent to the measure of the area enclosed by the hysteresis loop in the *H–B* plane. Thus, its area is proportional to the energy per unit volume dissipated in each magnetization cycle. The value of this energy represents *hysteresis losses*, the meaning of which will be described in Section 2.3.6.

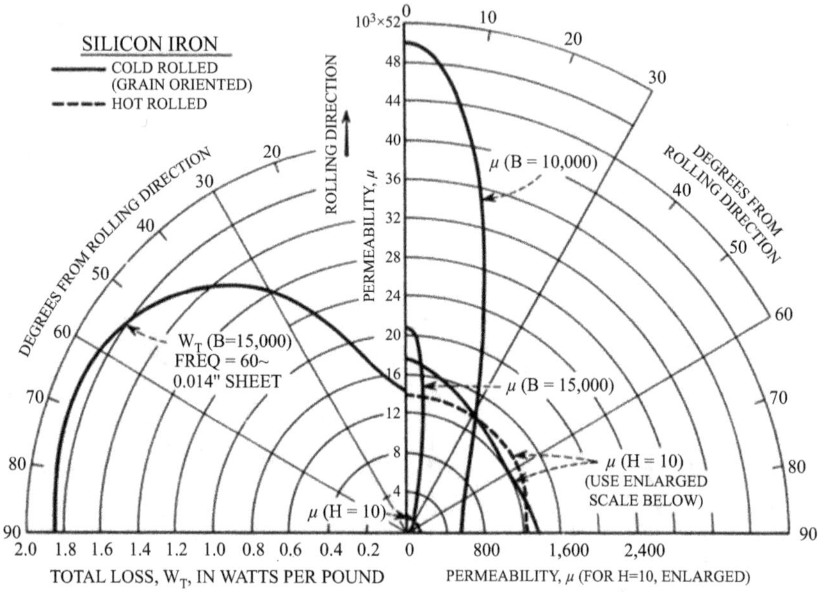

Figure 2.4 Specific losses and permeability as a function of magnetization direction

In addition to the above losses, there are also the additional ones in the magnetic core:

- *eddy current losses* (or *eddy losses* for short), due to the currents induced in ferromagnetic steel (which is an electric conductor, too) by the Faraday effect consequent to the variable magnetic field;
- *anomalous losses,* due to the creation and modification of *domain walls* in the ferromagnetic steel, the meaning of which will be explained in Section 2.3.6.

These losses are also present in the other ferromagnetic materials of the transformer, such as the clamps and tank (normally made of nonferromagnetic steel), and the magnetic flux shields that can be present in the walls inside the tank.

Hysteresis losses depend not only on the magnetization cycle, but also on the frequency of that cycle and the direction of the magnetization force with respect to any *easy* magnetization axis of the material that is in the *rolling* direction (Figure 2.4).

In order to make comparable the values of specific losses (i.e., losses per unit of mass) declared by different manufacturers of magnetic materials, two measurement procedures have been standardized, namely (Figure 2.5):

1. the *Epstein frame* [3];
2. the *Single Sheet Tester* (SST) [4].

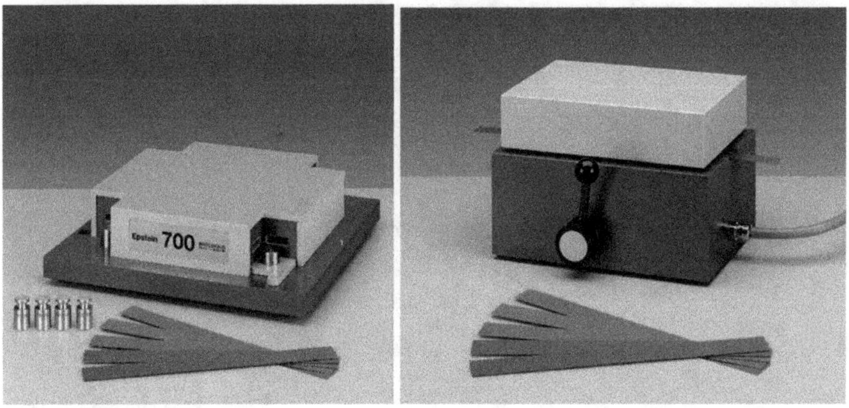

Figure 2.5 An Epstein frame (left) and a single sheet tester (right)

The Epstein frame is a circuit consisting of a series of two concentric windings arranged to form a square, inside which standardized stripes of magnetic sheets are arranged, stacked in multiples of 4. In contrast, the SST allows a single stripe to be placed inside two concentric windings, within a magnetically shielded container.

In both methods, a known current is injected into one winding, and the no-load voltage is measured at the ends of the other winding. Based on the knowledge of the standardized geometry of the circuit, it is possible to trace back the magnetizing force and the magnetic induction, as well as to measure the losses given by the active power absorbed by the supply circuit.

Figure 2.6 depicts a typical diagram of measured losses for a magnetic material by the traditionally used Epstein frame method.

In the absence of more precise information, it can be assumed that the specific hysteresis losses in a magnetic material can be calculated with the relation:

$$P_{V,hyst} = k_h B_m^2 f \tag{2.7}$$

and that the specific abnormal losses are calculable according to the relation:

$$P_{V,anom} = k_c (B_m f)^{3/2}, \tag{2.8}$$

where k_h, k_c are empirical coefficients [5].

Eddy current losses are of two types (Figure 2.7) and are squarely dependent on frequency, as can be easily demonstrated by calculating the induced voltage due to the Faraday effect. The losses per unit volume are calculated as [6] (10.7)

$$P_{V,eddy} = \left(\frac{\pi^2}{6}\right) \frac{f^2 B_m^2 w^2}{\rho}, \tag{2.9}$$

where f is the frequency of the current, w is the length of the path travelled by the magnetic flux, and ρ is the resistivity of the magnetic core (average resistivity of steel and insulating varnish, for interlaminar currents).

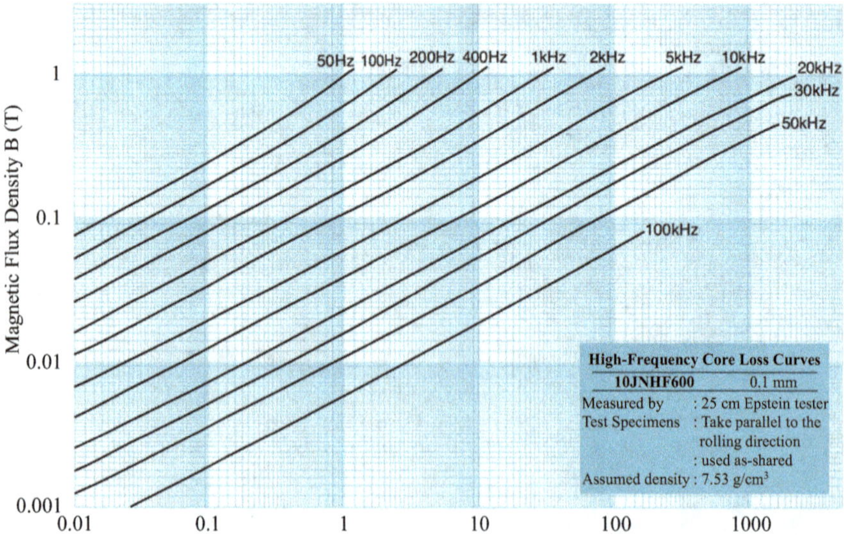

Figure 2.6 Loss curves in a magnetic material

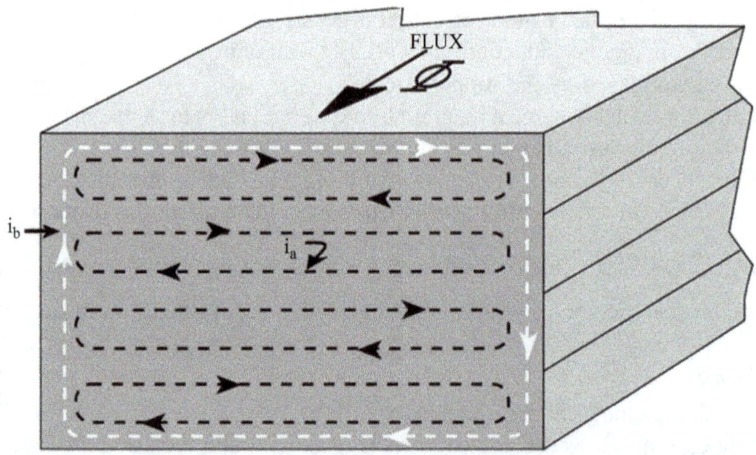

Figure 2.7 Intralaminar (i_a) and interlaminar (i_b) eddy currents

2.2.4 Construction of the magnetic core

The magnetic cores of transformers are formed by one or more *windows*, i.e., rectangular shapes made of ferromagnetic steel blocks (Figure 2.8), called *yokes* (horizontal) and *limbs* (vertical).

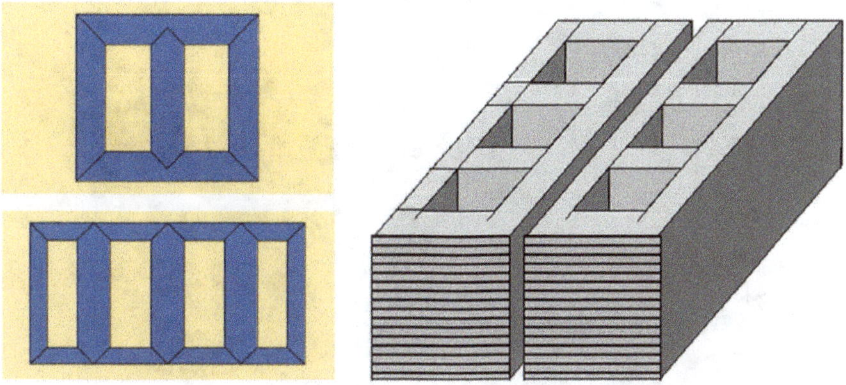

Figure 2.8 Magnetic cores for three and five limbs core-type (left) and for shell-type (right) transformers

Figure 2.9 Core of a five-limb transformer

The ferromagnetic material consists of thin laminated layers, called *sheets*, stacked in the plane parallel to the direction of magnetic flux and bounded by sheaths of nonmagnetic insulating material for limbs, and by nonmagnetic iron clamps for yokes (Figure 2.9).

Figure 2.10 Overall view of an automatic cutting machine and details of the operations of sheet unrolling, mechanical cutting, shaping, and laser cutting

The use of sheets coated with insulating varnish makes it possible to limit the circulation of eddy currents in the core, thus reducing the no-load losses.

Cutting operations of magnetic sheets are carried out by high-precision automatic machines (Figures 2.10 and 2.11), which minimize the mechanical stress applied to the sheet, and blurs at their edges, which are causes of additional losses.

From the cutting of a ferromagnetic steel sheet, lozenges and trapezoids are obtained that make up the core windows (Figures 2.11 and 2.12).

The connections among the different geometric figures that make up the windows, called *joints*, are obtained by stacking sheets of slightly different lengths alternately so that the surplus part of even-numbered ones fits into the space vacated by the defective part of odd-numbered ones (Figure 2.13). There are also some spaces inside the packs that allow the circulation of coolant. The *stacking factor* represents the ratio of the volume of the sheets to the volume occupied by the core, which includes the interlaminar space and the thickness of the varnish.

To improve the passage of magnetic flux, the joints are not all placed on the same axis but are staggered with each other (*step lap*) so as not to reduce the

Figure 2.11 Left, sheets packed with automatic machine; right, detail of some packs with staggered joints

Figure 2.12 Operations of manually assembling the sheets to form the core

equivalent cross section too much. In addition, in the corners, the joints are diagonal (*mitered*) to reduce the length of the sections where the flux is not aligned with the easy magnetization direction of the material (Figure 2.14).

In shunt reactors, the design of the core must consider the fact that there is no flux to compensate for that produced by the energized winding, so the magnetic circuit must have a sufficiently high reluctance to prevent the saturation of the magnetic material. To achieve this, sections composed of ferromagnetic steel are alternated in the wound limb with sections composed of paramagnetic insulating material, usually ceramic or steatite, to obtain an air gap capable of limiting the magnetic flux (Figure 2.15). In this case, the arrangement of the ferromagnetic sheets in the limb is done radially.

Figure 2.13 Corner of a window with joints and a cooling channel in the foreground

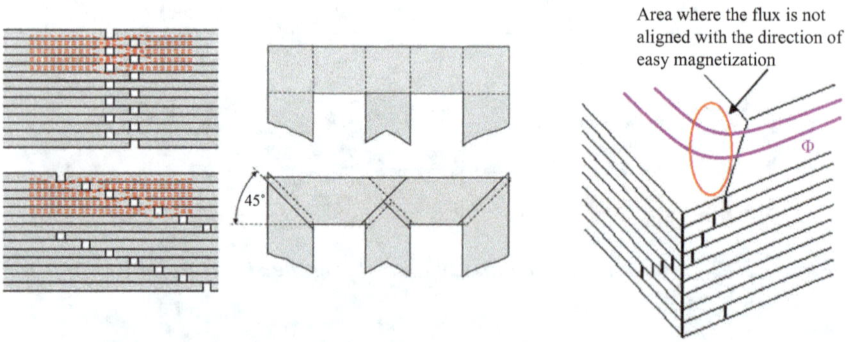

Figure 2.14 Single and multiple staggered joint sheet metal (left) and diagonal cut in the corner (right)

2.2.5 *Magnetostriction and noise*

Magnetic materials exhibit the phenomenon of *magnetostriction*, which consists in the change of size in proportion to the magnetization undergone, according to a square relationship (see Section A.4.3). The proportionality factor is called the *magnetostriction coefficient*, and it can be positive (elongation) or negative (contraction). Magnetostriction changes the dimensions of the body along all axes, although normally the predominant effect is in the direction of magnetization.

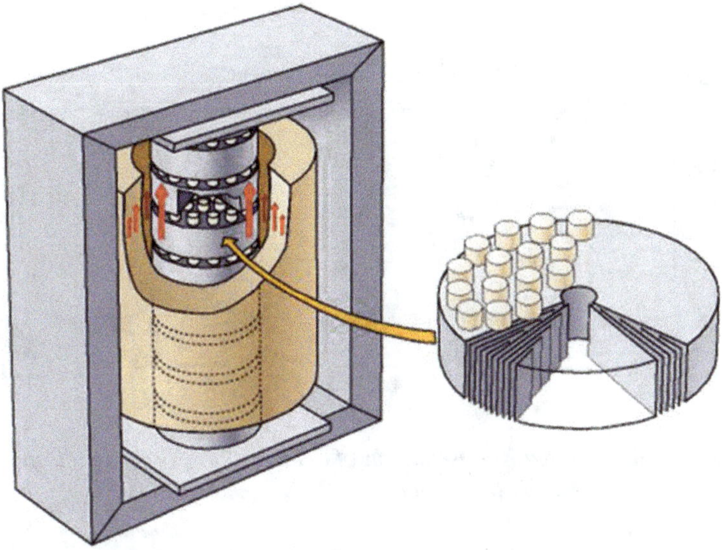

Figure 2.15 Schematic diagram of a single-phase reactor with a section of the column highlighted

In transformer applications, since the magnetizing force is alternating, the magnetostriction has a periodic trend with twice the frequency of the magnetizing force, and this happens because of the aforementioned square dependence.

Magnetostriction, in addition to creating mechanical strains in the core sheets, generates noise at frequencies equal to that of magnetization and its integer multiples, because the relationship is nonlinear and breaks down into harmonic components. The noise level depends on: the magnitude of the strains and consequently on the maximum magnetic induction and the direction of the magnetic flux; the stiffness of the core's mechanical supports, consisting of the clamps and tie sheaths; and the stacking of the sheets.

Therefore, to contain the noise produced by the transformer, it is necessary: to use high-quality steel sheets, which have less magnetostriction effects; an optimized magnetic core design, which reduces the areas where the magnetic flux is not aligned with the direction of the sheets (causing additional noise due to mutual collision between adjacent sheets); adequate clamping of the sheets.

Optimized design to reduce magnetostriction vibration is particularly important for reactors, where the presence of air gaps increases the possibility of deformation of the magnetic sheet (Figure 2.16).

2.3 Magnetic materials

2.3.1 Magnetic classification of materials

Materials can be classified according to their reaction to the presence of a magnetic field as

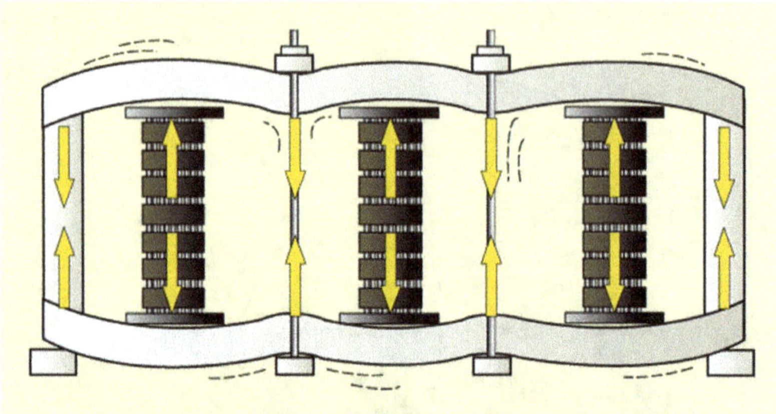

Figure 2.16 Accentuated representation of deformation from magnetostriction and vibration in a reactor

- *diamagnetic*: such materials exhibit a small negative susceptibility $\chi \simeq -10^{-5}$ not persistent. Diamagnetism is produced by a change in the orbital motion of electrons due to the applied magnetic field (*Larmor precession*). It is observed in all materials but—being weak—is only detected when other types of magnetism are totally absent. This form of magnetism has no practical importance.
- *paramagnetic*: such materials exhibit a small positive susceptibility $\chi \simeq 10^{-4}$ that is nonpersistent. In these materials, each atom possesses a permanent dipole moment by virtue of the incomplete cancellation of the orbital or spin magnetic moments of the electron (i.e., if $J \neq 0$, see Section 2.3.2). In the absence of an external magnetic field, the magnetic moments of these atoms have random orientations, so each elementary volume element possesses no macroscopic magnetization. However, in the presence of an external magnetic field, these atomic dipoles are free to rotate and can give rise to paramagnetism when, by rotation, they preferentially align with the field. The value of paramagnetism is derived from the Fermi statistics for electrons and depends on the overall azimuthal quantum number J, according to the formula ([7] Section 14 formulas 19, 20):

  $$M = N g_J J \mu_B \mathcal{B}_J(x),$$

  where
  - $x = \frac{g_J J \mu_B B}{k_B T}$, with k_B Boltzmann constant and T absolute temperature,
  - $\mathcal{B}_J(x) = \frac{2J+1}{2J} \coth \frac{(2J+1)x}{2J} - \frac{1}{2J} \coth \frac{x}{2J}$ (*Brillouin function*).

It can be seen that the Brillouin function (used to draw the magnetization curves $\mathfrak{m}=M/N$ in Figure 2.17) tends asymptotically to 1 for $x \gg 1$, while for low values of x, the slope of the curve $M/\frac{B}{T}$ is called the *Curie constant*.

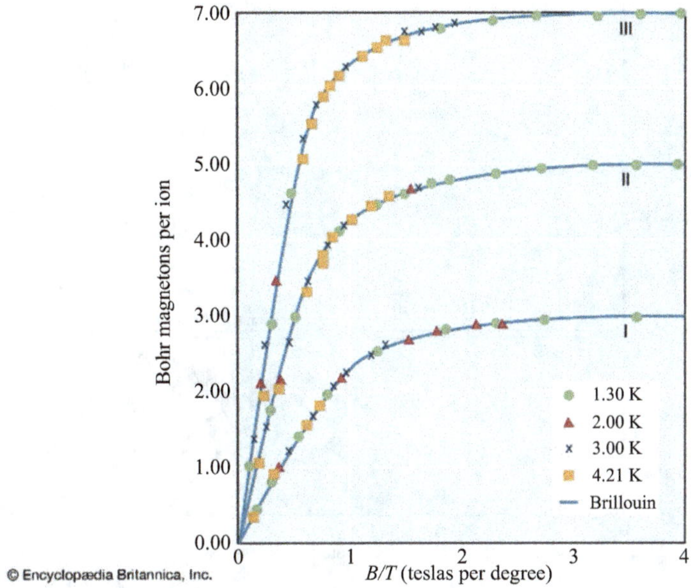

Figure 2.17 Trend of magnetization $\mathfrak{m}$ *for three substances (with* $g_J = 2$ *and* $J = 3/2, 5/2, 7/2$*)*

- *ferromagnetic:* such materials possess a permanent magnetic moment in the absence of an external field and have very intense and permanent magnetization. The characteristics of *ferromagnetism* are shown by transition metals: iron, cobalt, nickel, and some rare-earth metals such as gadolinium. Magnetic susceptibilities up to 10^6 are possible for ferromagnetic materials.
- non-magnetic: such materials do not exhibit magnetization $\chi = 0$. Some gases belong to this group, which are actually diamagnetic but, given their low density in the gaseous state, weakly detectable as such. For a meaningful comparison between elements in different physical states, it is more meaningful to consider molar susceptibility, i.e., the susceptibility multiplied by the molar mass of the element (Figure 2.18).

Table 2.1 lists the magnetic susceptibilities of some nonferromagnetic materials.

Both diamagnetic and paramagnetic materials are not considered "magnetic" because magnetization appears only in the presence of an external field. Moreover, for both, the magnetic induction $\boldsymbol{B}$ is approximately the same as in vacuum.

Figure 2.19 represents the microscopic behavior of diamagnetic and paramagnetic materials and their magnetization characteristic compared with that of a vacuum.

Permanent magnetic moments in ferromagnetic materials are due to spin magnetic moments that are not cancelled by compensation as a consequence of the particular electronic structure, due to mutual interaction (so far, the atoms have

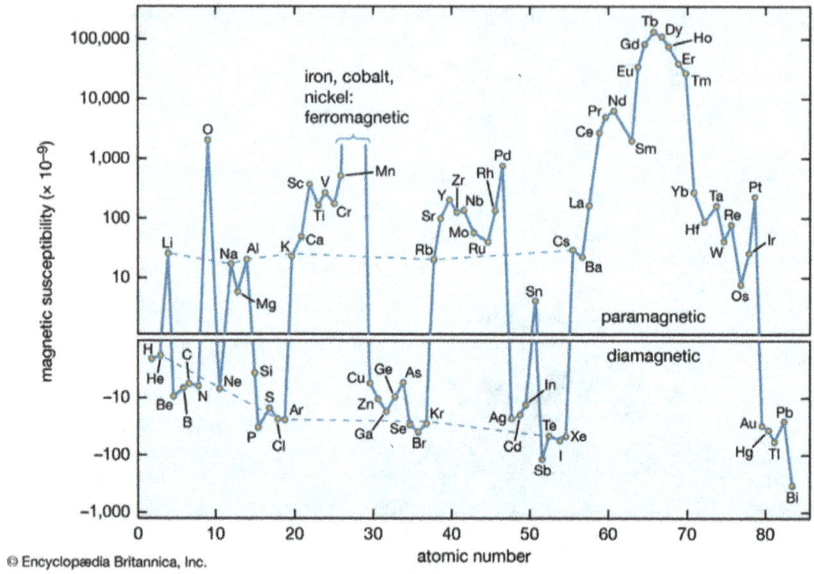

Figure 2.18 Molar susceptibility (in mol/kg) of some elements at room temperature

Table 2.1 Susceptibility of some diamagnetic and paramagnetic materials

Diamagnetics		**Paramagnetics**	
Material	**Susceptibility**	**Material**	**Susceptibility**
Silver	-2.38×10^{-5}	Aluminum	2.07×10^{-5}
Sodium chloride	-1.41×10^{-5}	Chromium chloride	1.51×10^{-3}
Mercury	-2.85×10^{-5}	Chrome	3.13×10^{-4}
Gold	-3.44×10^{-5}	Molybdenum	1.19×10^{-4}
Aluminum oxide	-1.81×10^{-5}	Sodium	8.48×10^{-6}
Copper	-0.96×10^{-5}	Manganese sulfate	3.70×10^{-3}
Silicon	-0.41×10^{-5}	Titanium	1.81×10^{-4}
Zinc	-1.56×10^{-5}	Zirconium	1.09×10^{-4}

been considered noninteracting, but they actually interact). What differentiates ferromagnetic materials from paramagnetic ones is the *exchange interaction*, a purely quantum-mechanical phenomenon between spins of adjacent atoms, which makes energetically convenient in some cases to align spin magnetic moments (the alignment of orbital magnetic moments is instead negligible), even in the absence of an external field (Figure 2.20).

Verification of the convenience of reciprocal alignment is provided by *Stoner's criterion*, which compares the density of states at Fermi energy $D(\mathcal{E}_F)$

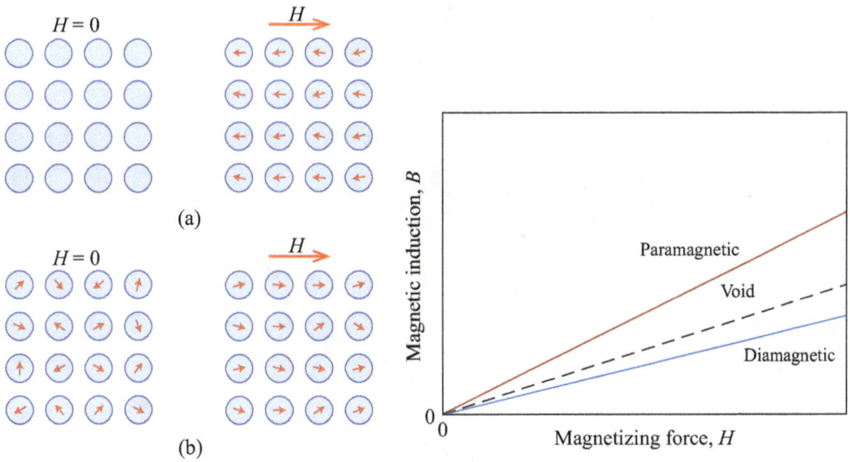

Figure 2.19 Magnetization in diamagnetic (a) and paramagnetic (b) materials and their B–H *curves*

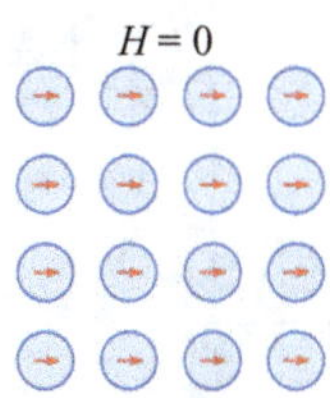

Figure 2.20 Magnetization in ferromagnetic materials

with the exchange integral $\mathcal{J}$ which represents the intensity of the exchange interaction. If their product is greater than 1, it is energetically convenient to align the spins, because as the energy level (hence kinetic energy) increases, there is a reduction in the exchange energy, and thus, the system arrives at a state with lower overall energy, hence more stable. This criterion accounts well for the fact that transition elements such as iron, nickel, and cobalt are ferromagnetic, unlike other metals (Figure 2.21).

The maximum possible magnetization, or saturation magnetization M_S, of a ferromagnetic material represents the magnetization that can be obtained when all magnetic dipoles are aligned with the external field; B_S is the corresponding saturation flux density. The saturation magnetization is equal to the product of the useful magnetic moment of each atom times the number of atoms present.

In nonferromagnetic materials, magnetic moment coupling phenomena can occur between adjacent ions and atoms. These couplings, if the exchange integral is negative (Figure 2.22), lead to favoring antiparallel alignments. When the alignment of spin moments of neighboring atoms (or ions) is in exactly opposite directions, the phenomenon is called *antiferromagnetism*.

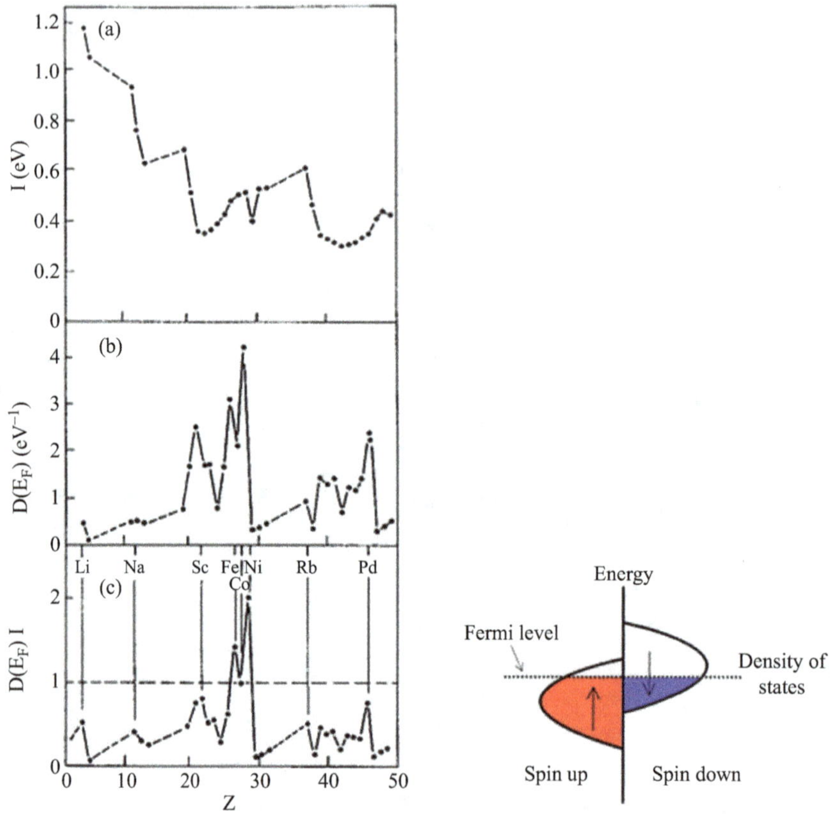

Figure 2.21 Left, calculation of the Stoner criterion for some elements: (a) exchange integral; (b) density of states; and (c) product of the aforementioned quantities; right, the representation of the imbalance between density of states for electrons with spin up and with spin down, due to exchange interaction

One material that exhibits this behavior is manganese oxide (MnO). It is an ionic ceramic material with Mn^{2+} and O^{2-} ions. In O^{2-} ions, both orbital and spin moments cancel each other so that the resulting magnetic moment is zero. In contrast, Mn^{2+} ions have a resultant magnetic moment due essentially to spin. These Mn^{2+} ions are arranged in the crystal structure in such a way that the moments of adjacent ions are antiparallel. Such an arrangement is depicted schematically in Figure 2.23. Obviously, the opposing magnetic moments cancel out and, as a result, the solid as a whole has no resultant magnetic moment.

Some ceramic materials also exhibit permanent magnetization, called *ferrimagnetism*. The macroscopic magnetic characteristics of ferromagnetic and ferrimagnetic materials are similar, and the difference lies in the origin of the resulting magnetic moments. The principles of ferrimagnetism can be illustrated by considering cubic ferrites. These ionic materials are represented by the chemical formula MFe_2O_4, where M represents any metallic element.

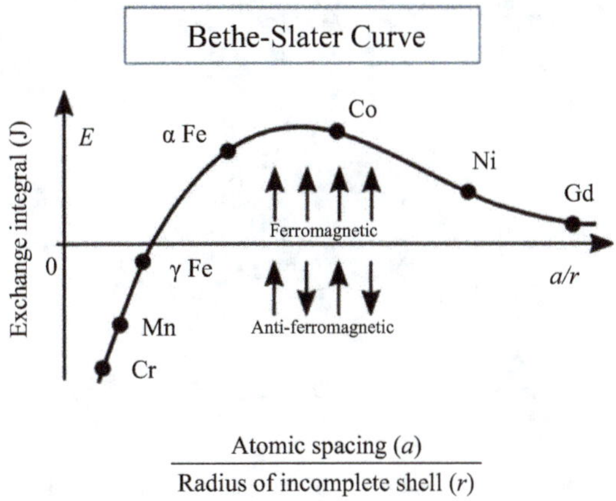

Figure 2.22 Bethe–Slater curve shows the dependence of the exchange integral on the ratio of the interatomic distance to the radius of the 3d *valence electron orbital*

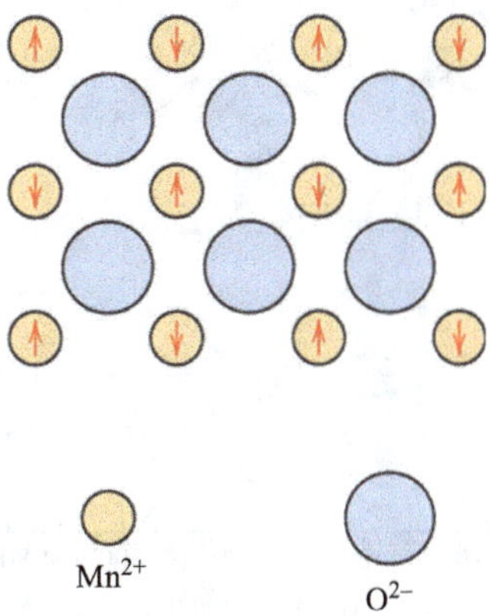

Figure 2.23 Magnetic behavior of manganese oxide

The exemplary ferrite is the mineral magnetite, Fe_3O_4, usually called simply magnetite. The formula Fe_3O_4 can be rewritten as $Fe^{2+}O^{2-}$–$(Fe^{3+})_2(O^{2-})_3$ where Fe ions exist in both +2 and +3 valence states in the ratio of 1:2. For each Fe^{2+} and Fe^{3+} ion, there is a resulting spin magnetic moment that corresponds to four and five Bohr magnetons, respectively, for the two types of ions. In contrast, the O^{2-} ions are magnetically neutral. Antiparallel spin coupling interactions with characteristics similar to those of antiferromagnetism appear between Fe ions. Consequently, the ferrimagnetic moment that may result originates from the incomplete cancellation of spin moments (Figure 2.24).

Saturation magnetization for ferrimagnetic materials is not as high as for ferromagnetic materials. On the other hand, ferrites, like ceramic materials, are

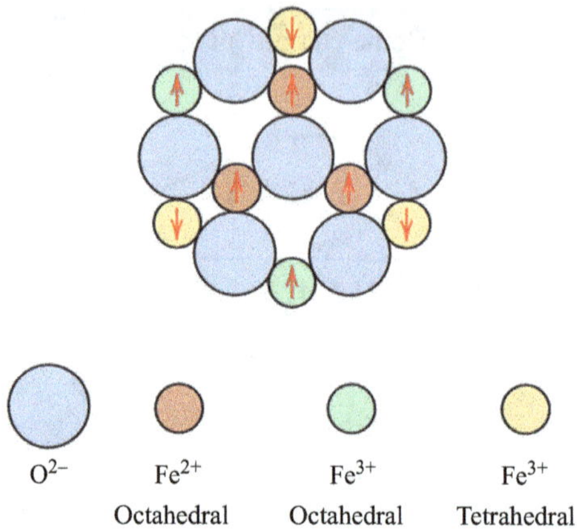

Figure 2.24 Magnetic behavior of magnetite

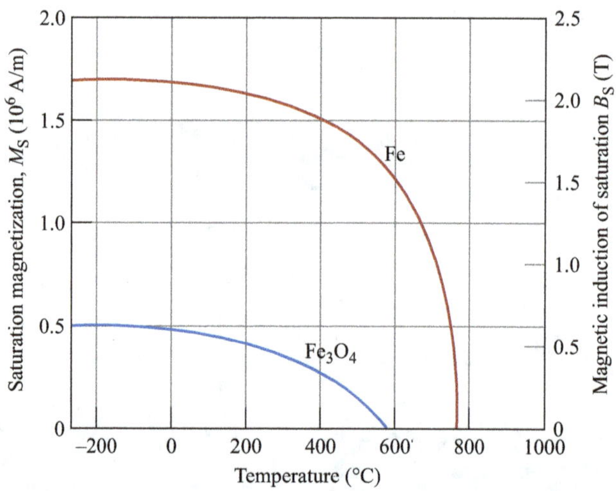

Figure 2.25 Diagram of saturation magnetization of iron and ferrite as a function of temperature

good electrical insulators, and for various magnetic applications, as in the case of high-frequency transformers, low electrical conductivity is highly coveted.

In ferromagnetic, antiferromagnetic, and ferrimagnetic materials, thermal agitation of atoms hinders the coupling forces between the dipole moments of adjacent atoms, causing loss of alignment even in the presence of an external field. This effect leads to decreased saturation magnetization for both ferromagnetic and ferrimagnetic materials. Saturation magnetization is highest at 0K (absolute zero), at whose temperature thermal vibrations are the lowest. As temperature increases, saturation magnetization gradually decreases and then falls abruptly to zero at the *Curie temperature* T_C (Figure 2.25), beyond which the material becomes paramagnetic.

Antiferromagnetism is also affected by temperature and disappears at a value called *Néel temperature* T_N. At higher temperatures, antiferromagnetic materials become paramagnetic, too.

Figure 2.26 summarizes the magnetic characteristics of pure elements; of course, molecular or ionic compounds containing several elements may have different characteristics from those of their individual constituent elements.

2.3.2 Magnetization in materials

In magnetostatics, magnetic effects are due to moving charges, represented by a current *I*, measured in amperes (A). In the approximation of a continuous distribution of moving charges in space, they are represented by a current density $\boldsymbol{J}$, measured in amperes per square meter (A/m^2). If the current flows in a circuit of negligible size relative to the distance from the observer, its effect can be described by means of the *magnetic dipole moment* ([8] Section 7.2):

$$\mathfrak{m} = \frac{1}{2}\int_{\mathcal{V}} \boldsymbol{r} \times \boldsymbol{J}(\boldsymbol{r})dV, \tag{2.10}$$

where the volume integral is extended to the entire portion of space $\mathcal{V}$ in which the current density $\boldsymbol{J}$ is present.[3] From (2.10), it can be deduced that the units of magnetic moment are amperes times square meter (A m^2) often referred to as joules per tesla (J/T).

The magnetic moment induces a *vector potential* $\boldsymbol{A}$, measured in weber per meter (Wb/m), equal to:

$$\boldsymbol{A} = \frac{\mu_0}{4\pi}\mathfrak{m} \times \frac{\boldsymbol{r}}{r^3}, \tag{2.11}$$

where $\boldsymbol{r}$ is the distance between the point where the vector potential is calculated and the position of the magnetic dipole. The vector potential determines the magnetic induction according to the following relation:

$$\boldsymbol{B} = \text{rot}\boldsymbol{A}. \tag{2.12}$$

Bodies are composed at the microscopic level by atoms, whose constituent particles are endowed with magnetic moments. The magnetization of the material is the sum of these microscopic magnetic moments:

$$\boldsymbol{M} = N\mathfrak{m}, \tag{2.13}$$

where *N* is the number of atoms per unit volume, with a unit of measurement m^{-3} (from which it follows that magnetization has the correct unit of measurement A/m).

[3]The expression (10) is independent of the choice of the origin of the reference system because any translation of it does not change its value: $\boldsymbol{r} = \boldsymbol{r}' + \boldsymbol{s} \Rightarrow \mathfrak{m}' = \mathfrak{m} - \boldsymbol{s} \times \int_{\mathcal{V}} \boldsymbol{J}(\boldsymbol{r}')dV = \mathfrak{m} - \boldsymbol{s} \times \boldsymbol{0} = \mathfrak{m}$, since the integral over all space of the current density cancels, as a consequence of the fact that its divergence is zero in magnetostatics.

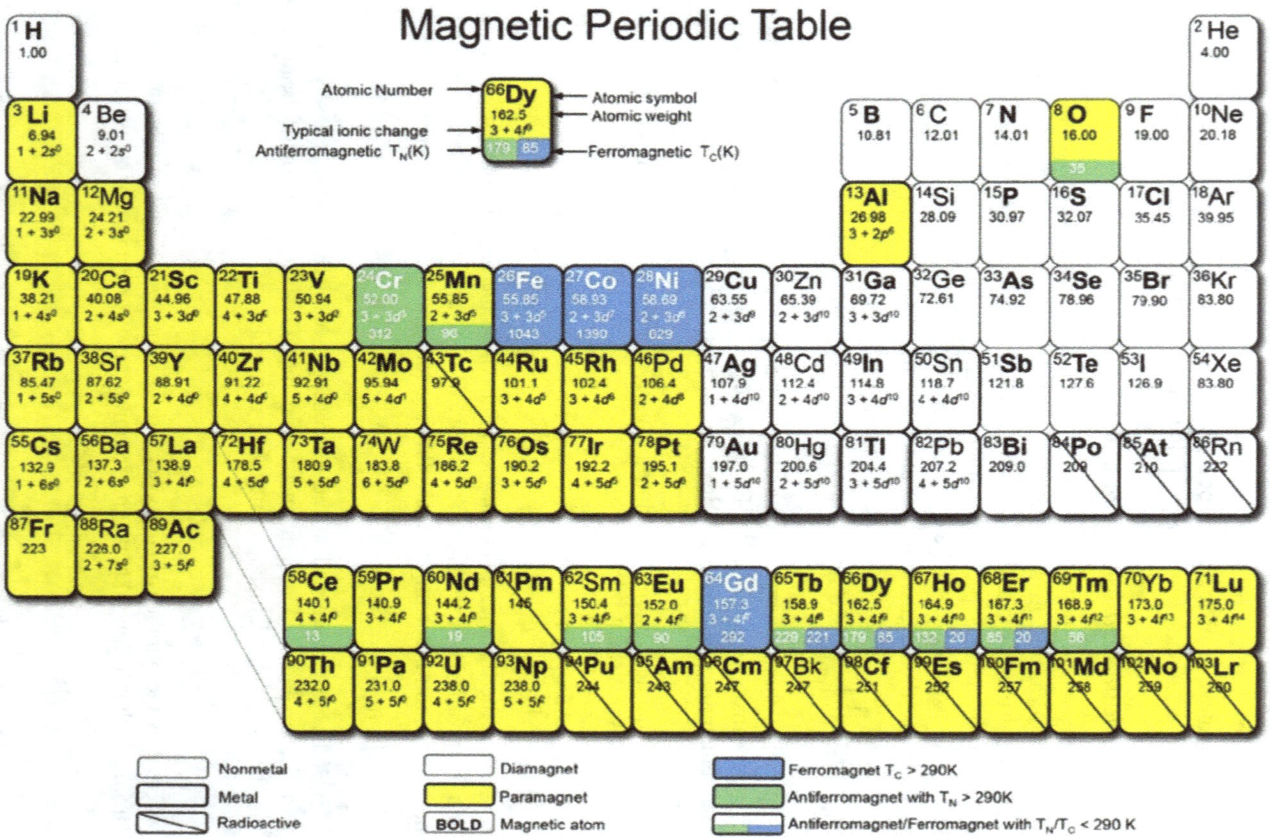

Figure 2.26 Periodic table with the magnetic classification of elements at room temperature

It can be demonstrated ([8] Section 7.2) that:

$$A(r) = \frac{\mu_0}{4\pi}\int_{\mathcal{V}} \frac{M(s)\times(r-s)}{\|r-s\|^3}dV_s$$

$$= \frac{\mu_0}{4\pi}\int_{\mathcal{V}} \frac{\mathrm{rot}M(s)}{\|r-s\|}dV_s + \frac{\mu_0}{4\pi}\int_{\partial\mathcal{V}} \frac{M(s)}{\|r-s\|}\times dS_s, \tag{2.14}$$

where ∂V is the frontier of V, i.e., the surface of the material. The second term represents the surface currents j_m of the magnetic material:

$$j_m = M \times n, \tag{2.15}$$

while the first term represents the magnetization current density J_m within the material:

$$J_m = \mathrm{rot}M \tag{2.16}$$

Therefore, Maxwell's equation related to the rotor of magnetic induction is modified in the presence of a magnetic material by the addition of a term (excluding the boundary effects due to the surface currents):

$$\mathrm{rot}B = \mu_0(J_c + J_m), \tag{2.17}$$

where J_c represents the density of conduction currents, which defines the magnetizing force:

$$J_c = \mathrm{rot}H \tag{2.18}$$

Substituting (2.18) and (2.16) in (2.17), we find the rotor of formula (2.1).

2.3.3 Origin of magnetic momentums

The angular momentum of atoms is given by three contributions:

- the *orbital angular momentum* of electrons due to their motion around the nucleus and denoted[4] by $\vec{L}$;
- the intrinsic angular momentum of electrons, called *spin* and denoted by $\vec{S}$;
- the *angular momentum of the nucleus*, given by the sum of the intrinsic angular momentums of the protons and neutrons and their angular momentums with respect to the center of mass, denoted by $\vec{I}$.

The *overall angular momentum* is defined as the vector sum $\vec{J} = \vec{L} + \vec{S}$. The magnetic momentum of the atom is derived from these angular momentums. In fact, for a point charge, the current density is written:

$$J(r) = qv\delta(r - r_0) \tag{2.19}$$

[4]The symbols of angular momentum contain the above arrow to avoid confusion with other symbols that use the same letters.

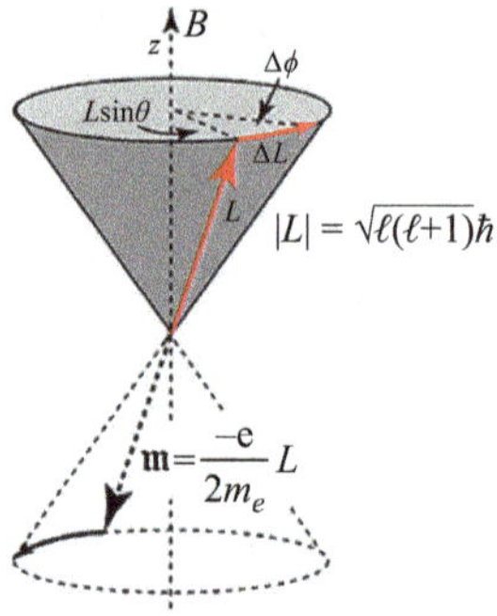

Figure 2.27 Precession of angular momentum

where q is the charge, measured in coulombs (C), $\boldsymbol{v}$ the velocity of the particle, and δ the three-dimensional *Dirac delta* function. Accordingly, (2.10) reduces to:

$$\mathfrak{m} = \frac{1}{2}\boldsymbol{r} \times \boldsymbol{v}q = \frac{1}{2}\vec{\boldsymbol{l}}\frac{q}{m}, \tag{2.20}$$

where $\vec{\boldsymbol{l}}$ represents the angular momentum of the particle[5] and m its mass.

According to quantum theory, the orbital angular momentum of electrons bound to an atomic nucleus is quantized according to a number ℓ, called the *azimuthal quantum number,*[6] which can take integer values between 0 and $n - 1$, where n is the *radial quantum number,* which defines the energy level of the electron in the unperturbed atom. According to this scheme, the modulus of the orbital angular momentum of the electron holds:

$$l = \hbar\sqrt{\ell(\ell + 1)}, \tag{2.21}$$

where $\hbar$ is the reduced Planck constant ($\hbar \simeq 1.055 \times 10^{-34}$Js). However, this value is of little meaning for the magnetic effects, as the orbital momentum makes a precession with respect to an axis[7] (Figure 2.27).

Therefore, what matters is the value of the projection of angular momentum on the precession axis (usually denoted by z), which constitutes the average value of angular momentum. This angular momentum is also quantized, according to a number m_ℓ, called the *azimuthal magnetic quantum number*, which can take integer values between $-\ell$ and $+\ell$.

[5]Since this is a single particle and not a current in a circuit, the origin with respect to which angular momentum is calculated must be specified; for atoms, the origin is fixed in the atomic nucleus.
[6]The designation "orbital" is used here in reference to the angular momentum $\vec{l}$ but not in reference to the quantum number ℓ, because it does not represent the magnitude of the momentum, and there is an analogous quantum number for spin, which has no analogy to an orbit of the particle.
[7]In the absence of an external magnetic field, the direction of the precession axis is not fixed a priori; in the presence of an external magnetic field, the direction of the axis aligns with that of the magnetic field.

Accordingly, the electron's elemental magnetic moment, or *Bohr's magneton,* is defined:

$$\mu_B = \frac{\hbar \mathsf{e}}{2m_\mathsf{e}}, \tag{2.22}$$

where e is the elementary charge ($\mathsf{e} \simeq 1.602 \times 10^{-19}$C) and m_e the mass of the electron ($m_\mathsf{e} \simeq 9.109 \times 10^{-31}$kg), from which $\mu_B \simeq 927.4 \times 10^{-26}$J/T.

Resuming the (2.20), the result is as follows:

$$\mathfrak{m}_z = -m_\ell \mu_B \tag{2.23}$$

where the minus sign considers that the charge of the electron is negative.

Electrons, like all subatomic particles in the fermion family, are also endowed with an intrinsic angular momentum $\vec{s}$, that is unrelated to their own motion, quantized according to a fixed *spin quantum number* $s = \frac{1}{2}$. This angular momentum also performs a precession motion around an axis, and therefore to the magnetic effects what counts is the *spin magnetic quantum number* m_s which can take only the values −½ and +½.

The relation (2.20) is not valid for the magnetic moment due to intrinsic angular momentum. To obtain an equivalent relation, we must introduce a coefficient, called the *g-factor*, which is worth[8] $g_s = 2$, and write:

$$\mathfrak{m}_z = -g_s m_s \mu_B. \tag{2.24}$$

The overall angular momentum of each electron $\vec{j}$ is similarly associated with a magnetic quantum number $m_j = m_\ell + m_s$ and an azimuthal quantum number j, so that $-j \le m_j \le j$. However, j is not the sum of ℓ and s, because its actual value depends on the vector sum of $\vec{l}$ and $\vec{s}$, and it gives the modulus of the overall angular momentum as $|\vec{j}| = \hbar\sqrt{j(j+1)}$.

The azimuthal and spin magnetic quantum numbers of the several electrons add up to give the total magnetic numbers:

$$M_L = \sum_i m_{\ell,i}, \quad M_S = \sum_i m_{s,i}. \tag{2.25}$$

The total magnetic quantum number is the sum of the total azimuthal and spin magnetic quantum numbers (*Russell–Saunders coupling,* [9] Section 5-7.B):

$$M_J = M_L + M_S. \tag{2.26}$$

The azimuthal and spin quantum numbers l and s, on the other hand, do **not** sum algebraically, since they do not represent projections of angular momentums—as shown in (2.21). Instead, they depend on the possible combinations of the

[8]Neglecting radiative corrections.

magnetic numbers[9]: those resulting in the maxima of total magnetic numbers M_L and M_S give the values of L and S.

The total overall azimuthal quantum number J, similarly, is not the sum of L and S but can have any value between their sum and their difference, as long as it is not less than the absolute value of the overall magnetic number. That is, the overall azimuthal quantum number must respect the relations:

$$|L - S| \leq J \leq L + S, \quad J \geq |M_J|. \tag{2.27}$$

The actual value of J depends on the particular combination in the atom under consideration.

An atomic g-factor, also called *Landé factor* (see equation (5-60) in [9]), is defined as:

$$g_J = \frac{3}{2} + \frac{S(S+1) - L(L+1)}{2J(J+1)}. \tag{2.28}$$

Then the following formula for the total overall magnetic momentum along the magnetization axis holds:

$$\mathfrak{M}_{J,z} = -g_J M_J \mu_B. \tag{2.29}$$

More details are described in Section A.1.

To determine the value of the actual quantum numbers, it must be considered that by the *Pauli exclusion principle*, no two electrons with all coincident quantum numbers can be present in the same atom.

Electrons occupy orbitals (with some exceptions) in increasing order of the sum $n + \ell$, and with the lower n among orbitals having the same sum. *Hund's rule* states that electrons are arranged within the same orbital so that the following criteria are met, in descending order of importance ([7] p. 424):

- Maximize the total spin quantum number S;
- Maximize the total azimuthal quantum number L;
- If the orbital is less than half full, then $J = |L - S|$; if it is more than half full, then $J = L + S$, if it is half full $J = S$.

Figure 2.28 represents some examples of the application of Hund's rule.

The angular momentum of the nucleus can also take on only half-integer multiples of the reduced Planck constant. However, since the nuclear magneton has a much smaller value than the electrons,[10] its effect on the atomic magnetic moment can be neglected.[11]

[9]Normally the maximum value of m_ℓ is ℓ, but the Pauli exclusion principle can restrict the choice of possible values when dealing with several electrons.

[10]Since in the magneton formula, the mass of the particle appears in the denominator, and the masses of the protons and neutrons are about 1836 and 1839 times that of the electron, respectively.

[11]Indeed, the spin g-factors of the proton and neutron are in absolute value higher than that of the electron: 5.59 and −3.83, respectively. However, this does not compensate for the small value of the nuclear magneton.

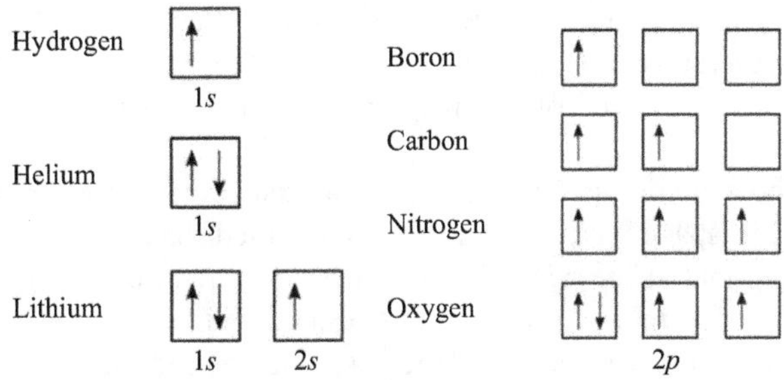

Figure 2.28 Electron arrangement and spin orientation in some atoms

2.3.4 Crystal structure of solids

The preceding paragraphs show that, to understand magnetization in solids, it is essential to know the arrangement of atoms in space.

As it is well known, atoms in solids form molecular or ionic bonds. The main difference between the two bonds, at the electronic level, is that in molecules, the electronic orbitals are shared between two or more atoms (*covalent bonding*), while in ions, an atom—called *anion*—takes on a negative charge, by adding to one of its own orbitals an electron given up by another atom—called *cation*—which loses the electron and takes on a positive charge. Thus, an electrostatic attractive force (*ionic bond*) is established between these two ions.

At the higher structural level, in *molecular solids*, molecules bind together through intermolecular forces, which can be of the following types: *dipolar forces* in cases where the electric charge is not uniformly distributed in the molecule, and then it has an electric dipole moment, *van der Waals forces* in which there is a residual electrostatic attraction force between neutral molecules and atoms.

Instead, ionic bonds are also responsible for the macroscopic structure of *ionic solids* because ionic bonding is stronger than covalent bonding and, therefore, acts on all adjacent atoms, not only on a specific group as in molecules. There are also cases of large-scale molecular structures, called *covalent solids*, in which the covalent bonding extends to all the atoms in the solid. *Metal solids* represent an intermediate case between ionic and molecular solids, in which the *conduction electrons* in the upper orbitals of metal atoms are weakly bound and are free to move between adjacent atoms, allowing electrical conduction.

Ionic solids and covalent solids always assume a regular structure in space, called a *crystalline structure*. Metallic and molecular solids can assume either a crystalline structure or a non-regular arrangement in space, called an *amorphous structure*. The type of structure a solid takes depends on: its homogeneity, considering that bodies made of different materials or containing impurities tend to have a less regular microscopic structure; and the solidification process since liquids that cool quickly tend to form a less regular structure. The portion of space occupied by a regular arrangement in a crystalline or polycrystalline solid is called a *grain* (or *crystallite*).

The macroscopic magnetic moment depends on the sum of the atomic magnetic moments; therefore, it is essential to know the structure of the magnetic solid.

The crystal structure of grains constitutes the subject of study in *crystallography.*

Ideally, a crystal consists of a *base*, i.e., an arrangement of atoms or molecules or a set of ions in space with respect to a point called the *origin*. The structure is replicated throughout space according to a *crystal lattice*, consisting of a regular arrangement of points in space. At each point in the lattice, there is a replica of the base. In addition to the translation of the base with respect to its origin, rotations and inversions are also allowed.

The set of translation, rotation, and inversion operations and their combinations that bring the crystal into a replica of itself constitutes the *spatial symmetry group*, or simply *spatial group* of the crystal. It has been calculated that there is a total of 230 possible spatial groups in three dimensions ([10] Section 3.5), and thus as many different types of crystals. Section A.3 provides more details about spatial groups.

It is possible to categorize space groups into *crystal systems*, which in turn depend on *lattice systems*, that include all crystals sharing the same point symmetries (i.e., the same rotations and inversions). To each lattice system corresponds a *unit cell*, which, when replicated, covers all space (Table 2.2). The different lattices (called *Bravais lattices*) differ in the arrangement of the base in the unit cell. Section A.3 provides more details on the point symmetries of lattice systems.

The most recurrent structures in metallic solids are *body-centered cubic* (*bcc*) and *face-centered cubic* (*fcc*) lattice systems and, to a lesser extent, *hexagonal close packed* (*hcp*), which has a base consisting of two atoms and allows for the greatest possible compactness (Figure 2.31).

Specifically, iron assumes three crystal structures[12] at ambient pressure, depending on the temperature ([11] Section 2.2.3, Figure 2.32):

- up to 910°C, iron-α has *ccc* structure;
- 910°C to 1394°C, iron-γ has *cfc* structure;
- 1394°C to 1538°C, iron-δ has *ccc* structure, but different from that of iron-α.

Iron-β is the name, now obsolete, that had been assigned to iron-α above its own Curie temperature of 771°C.

2.3.5 Crystal structure formation in metals

The formation of a crystalline structure in metallic solids is a consequence of the lower energy available in atoms in the solid state compared to the liquid state. As it is well known, physical systems at equilibrium arrange themselves in the lowest potential energy configuration. Since the interactions between atoms are mainly electrostatic, the lowest energy configuration corresponds to the one in which the distances between heterologous charges are minimized, while the distances between

[12]The different structures that a material can assume are called *allotropes*.

Table 2.2 Unit cells

Lattice system and solid	**Sides**	**Angles**	**Image**
Triclinic—parallelepiped	$a \neq b \neq c$	$\alpha \neq \beta \neq \gamma$	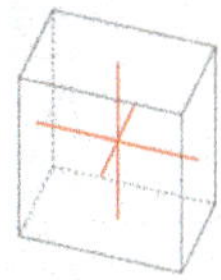
Monoclinic—oblique prism with a rectangular base, in which one side face is perpendicular to the base	$a \neq b \neq c$	$\alpha = \gamma = 90°, \beta \neq 90°$	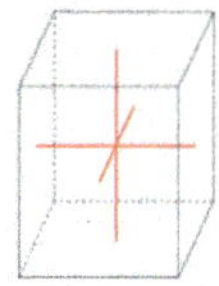
Orthorhombic—straight prism with rectangular base	$a \neq b \neq c$	$\alpha = \beta = \gamma = 90°$	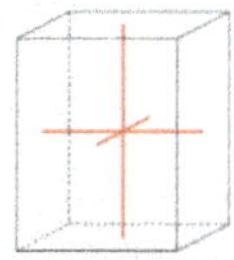
Tetragonal—right prism with a square base	$a = b \neq c$	$\alpha = \beta = \gamma = 90°$	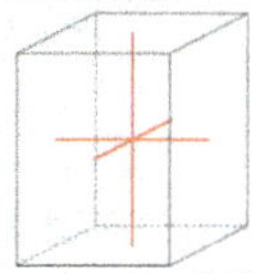
Rhombohedral[a]—equal-sided rhombohedron	$a = b = c$	$\alpha = \beta = \gamma \neq 90°$	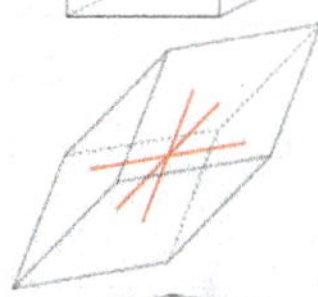
Hexagonal—right prism with a rhombic base, consisting of two equilateral triangles[b]	$a = b\ (\neq c)$	$\alpha = \beta = 90°, \gamma = 120°$	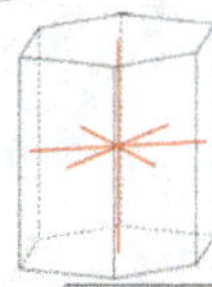
Cubic—cube	$a = b = c$	$\alpha = \beta = \gamma = 90°$	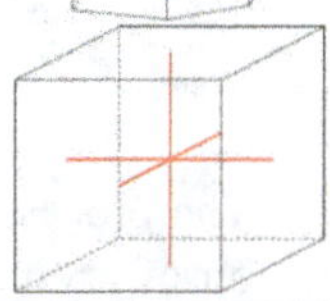

[a]The rhombohedral and hexagonal lattice systems share some symmetries. In fact, with a different choice of axes, it is possible to transform a rhombohedral system into a hexagonal system (Figure 2.29). The crystal system of symmetries common to the two lattice systems is called the *trigonal system.*

[b] Although the basic solid of the hexagonal system is a prism with a rhombic base, it is more straightforward to visualize the symmetry by considering a hexagon, which contains three basic prisms: two integers, and two in halves (Figure 2.30).

homologous charges are maximized. Therefore, it is not surprising that solidification occurs by the creation of regular structures that replicate the equilibrium configuration between adjacent atoms. The specific arrangement is determined by the shape of the outer electronic orbitals of the atoms in the metal or metal alloy.

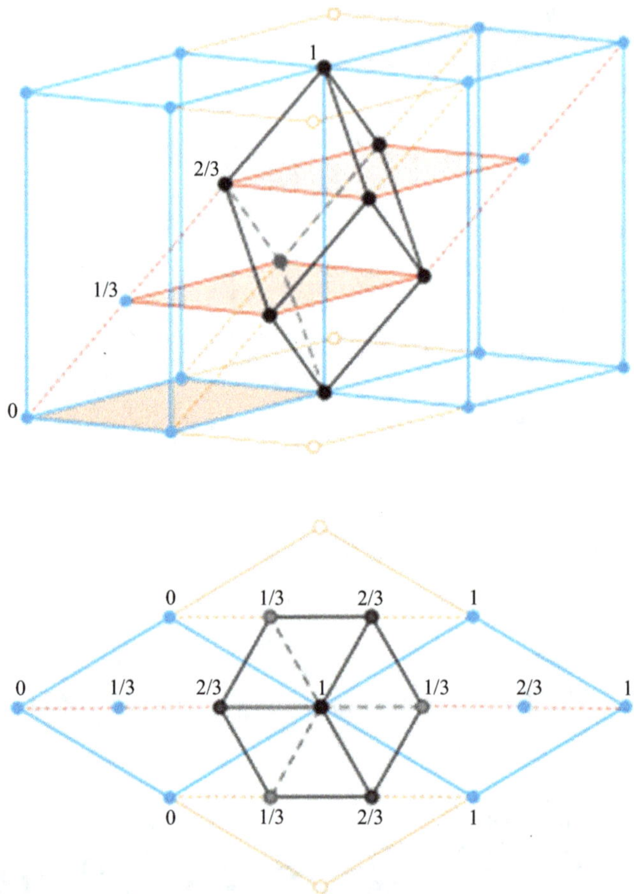

Figure 2.29 Relationship between hexagonal and rhombohedral systems

The process of crystal structure formation (Figure 2.33) begins with the formation of crystal germs (*nucleation*) at catalysts, such as impurities in the liquid or lower temperature parts. Next, the crystal grows to form a grain (*accretion*), which reaches up to the limit of the growth of adjacent grains, delineating a boundary edge.

Consequently, grain boundaries—which constitute a discontinuity in the crystal structure—must be reduced in order to achieve a more homogeneous structure and so a better crystal.

To promote the growth of larger grains, it is necessary to limit the presence of impurities and slowly cool the liquid. In fact, a rapid decrease in temperature, especially when combined with mechanical stresses (such as vibration or agitation of the liquid), promotes a more disordered solidification and, at the limit, the metallic solid may assume an amorphous configuration. At the opposite limit, in the case of slow temperature decrease and absence of liquid's motions, *undercooling*

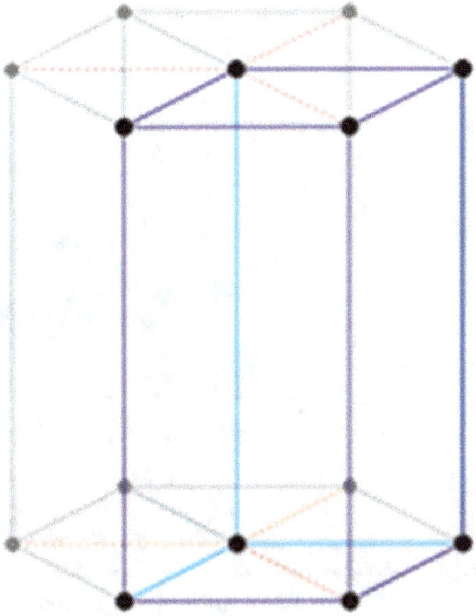

Figure 2.30 Primitive cell of a hexagonal system

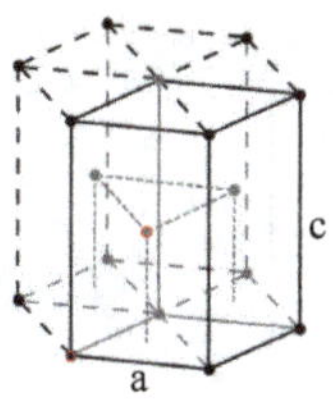

Figure 2.31 Close compact hexagonal crystal with base atoms highlighted in red

can occur, that is, a reduction of the liquid temperature below its solidification temperature without phase change. In this situation, a minimal perturbation of the system leads to solidification with rapid crystal growth, as this configuration is energetically favored.

The thermal gradient also influences crystal formation. In the case of a directional gradient (e.g., a top surface temperature higher than the bottom surface temperature), growth on planes perpendicular to the gradient, called *terraced* growth, is favored. In the case of a uniform gradient and slow decrease in temperature, crystallization occurs through the growth of *dendrites* (Figure 2.34), branched microcrystalline structures that extend in one main direction and then grow evenly in the other two directions until they fill the entire space.

For polymorphic materials, such as iron, crystallization occurs according to the most stable allotrope just below the solidification temperature. Further cooling

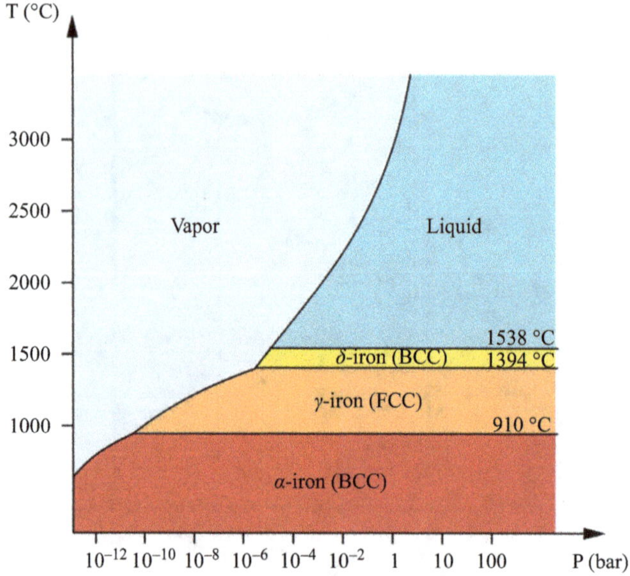

Figure 2.32 Phase diagram of pure iron

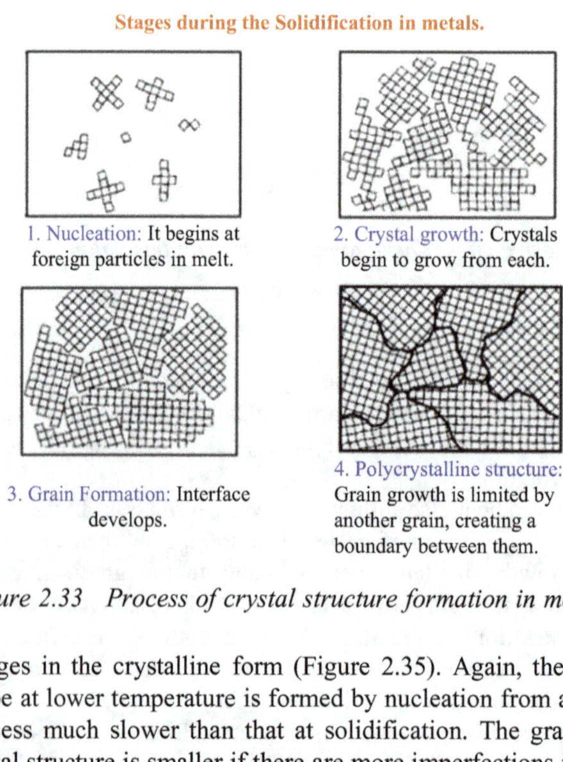

Figure 2.33 Process of crystal structure formation in metals

leads to changes in the crystalline form (Figure 2.35). Again, the crystal of the stable allotrope at lower temperature is formed by nucleation from a few accretion points, a process much slower than that at solidification. The granularity of the resulting crystal structure is smaller if there are more imperfections in the lattice of

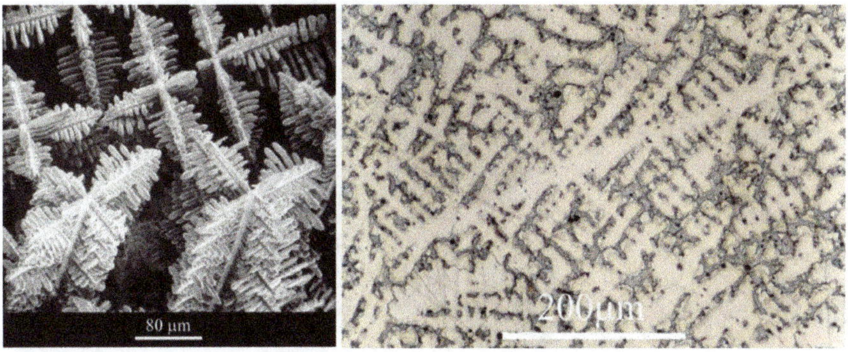

Figure 2.34 Micrograph of a growing dendrite (left) and a surface with signs of crystalline growth dendrites (right)

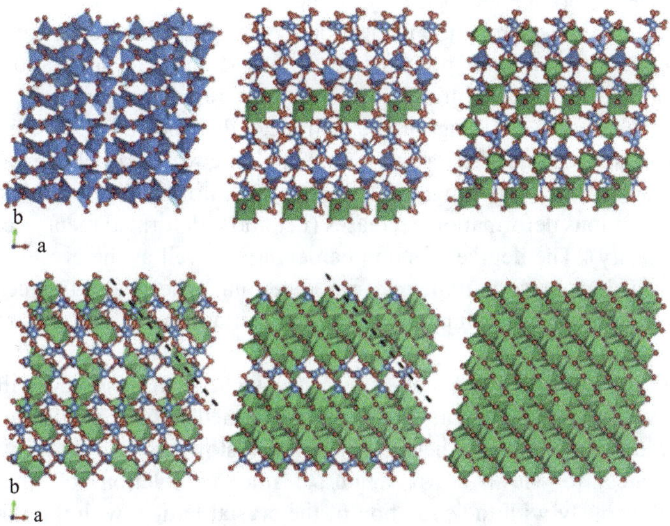

Figure 2.35 Simulation of a polymorphic transition

the crystal structure of the previous allotrope. Normally grain boundaries, inclusions, and internal stresses constitute preferential accretion points. The crystal structure that is formed thus bears the imprint of the previous structure.

To obtain a more homogeneous crystal structure in low-temperature allotropes, the process of *annealing* is employed, that is, heating to an appropriate temperature in the range of 1000°C–1500°C for steel, maintained for a sufficiently long time. At the microscopic level, three phenomena occur during annealing:

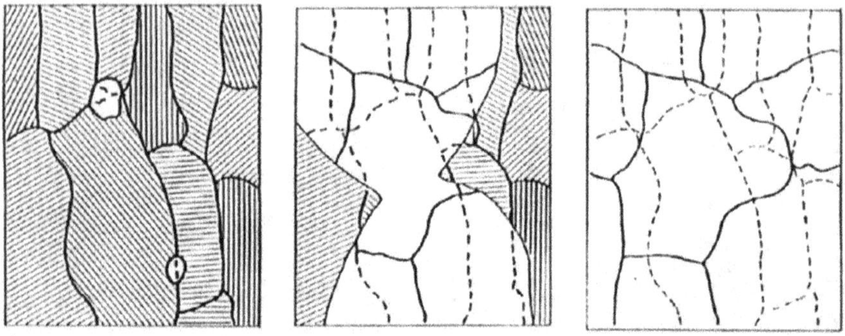

Figure 2.36 Schematic drawing of the recrystallization process

- *recovery*, which is the relaxation of crystalline deformations caused by internal material stresses created during cold processing such as *work hardening* or other metalworkings beyond its yield point;
- *recrystallization*, that is, the nucleation of new crystalline grains, which grow at the expense of those pre-existent (Figure 2.36). The nucleation proceeds until all the expanding grains, whose edges are stress-free, come in contact with each other, resulting in the complete disappearance of the deformed previous structure. The temperature to which the work-hardened material must be brought in order to allow the start of recrystallization increases as the grade of the previous deformation decreases (i.e., more deformed metal recrystallizes more easily). The degree of strain hardening, as well as the crystal lattice and any inclusions present in the material, determines the size of the recrystallized grain, which is also a function of the temperature at which the treatment is carried out;
- *grain growth*, which occurs after the initial incubation period described above, during which recrystallization nuclei are formed only at particular locations in the deformed grain. There is an initially slow growth phase, which then accelerates to slow down again at the completion of the process. Simultaneously with the evolution of the crystal lattice, with the formation of new stress-free grains there is a change in mechanical properties, which essentially results in progressive softening of the material.

When the growth phase is completed, the crystal structure of the metal has assumed a new arrangement, free of stresses associated with the previous deformation and with small grain sizes that are a function of: the rate of work hardening previously induced in the metallic material, the annealing temperature, and the chemical composition.

Further permanence at the annealing temperature results in an increase of the grains' size. During the recrystallization phase, the driving force for the process comes from the internal energy associated with the deformed state, while in the growth phase it depends on the interface energy existing at the edge of the grains.

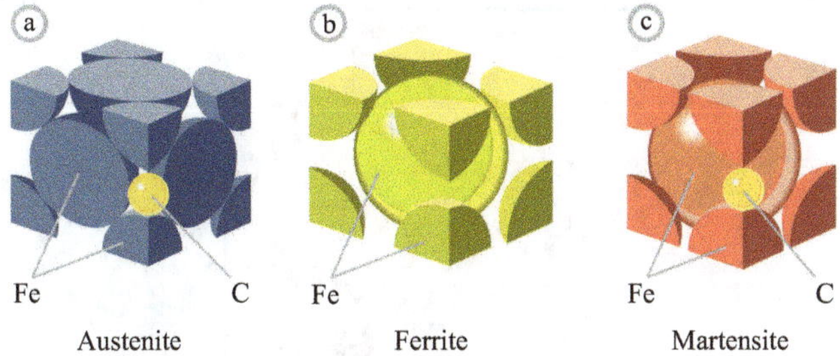

Figure 2.37 Crystal structures of steel compared with those of iron-α (ferrite)

Therefore, the final accretion rate of stress-free grains is much lower than the growth rate in the recrystallization phase.

To make iron mechanically stronger, carbon is added to the liquid phase; carbon atoms end up positioned into the gaps left by the crystalline structure of iron, making the crystal more robust (Figure 2.37). If the mass percentage of carbon is between 0.008% and 2.14%, the resulting alloy is called *steel*; if it is between 2.14% and 6.7%, it is called *cast iron*. Above that percentage, it is called *cementite*. For electrotechnical uses, only steel is of interest since other alloys are too brittle. The crystal structure of steel is *fcc* type for temperatures ranging from the melting temperature of 1495°C (with a carbon content of 0.17%) down to 723°C, and it is called *austenite*. For temperatures below 723°C, the crystal structure is *bcc* type and is called *martensite*.

However, since low electrical conductivity is desirable in steels for magnetic applications, silicon instead of carbon is added to iron for transformer cores, with contents up to 4% by mass, since higher contents would make the metal alloy too brittle. This makes possible to reduce conductivity by 75% as well as magnetostriction; conversely, the saturation induction is reduced from 2.15 to 2.03 T.

2.3.6 Magnetic domains

Every ferromagnetic material, below its Curie temperature, consists of a set of small-volume regions within which all magnetic dipole moments are aligned in the same direction, as shown in Figure 2.38. Such a region is called a *domain* and is magnetized up to its saturation. Adjacent domains are separated by edges called *Bloch walls*, in which the direction of magnetization varies gradually from that of one domain to that of the other (Figure 2.39, left).

Each crystalline grain may consist of more than one domain. Thus, a macroscopic volume of material may have a large number of domains, all with different magnetization orientations. The value of the $\boldsymbol{M}$ field for the entire solid is given by the vector sum of the magnetizations of all the domains, weighing the contribution of each domain with its volume fraction. For an unmagnetized material, the vector sum of the magnetization of all domains is zero.

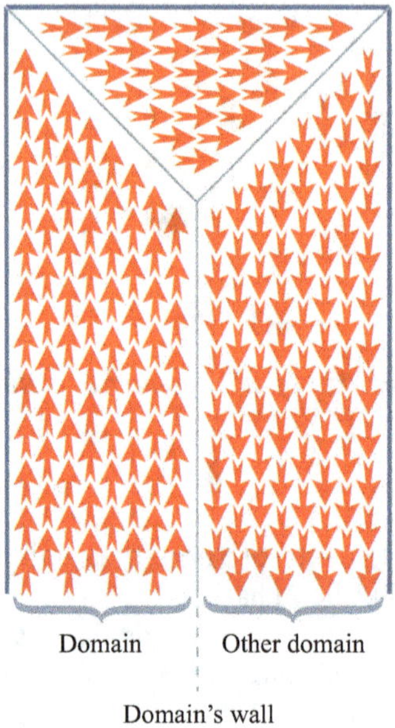

Figure 2.38 Magnetic domains

The explanation for the separation of each grain into more than one magnetic domain is given by energy considerations. In fact, the magnetic energy is given by the following formula [8] (M-12b)[13]:

$$\mathcal{E} = \frac{1}{2\mu}\int_{\mathcal{V}} B^2 dV. \tag{2.30}$$

Therefore, in order to minimize the magnetic energy of each grain, it is convenient for it to be divided into several portions with opposite magnetization (Figure 2.39, right), which compensate for each other. The presence of the triangular domains at the top and bottom edges, called *closure domains*, allows the surface currents (2.15) to be generated which compensate for the magnetic field at the edges.

Since a certain amount of energy is required to introduce a Bloch wall, an equilibrium position is reached when the energy required to form an extra wall is equal to the reduction in the energy of the external field.

Bloch walls are not fixed like grain separations but vary as a function of the magnetizing force applied (Figure 2.40). Specifically, the walls first deform to increase the surface area of the domains oriented toward the direction of the magnetizing force, and then the domains change the alignment of the magnetization

[13]This formula should not be confused with (2.6), which concerns the *work* done by the magnetic field for a change in the intensity of the magnetizing force. The (2.30), on the other hand, concerns the *energy* stored in space due to the presence of the magnetic field and is valid even when there is no change of the magnetizing force.

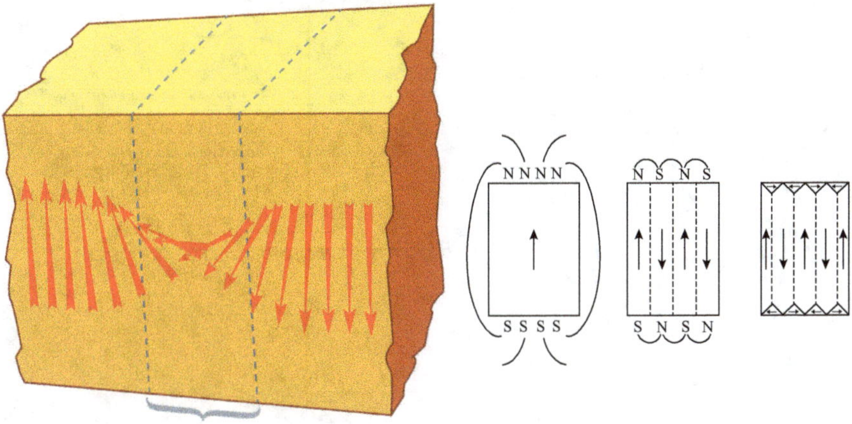

Figure 2.39 Domain separation wall (left) and wall formation (right)

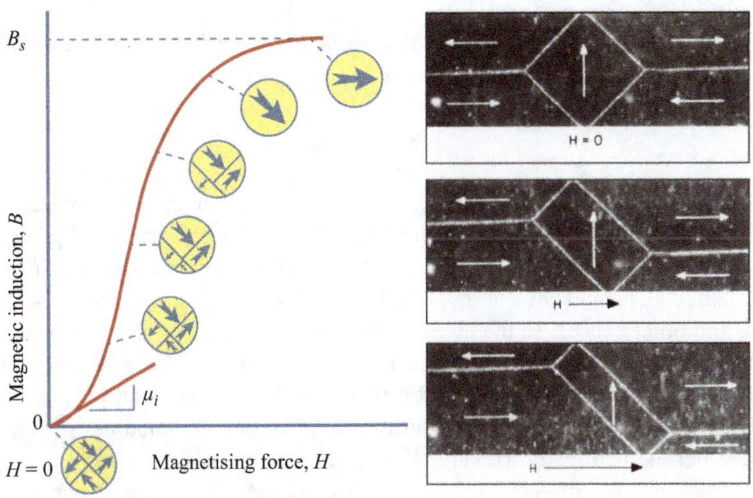

Figure 2.40 Movement of Bloch walls with increasing magnetizing force

direction to coincide with that of the magnetizing force. At saturation, each grain consists of a single domain that has magnetization facing the direction of the magnetizing force (Figure 2.41). At this point, the differential magnetic permeability, i.e., the derivative of the curve $B(H)$, has the same value as the magnetic permeability of vacuum, and the material has no longer ferromagnetic behavior.

The energy required for the creation or annihilation of Bloch walls is the main cause of anomalous losses ([12] Section 7.8.2), whereas simple wall motion is reversible and, therefore, does not create energy dissipation.

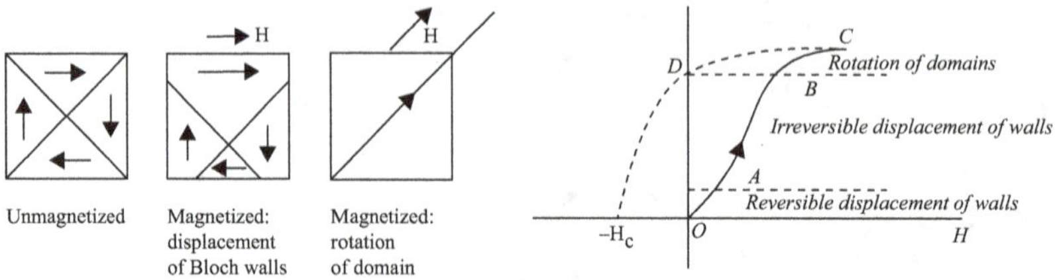

Figure 2.41 Bloch wall motion

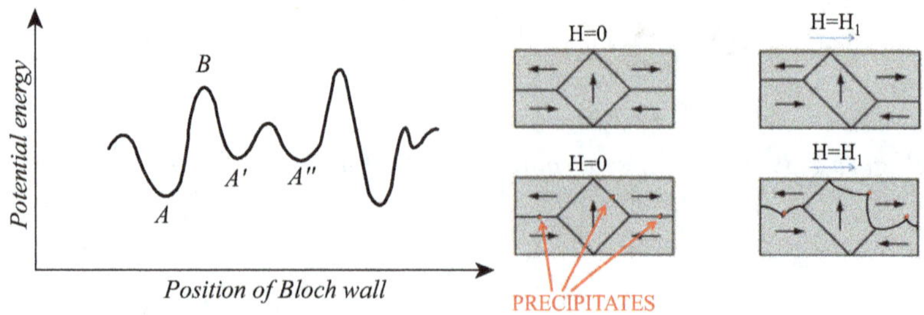

Figure 2.42 Potential energy associated with Bloch wall position and pattern of wall motions in the presence of impurities

Changes in magnetization can be reversible or irreversible in character. This phenomenon can be explained by considering the energy diagram in Figure 2.42, which shows the potential energy of a Bloch wall as a function of its position in the crystal lattice. Changes in energy result from local stresses, impurities, lattice defects, and other causes. In the absence of external magnetic field, the position of the wall will be so to minimize the energy, e.g., point A in the diagram. When a magnetic field is applied, magnetization is reversible until the wall position passes point B, after which there are new equilibrium positions (denoted as A′, A″, etc.), each of which is manifestly irreversible in character.

The presence of irreversible Bloch wall displacements is the explanation for the hysteresis curve and the corresponding losses.

However, it should be considered that magnetization does not occur in the same way in all the directions. In fact, the crystal structure endowes the material with preferential directions. The directions along which saturation is reached with a lower magnetizing force are called *easy directions*, those along which it is reached with a higher field are called *hard directions*. In Figure 2.43, it is observed that in iron, the easy direction is denoted by [1 0 0] (see Section A.3.4 for the definition of *Miller's indices*), while the hard directions are indicated with [1 1 0] and [1 1 1].

Therefore, it is convenient to obtain magnetic sheets that have the easy magnetization direction corresponding to the rolling direction, since they are invested by the magnetic flux in that direction. Magnetic steels rolled according to the easy

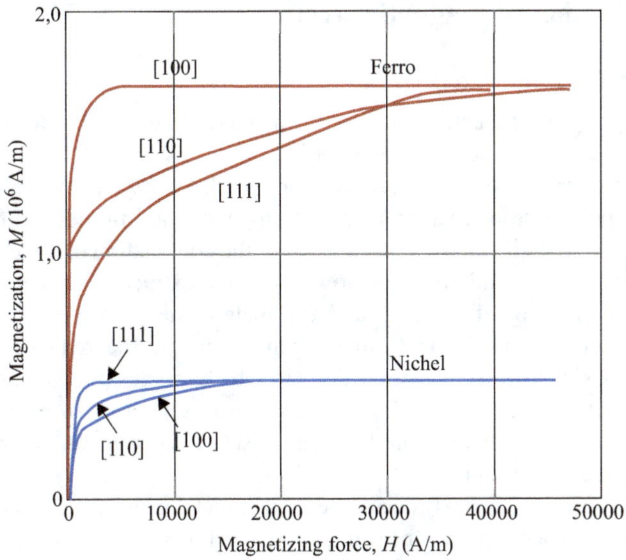

Figure 2.43 Magnetization along the crystalline directions of iron and nickel

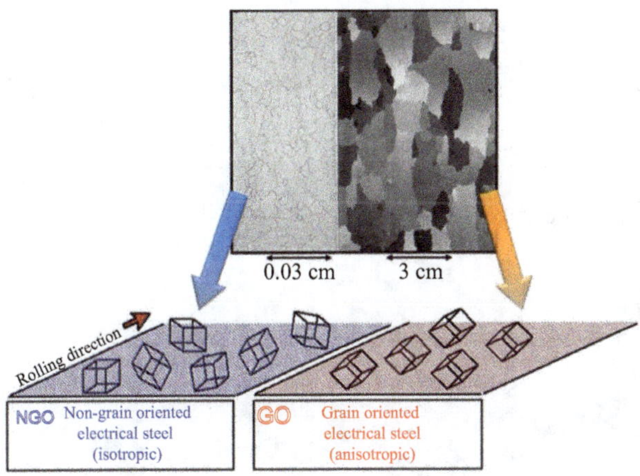

Figure 2.44 Grain-oriented steel sheet and Goss texture

magnetization direction are called *grain oriented* (*GO*) to distinguish them from *non-grain oriented* (*NGO*) steels.

Grain-oriented steels subjected to hot rolling, which allows grain alignment, are called *conventional grade* (*CGO*), and the grain configuration is called *cube texture*. Steels subjected to a further annealing process, which allows more grain growth and better alignment, are called *high permeability grade* (*HGO*), and the grain configuration is called *Goss texture*, as depicted in Figure 2.44.

2.4 Production of magnetic steel

2.4.1 Smelting and casting

The first step in the production process of magnetic steel sheets, made with iron–silicon alloy, is to melt the raw materials in a blast furnace.

The *blast furnace* is a shaft furnace, into which feedstock materials, including iron oxides and limestone that act as a cleaning agent, are introduced from above. Through nozzles hot air is blown, necessary for the combustion of coke coal used to heat the blast furnace, while from the lower part, the extraction of molten materials takes place, consisting of pig iron and the melt waste, called *slag*. In the blast furnace, the separation of iron from oxygen occurs by the reduction of carbon binding with oxygen to form carbon dioxide which, being gaseous, escapes from the mouth of the blast furnace. The molten iron is separated from the slag by a refining process and is then poured into a vessel called a *ladle*, where silicon is added in the required proportion.

In the *continuous casting* process, the molten metal is drawn through rollers and simultaneously cooled with jets of water spray until it is completely solidified. It is then cut into slabs of the desired size (Figure 2.45). The resulting product of solidification is a semi-finished product, on which *finishing* operations are performed.

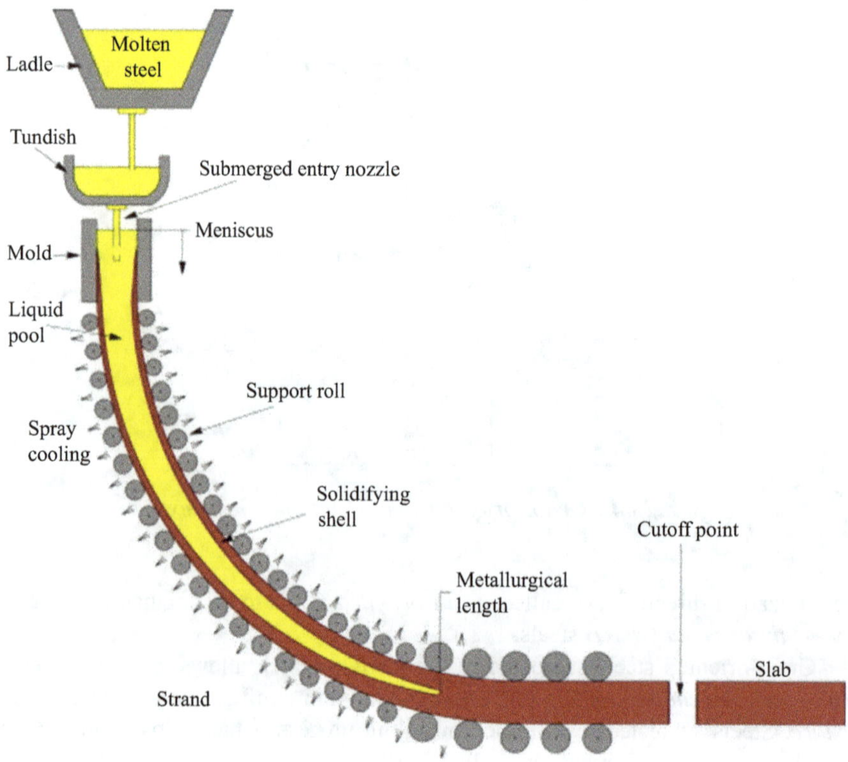

Figure 2.45 Continuous casting process

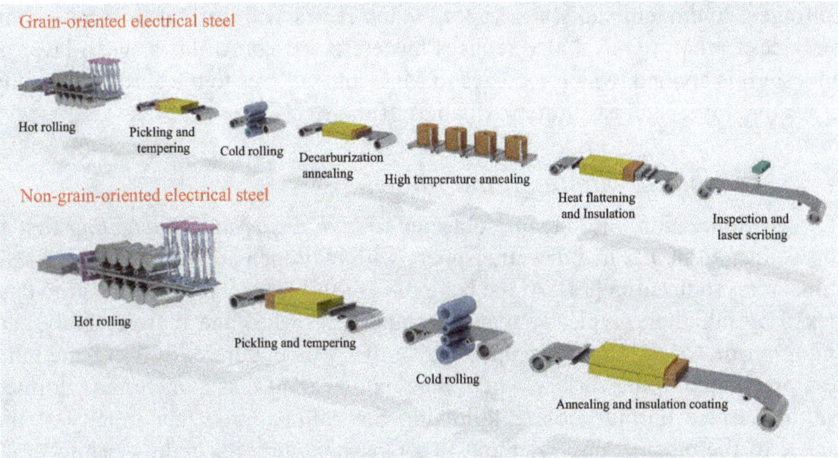

Figure 2.46 Diagram of the production steps of magnetic sheets

The production of magnetic steel is a complex process structured in several steps described in the following paragraphs and briefly illustrated in Figure 2.46.

2.4.2 Hot rolling

Hot rolling follows solidification and occurs in heated furnaces, where the steel sheet is pressed by rollers at austenite-forming temperatures (~1200°C), reducing the thickness from 250 to ~2.3 mm. Hot rolling allows the austenite to crystallize with crystals ~100 μm in size in the rolling direction, which are partially preserved in the subsequent cooling and transition into the martensite allotrope. After the hot rolling process, the steel shrinks slightly as it cools, making shape control less precise and resulting with a rough surface, rounded edges, and small distortions.

2.4.3 Tempering and pickling

The cooled sheet after hot rolling is subjected to a *tempering* process, that is, slow heating to about 700°C, so as to reduce the internal stresses in the material created during cooling shrinkage. At this stage, the grain size grows up to 5 mm.

In addition, the surface of the sheet is treated with a *pickling* process, which aims to remove the oxide scales that form on an artifact when it undergoes heating at a high temperature in the presence of atmospheric oxygen. The pickling techniques can be mechanical, e.g., sandblasting with silica sand, or chemical, with baths in acidic solutions that remove oxides.

2.4.4 Cold rolling and decarburization

After the tempering and pickling operations, the sheet undergoes further *cold rolling* to reduce the thickness to ~0.30 mm, followed by a *decarburization* process, which removes surface carbon impurities by heat treatment in an atmosphere of nitrogen, hydrogen, and water vapor, which reacts with carbon to form carbon monoxide. Carbon removal also reduces hysteresis losses. At this stage, a layer of magnesium is applied, which oxidizes to MgO and adheres to the surface, forming a protective layer against further oxidation of the steel.

2.4.5 Annealing, painting, and flattening

After decarburization, the sheet is subjected to *high temperature annealing* first at 1200°C, then at 800°C in a dry atmosphere with hydrogen and nitrogen to achieve optimal recrystallization [13]. At the beginning of this stage, normal grain growth is blocked by inhibitors, typically manganese sulfide (MnS) and more recently aluminum nitride (AlN). However, upon reaching a temperature of 1000°C, the inhibitors precipitate to the surface where they oxidize, and thus some grains, formed near the surface during cooling following hot rolling, grow abnormally at the expense of the other grains, resulting in a Goss texture with grain sizes up to 20 mm, although this results in an angle 3°–7° of mismatch between the rolling and texturing directions.

Next, a layer of protective insulating varnish is applied. In fact, the magnesium oxide layer already applied and supplemented by inhibitor precipitation is too abrasive to be left exposed on the surface. Various types of coatings can be applied, and many solutions are patented by manufacturers. The most used coatings are enamels, phosphate-based varnishes, and ceramic fillers used to increase mechanical strength.

Finally, the sheet is further passed between rollers to flatten it uniformly (*flattening*) so to reduce its size to ~0.23 mm.

2.4.6 Laser scribing

Since the 1960s, several *domain refinement* techniques have been developed, employed only for high-performance products.

This step, which occurs after flattening, involves making microscopic incisions on the surface of the sheet. In 1982, the laser engraving technique, called *laser scribing*, was developed [14]. Other engraving techniques, such as chemical *etching* or mechanical engraving, are now considered outdated. Because the intensity of the laser beam can be calibrated to achieve microfusion of the metal layer underneath the surface without affecting the coating, the technique provides very precise and minimally invasive results that do not alter the surface treatments of the sheets. However, sometimes an additional layer of paint is applied at the end of the treatment.

The incisions are made perpendicular to the rolling direction (Figure 2.47) and have the effect of decreasing the length of the domains and increasing the resistance along them; since the anomalous losses depend nonlinearly on the length of the Bloch walls, this reduces them by 10% and more.

The major disadvantage of sheets subjected to domain refining is that the effects of the treatment are lost if the material is brought to high temperatures, as grain relaxation can occur with the loss of the refined domains. Therefore, the

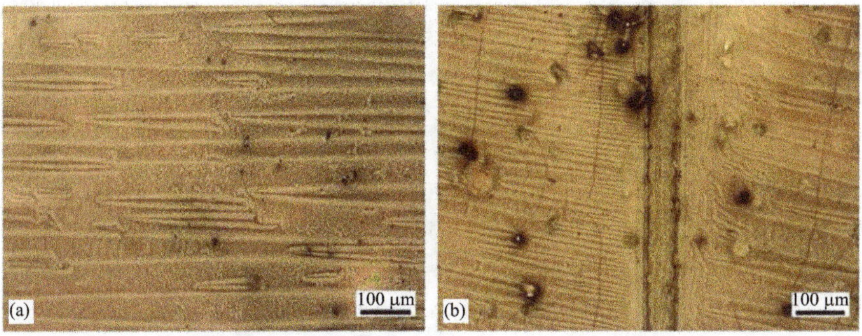

Figure 2.47 Sheet metal surface before (a) and after (b) laser scribing treatment

performance of laser-scribed sheets should be verified after the heat treatment of the active part of the transformer.

Since the Bloch walls of the domains are parallel to the rolling direction, this process does not affect their movement during the magnetization of the material, so hysteresis losses and saturation point are not affected.

On the contrary, the process results in the formation of additional closure domains at the incisions, which are a source of magnetostriction and thus of additional noise [15]. However, by properly calibrating the power and collimation of the laser beam and the application time, uniform incisions can be created that achieve optimal magnetostriction performance without sacrificing loss reduction (Figure 2.48).

2.5 Technological developments

2.5.1 State of the art

The production process of both conventional and high-permeability oriental crystal magnetic steel has achieved sufficient standardization so that the quality of the product can be regulated by international standards.

The standard IEC 60404-8-7 [16] specifies the different characteristics of oriented crystal sheets and establishes a designation convention, through an acronym consisting of:

- the letter M, for magnetic steel;
- the specific losses at 1.7 T of magnetic induction with frequency 50 Hz, in W/kg × 100;
- a dash followed by the nominal thickness in mm × 100;
- the letter S for CGO, P for HGO;
- the frequency in Hz/10, e.g., 5 for 50 Hz.

Table 2.3 shows the characteristics of conventional and high-permeability steels with the lowest losses among those in the standard.

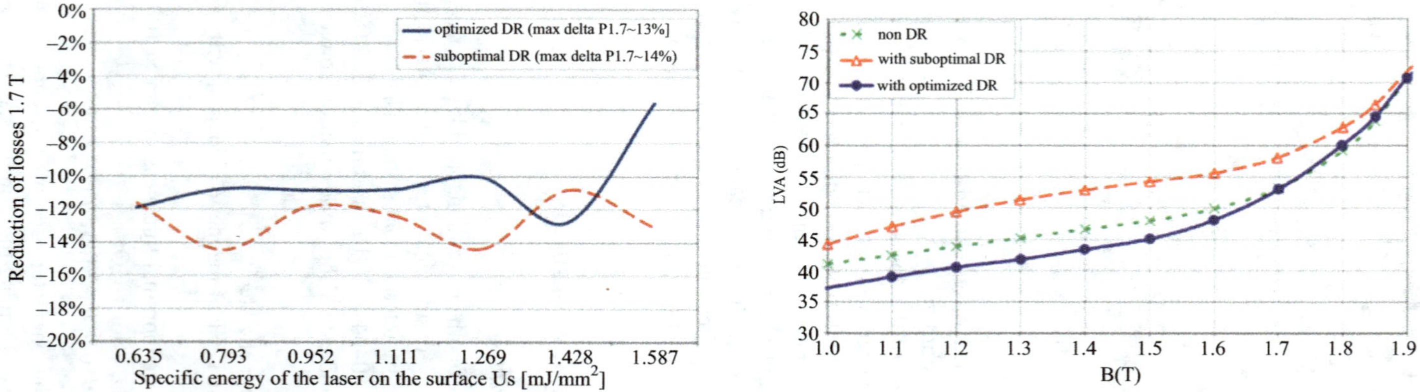

Figure 2.48 *Loss reduction and noise with optimized and suboptimal domain refining (DR) technique*

Table 2.3 Characteristics of the best GO steels according to IEC 60404-8-7

Designation	**Nominal thickness (mm)**	**Maximum losses at 1.5 T 50 Hz**	**Maximum losses at 1.7 T 50 Hz (W/kg)**	**$\mu_0 M$ minimum at H=800 A/m (T)**
M110-23S5	0.23	0.73 W/kg	1.10	1.78
M85-23P5	0.23	Nonsignificant	0.85	1.85

Table 2.4 Characteristics of some top-quality GO steels on the market

Manufacturer	**Commercial designation**	**Nominal thickness (mm)**	**Maximum losses at 1.7 T 50 Hz (W/kg)**	**$\mu_0 M$ minimum at H=800 A/m (T)**
ThyssenKrupp	PowerCore® H 075–23 L	0.23	0.75	1.88
AK Steel	TRAN-COR® H–OCAR-LITE DR	0.23	0.80	1.88
JFE	G-CORE 23JGSD075	0.23	0.75	1.87
Nippon steel	Orientcore HI-B™ 23ZDKH75	0.23	0.75	1.88
POSCO	PHD-Core 23PHD080	0.23	0.80	1.88

The standard specifies that for HGO steels with lowest losses, domain refining treatment is allowed.

However, there are already better qualities of sheets on the market than those indicated in the standard. Table 2.4 lists the characteristics of low-loss sheets taken from the catalogs of some of the major manufacturers.

Because the cutting operations required to produce the steel strips used in the Epstein frame introduce significant deformation at the edges, resulting in deterioration of grain characteristics, the stripes are subjected to annealing to restore optimal grain conformation. Because this operation would eliminate the refining of domains in the laser-scribed sheets, these can be measured exclusively by the SST, which—since it allows a sheet with larger surface area—makes less influential the presence of edge deformations not corrected by annealing.

However, it has been found that specific loss measurements made according to the two normative methods of Epstein frame and SST do not give perfectly comparable results. Therefore, a normative update was undertaken to clarify what is the official conversion factor between the two methods; previously, there was only a nonbinding conversion criterion described in Annex C of IEC 60404-3 [4]. In general, it appears that the Epstein frame measures lower losses than the SST.

In addition, the update of IEC 60404-8-7 [17] introduced sheets with a thickness of 0.20 mm and maximum specific losses of 0.70 W/kg at 1.7 T 50 Hz.

Figure 2.49 depicts the performance evolution of silicon steel sheets.

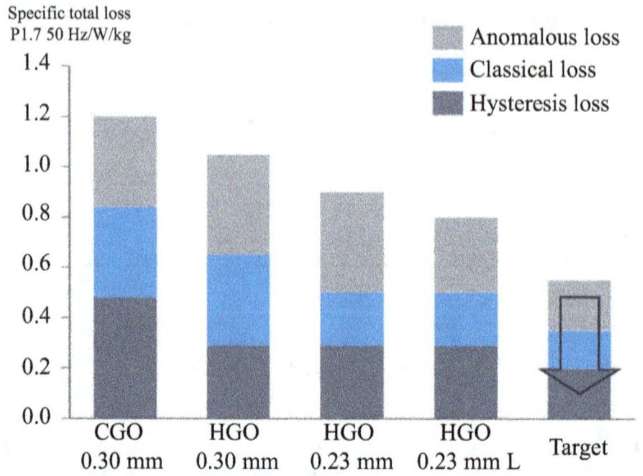

Figure 2.49 Composition of losses in silicon steel sheets

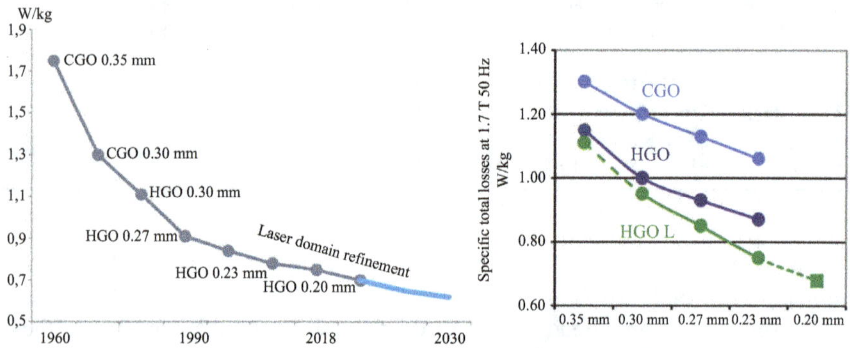

Figure 2.50 Perspectives on the evolution of specific losses of grain oriented sheets

2.5.2 High permeability steels

A regulatory proposal circulated by the European Commission [18] to update the Ecodesign Regulation on transformers calls for requirements both on overall transformer efficiency and on specific losses of the sheets to be used, at least in some cases. A similar approach has been taken by other regulators (see, e.g., "2016 U.S. Dept. of Energy efficiency standard").

Therefore, there has been intense technological research for crystal-oriented sheets with low losses, especially through an optimization of the laser scribing technique. For example, Figure 2.50 shows two graphs prepared by ThyssenKrupp regarding the prospects for further reduction of guaranteed specific losses on sheets with a thickness of 0.20 mm. It is expected that sheets with a thickness of 0.18 mm will eventually be commercialized.

The production capacity of very thin sheets is limited by the precision of cold rolling, as even small inaccuracies result in production rejection. In addition, the

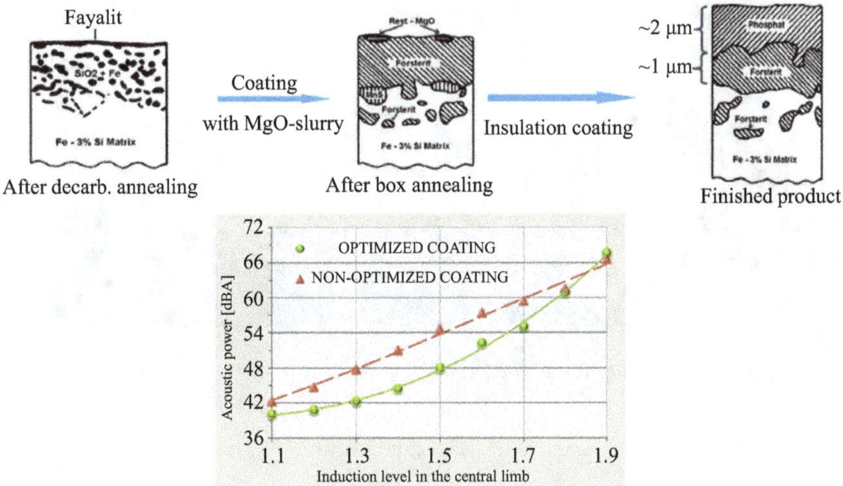

Figure 2.51 Optimized coating process and acoustic power comparison

use of thin sheets in the construction of cores implies increased attention, as cutting operations increase losses at the edges of the sheets. These problems, together with the criticality of domain refining treatment, cause higher costs and the risk that the technological advantage of thin sheets is lost during the transformer's core construction and treatment operations.

In addition, optimized protective coating techniques have been developed to achieve greater noise reduction (Figure 2.51).

2.5.3 Amorphous steels

The search for minimal loss steels has led to the production of sheets with very low thickness (~25 μm) and amorphous structure (like glass). The amorphous structure does not have crystals with defined edges; rather, the atoms are randomly arranged in space, and therefore the material does not possess a preferred direction but is isotropic.

Amorphous steel is made by a completely different process from that used for crystalline steel (Figure 2.52). A material that promotes glass solidification such as boron (B) is added to the ladle, and the molten mixture is sprayed onto a roller with high rotational speed, partially immersed in a cooling bath that allows cooling rates of 10^{6}°C/s, thus preventing the formation of crystalline grains. This produces strips with thicknesses of ~40 μm, which can be further hot rolled.

However, such rapid cooling introduces internal stresses into the material, which must then be subjected to annealing at 200°C–280°C.

Amorphous steel strips, given their small thickness and higher resistivity, have very low eddy current losses (<10 mW/kg compared with 200 mW/kg in the best HGO steels), so hysteresis losses make up about 75% of the total losses.

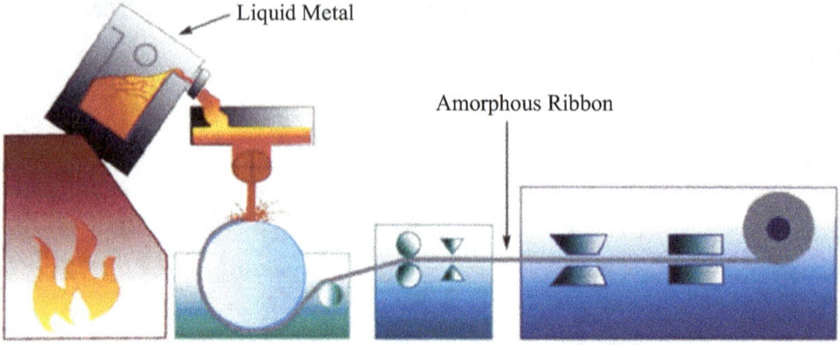

Figure 2.52 Diagram of the production process of amorphous steel strip

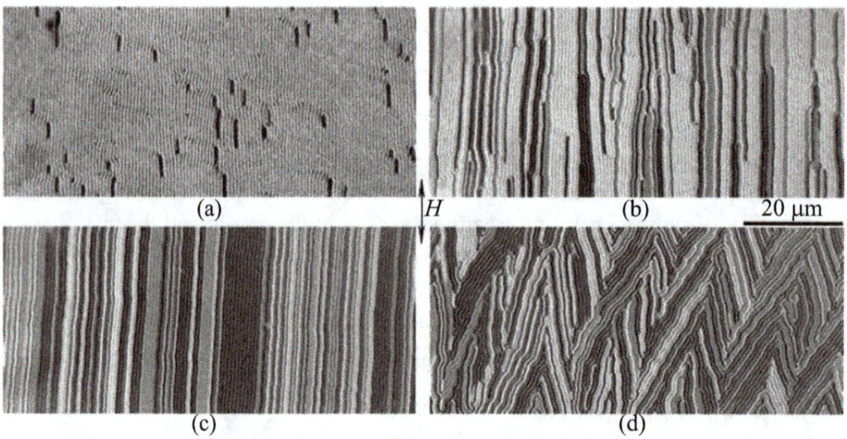

Figure 2.53 Stripe magnetic domains

Although amorphous steel has no defined grains, the arrangement of magnetic moments takes on a regular structure, forming *stripe magnetic domains*, arranged according to the rolling plane. In Figure 2.53 are depicted the domains under recombination for the application of a parallel magnetizing force (a, b), the arrangement in the absence of magnetizing force (c) and the rotation of the domains in the presence of a transverse magnetizing force (d).

The preferential direction of the stripe domains creates the so-called *rotatable anisotropy*, whereby the hysteresis curve depends on the direction of the applied magnetization force relative to the direction of the residual magnetization. The hysteresis loop is obviously larger if the direction of the new magnetization is different from that of the old magnetization. However, this is not a problem for transformers, where the direction of the magnetizing force is constant, at least under normal operating conditions.

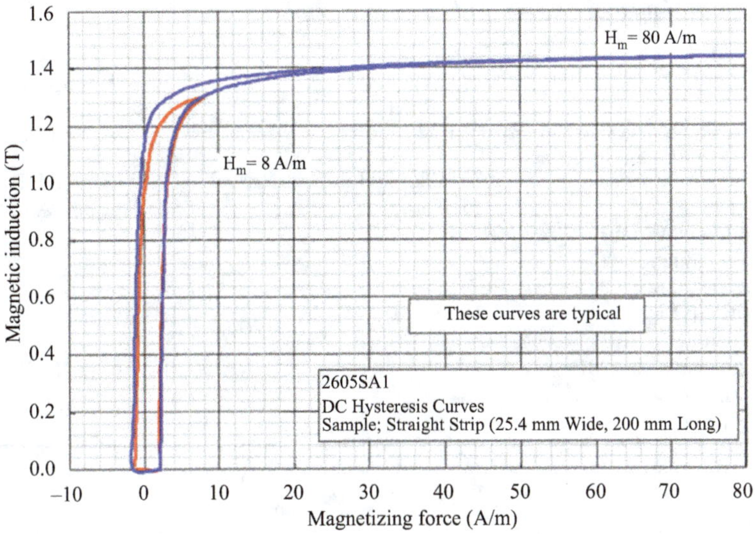

Figure 2.54 Magnetization curve of an amorphous steel

The apparent magnetic permeability of amorphous steels is higher than that of crystalline steels, due to the higher internal stresses and due to the surface deformation of bond energies (*surface pinning*), which is very pronounced given the reduced thickness. Thus, the typical magnetizing force is 80 A/m, instead of 800 A/m as in crystalline steels. In addition, the presence of vitrifying material reduces the saturation induction to ~1.5 T (Figure 2.54).

The low Curie temperature of amorphous steels, <400°C, compared to 745°C of crystalline steels, reduces the maximum use temperature of amorphous steels to ~120°C.

The amorphous structure makes the material harder but also more brittle, considering its thin thickness. Therefore, cutting operations must be carried out with special care.

Table 2.5 compares the salient features of a typical amorphous steel with an HGO-L type crystalline steel.

Amorphous steel has a strong dependence of losses on compressive stress (Figure 2.55). This implies that in a large core, which is held together by sheaths and clamps with significant compressions, the losses of amorphous steel can increase by as much as 50% above the nominal value. In addition, amorphous steel is produced in strips of limited width, making its use in large power transformers difficult.

Because of the abovementioned reasons, amorphous steel is more suitable to be wrapped rather than packed in parallel layers (Figure 2.56). This has limited its use so far to small distribution transformers ($S_r \leq 2500$ kVA).

Amorphous steel has a strong tendency to magnetostriction. For comparison, while the strain constant $\lambda = \Delta L/L$ (L being the length in the magnetization direction) is about -0.20 ppm at 1.5 T in crystalline steels, it is about -27 ppm for amorphous steels. As a result, the vibrations and resulting noise are much greater.

Table 2.5 Comparison of amorphous and crystalline steel

Material	Metglas® 2605HB1M	ThyssenKrupp H 075–23 L
Thickness (mm)	0.025	0.23
Losses	0.20 W/kg at 1.4 T 60 Hz	0.73 W/kg at 1.5 T 60 Hz
Magnetization	1.35 T at 80 A/m	1.88 T at 800 A/m
Maximum width (mm)	213.4	1000
Saturation induction (T)	1.63	2.03
Curie temperature (°C)	364	745
Resistivity (μΩ m)	1.3	0.48
Vickers hardness	900	175–195
Breaking tension R_m (MPa)	2100	330–370
Density (kg/dm^3)	7.33	7.65
Packing factor (%)	84	95.5

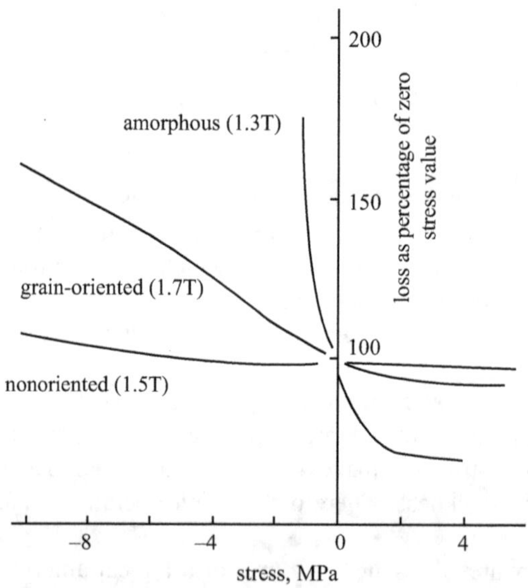

Figure 2.55 Dependence of losses on compression

Finally, the processing of amorphous steel for the construction of magnetic cores requires special care, lest the advantages of its use be nullified.

From the economic point of view, amorphous steel has a higher cost than crystalline HGO steel, which has to be compared with the savings for loss reduction.

2.5.4 Alternative magnetic materials

Silicon steel, crystalline or amorphous, satisfies most of the needs in magnetic applications with soft materials, that is, materials that do not require permanent

Figure 2.56 Wrapped core with amorphous steel

magnetization. However, there are some ferrous alloys that are used in special applications.

For example, iron powder cores are made by pressing a mixture of iron powder and epoxy resin and then baking the mixture at the curing temperature of the resin. This produces a monolithic block of iron particles isolated from each other, thus minimizing eddy current losses. In contrast, the particles are nonoriented crystals.

This technology is applicable only to very small cores suitable for measurement transformers. Such cores have highly variable differential permeability so, to achieve more stable behavior and to reduce losses, powders with up to 81% nickel content are used. This stabilizes the permeability but reduces saturation induction.

There are also iron/nickel alloys, known as *Permalloy,* and iron/cobalt alloys, which are produced in continuous loops wound over a mandrel. These materials can achieve very high differential permeabilities, in the presence of nickel, or high saturation inductions, in the presence of cobalt, but the manufacturing process limits their use to toroidal or otherwise wound cores. These materials are mainly used in high-frequency applications or where a specific magnetization characteristic is needed.

Figure 2.57 classifies the magnetic materials according to saturation induction and relative magnetic permeability.

2.6 Conclusions and suggestions

At present, the soft magnetic material with the largest market share remains silicon steel (Figure 2.58), followed by cobalt or nickel alloys, especially for measurement or high-frequency applications.

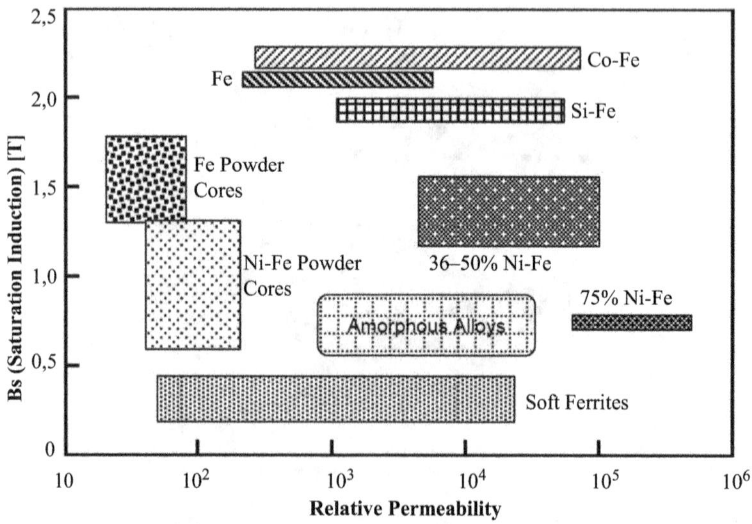

Figure 2.57 Characteristics of different magnetic materials

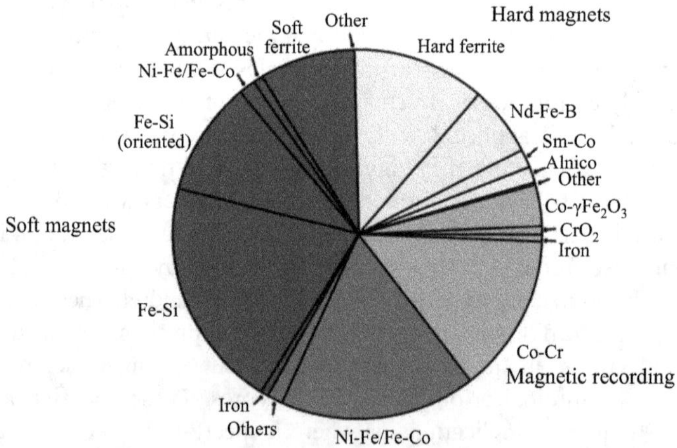

Figure 2.58 Market shares of magnetic materials

Amorphous steel is little used, both because of its high cost and the inherent disadvantages of the technology. Nonetheless, technological research in this field continues, along with the development of even better performing steels consisting of nanocrystals with very low hysteresis losses (Figures 2.59 and 2.60). However, microcrystalline steels are limited in size and therefore currently used in high-frequency applications in electronics. Foreseeably in the near-term future, steels with refined domains will continue to be the magnetic material for large powers with higher energy efficiency performance.

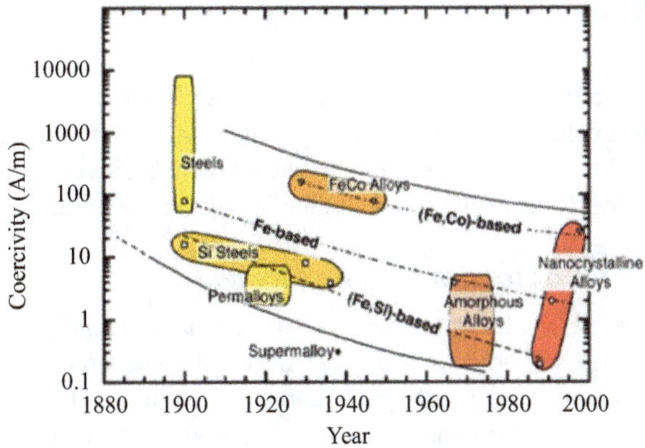

Figure 2.59 Evolution of soft magnetic materials

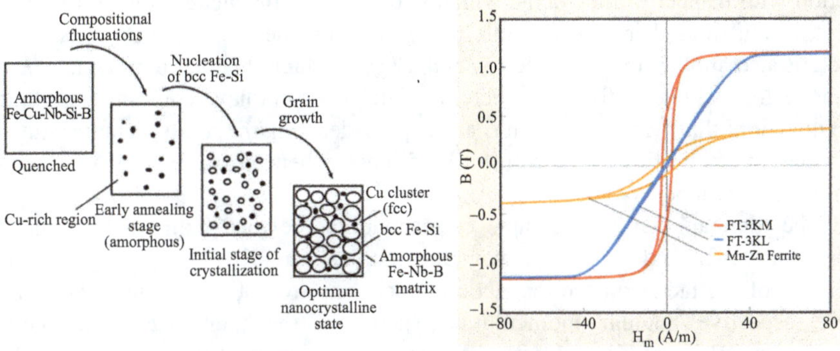

Figure 2.60 Manufacturing process of nanocrystals (left) and comparison of hysteresis curves (right)

To guarantee the quality of the material used in future power transformers, the generic power utility should prescribe the maximum specific loss value for the steel sheets used, for example, limiting the specific losses to 0.80 W/kg. In this way, all manufacturers would have to meet the same contractual conditions, and the guaranteed value of no-load loss could be more precisely defined. Otherwise, the choice is left to the generic manufacturer, who may use poor-quality steel sheets at the expense of transformer size while still meeting the no-load loss requirements.

The innovative materials previously described (alloys, amorphous steels, and nanocrystalline materials) might be a viable alternative in the future, if advances in the manufacturing process can overcome the drawbacks that currently discourage their use in large power transformers. Since research and development in the field are very active and could be further stimulated by regulatory measures, power utilities need to keep monitoring the development of this technology.

A Appendices

Appendices contain explanatory and theoretical insights of the topics discussed in the chapter. The explanations, especially the underlying mathematical theory, are by no means to be considered complete but provide an overview that helps to understand the physical reasons for the behavior of the magnetic materials.

A.1 Calculation of total magnetic momentum

According to Russell–Saunders coupling (also called *L-S coupling*), the azimuthal angular momentum $\vec{l}$ and spin $\vec{s}$ of each electron add up separately, resulting in the total azimuthal angular momentum $\vec{L}$ and total spin $\vec{S}$. These total angular momentums add up to form the total overall angular momentum $\vec{J}$. Both total angular momentums precede with the same velocity with respect to the axis of $\vec{J}$, so the modulus and direction of the latter remain constant, and it is possible to define an azimuthal quantum number J. The vector $\vec{J}$ in turn makes a precessional motion with respect to the z-axis, which is the axis of magnetic induction (where present), and it is therefore possible to define a magnetic quantum number M_J. There is also another model, called *j-j coupling*, in which the total momentums $\vec{j}$ of each electron add up to form the vector $\vec{J}$ which does not have constant amplitude and direction; this model, in which $\vec{L}$ and $\vec{S}$ precede along the z-axis, is better suited to strong magnetic fields and will not be discussed henceforth. Figure A.1 exemplifies the two models.

The total angular momentums correspond to the total azimuthal magnetic momentum $\mathfrak{M}_L$ and total spin magnetic momentum $\mathfrak{M}_S$; however, given the presence of the factor g_s, the magnetic momentums are not in the same ratio with their respective angular momentums. Therefore, the total overall magnetic momentum $\mathfrak{M}$ is not aligned with the axis of $\vec{J}$; however, it precedes around the same axis of $\vec{J}$, since it is the sum of magnetic momentums preceding around that axis. Consequently, it is meaningful to treat its projection $\mathfrak{M}_J$ as the sum of the projections of the azimuthal and spin magnetic momentums:

$$\mathfrak{M}_J = \mathfrak{M}\cos\left(\mathfrak{M}, \vec{J}\right) = \mathfrak{M}_L\cos\left(\vec{L}, \vec{J}\right) + \mathfrak{M}_S\cos\left(\vec{S}, \vec{J}\right). \tag{A.1}$$

Taking advantage of Carnot's theorem, we can write

$$\left|\vec{L}\right|^2 = \left|\vec{J}\right|^2 + \left|\vec{S}\right|^2 - 2\left|\vec{S}\right|\left|\vec{J}\right|\cos\left(\vec{S}, \vec{J}\right), \tag{A.2}$$

$$\left|\vec{S}\right|^2 = \left|\vec{J}\right|^2 + \left|\vec{L}\right|^2 - 2\left|\vec{L}\right|\left|\vec{J}\right|\cos\left(\vec{L}, \vec{J}\right), \tag{A.3}$$

from which

$$\mathfrak{M}_J = -\left|\vec{L}\right|\frac{\left|\vec{J}\right|^2 + \left|\vec{L}\right|^2 - \left|\vec{S}\right|^2}{2\left|\vec{L}\right|\left|\vec{J}\right|}\mu_B - g_s\left|\vec{S}\right|\frac{\left|\vec{J}\right|^2 + \left|\vec{S}\right|^2 - \left|\vec{L}\right|^2}{2\left|\vec{S}\right|\left|\vec{J}\right|}\mu_B. \tag{A.4}$$

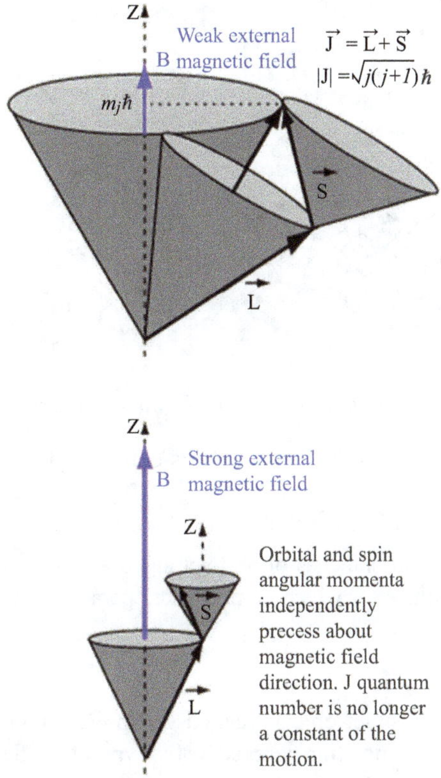

Figure A.1 Precessions of electron angular momentums

If we define the Landé factor g_J with the relation:

$$\mathfrak{M}_J = -g_J \left| \vec{\boldsymbol{J}} \right| \mu_B, \tag{A.5}$$

and since (2.21) holds for all total angular momentums, it results:

$$\begin{aligned} \mathfrak{M}_J = & -\frac{J(J+1) - S(S+1) + L(L+1)}{2J(J+1)}\mu_B \\ & - 2\frac{J(J+1) + S(S+1) - L(L+1)}{2J(J+1)}\mu_B, \end{aligned} \tag{A.6}$$

and collecting the terms it gives (2.28).

The same Landé factor can be used to define the value of the projection of the total overall magnetic momentum along the z-axis:

$$\mathfrak{M}_{J,z} = \mathfrak{M}_J \cos\left(\vec{\boldsymbol{J}}, \boldsymbol{B}\right) = -g_J \left| \vec{\boldsymbol{J}} \right| \cos\left(\vec{\boldsymbol{J}}, \boldsymbol{B}\right) \mu_B = -g_J M_J \mu_B. \tag{A.7}$$

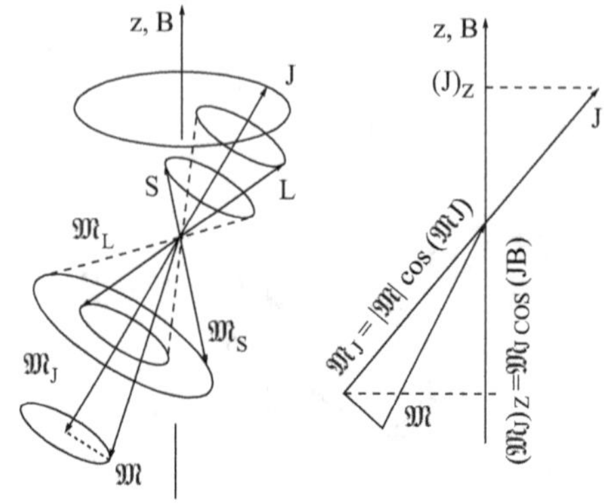

Figure A.2 Composition and projections of the atomic magnetic momentums

Figure A.2 represents the different vectors considered in the calculation.

A.2 Groups of point symmetry

Group theory is one of the branches of abstract algebra, the discipline that deals with nonnumeric abstract entities for which number-like operations and relationships can be defined.

A.2.1 Definitions

A *group* $\mathfrak{G}$ is a set of abstract objects (called group *elements*) between which it is defined the multiplication operation denoted by the symbol $\circ$ (often omitted) such that:

1. multiplication between two elements results in a group element:

$$g_i, g_j \in \mathfrak{G} \Rightarrow g_i \circ g_j = g_k \in \mathfrak{G}; \qquad \text{(A.8)}$$

2. multiplication enjoys the associative property:

$$g_i \circ (g_j \circ g_k) = (g_i \circ g_j) \circ g_k; \qquad \text{(A.9)}$$

3. there exists a neutral element g_1, called *identity*, such that:

$$g_1 \circ g_i = g_i \circ g_1 = g_i \quad \forall g_i; \qquad \text{(A.10)}$$

4. for each element g_i, there exists an inverse g_i^{-1}, such that[14]:

$$g_i \circ g_i^{-1} = g_i^{-1} \circ g_i = g_1. \qquad \text{(A.11)}$$

[14]The fact that the inversion operation is involutive, viz., $\left(g_i^{-1}\right)^{-1} = g_i$, and the commutativity of the identity are actually demonstrable given the definitions $g_i \circ g_1 = g_i$ and $g_i \circ g_i^{-1} = g_1$. In fact, multiplying to the left by g_i^{-1} the second relation:

$$g_i^{-1} \circ \left(g_i \circ g_i^{-1}\right) = g_i^{-1} \circ g_1 = g_i^{-1} \Rightarrow \left(g_i^{-1} \circ g_i\right) \circ g_i^{-1} = g_i^{-1} \Rightarrow g_i^{-1} \circ g_i = g_1,$$

and using the relationship found:

$$g_1 \circ g_i = \left(g_i \circ g_i^{-1}\right) \circ g_i = g_i \circ \left(g_i^{-1} \circ g_i\right) = g_i \circ g_1 = g_i \Rightarrow g_1 \circ g_i = g_i$$

If multiplication also enjoys the commutative property, the group is called *abelian*; otherwise, it is called *non-abelian*.

Groups are distinguished into *finite* and *continuous*. Finite groups have a fixed number of elements, called the group *order*, which is denoted as dim$\mathfrak{G}$. The elements of continuous groups depend on one or more parameters given by real numbers, and the number of such parameters is called the group order.

For symmetry operations on crystals, it is sufficient to deal with finite groups. Therefore, we will be referring exclusively to them henceforth.

Two groups having the same order are said to be *isomorphic* if it is possible to associate biunivocally each element of one group with one of the other group, so that each operation performed on the elements of one group gives the same result if performed on the corresponding elements of the other group. If the correspondence is not biunivocal, but each element in one group can be matched by more than one in the other group while maintaining the correlation between operations, the two groups are said to be *homomorphic*; in this case, one of the two groups is larger than the other and has an order that is a multiple of the order of the smaller group.

A *subgroup* is a subset of a group such that its elements form a group of their own. Evidently each group contains several subgroups (e.g., identity, an element and its inverse form a subgroup), most of which are of no specific interest. Subgroups that give additional information about the groups are the *invariant subgroups* (also called *normal subgroups*), whose elements are permutable with any other element in the group. An element of a subgroup $\gamma_j \in \Gamma$ is said to be *permutable* with an element g_i if the following relation holds:

$$g_i^{-1} \circ \gamma_j \circ g_i = \gamma_k \in \Gamma \tag{A.12}$$

The two elements γ_j, γ_k are said to be *conjugate* to each other. If this relationship is valid $\forall g_i \in \mathfrak{G}$, the subgroup $\Gamma \subset \mathfrak{G}$ is invariant.

An invariant subgroup may also contain smaller subgroups, so it is meaningful to consider only *maximum invariant subgroups* that are not contained in any other invariant subgroup.

A group is said to be *simple* if it contains no invariant subgroups, otherwise, it is said to be *non-simple*.

The *coset* Γg_x of a group with respect to a subgroup is a set of elements of the group in which each element of the coset is obtained by multiplying an element of the subgroup by a fixed element $g_x \notin \Gamma$ of the group: $\gamma_k g_x$. The subgroup itself Γ forms a coset; other cosets are obtained by multiplying on the right (*right cosets*) or on the left (*left cosets*) an element of the subgroup by an element of the group not included in the subgroup; they are not subgroups. Two right (or left) cosets are either identical or disjoint. If $\Gamma, \Gamma g_2, \Gamma g_3, \ldots, \Gamma g_q$ are disjoint cosets such that their union is equal to the group $\mathfrak{G}$, then they form the complete system of right cosets of $\mathfrak{G}$ with respect to Γ, in number q; same holds with the left cosets. This also shows that the order of a subgroup is a submultiple of the order of the group; the ratio between the two orders is called the subgroup *index*.

The *complementary* or *quotient group* of a subgroup Γ with respect to the group $\mathfrak{G}$ is the group $\mathcal{H} = \mathfrak{G}/\Gamma$ of the cosets of $\mathfrak{G}$ that form a complete system with respect to the Γ subgroup, i.e., $\Gamma, \Gamma g_2, \Gamma g_3, \ldots, \Gamma g_q \in \mathcal{H}$ and $q = \dim\mathcal{H}$ (similarly considering left cosets). The complementary group is in correspondence with the total group: if the conjugate subgroups $\Gamma, g_2^{-1}\Gamma g_2, \ldots, g_q^{-1}\Gamma g_q$ have no element in common except the identity, the correspondence is isomorphic ($q = \dim\mathfrak{G}{=}m$); otherwise, they coincide and the correspondence is homomorphic. Specifically, if Γ is non-invariant, $\mathcal{H}$ is isomorphic with $\mathfrak{G}$; if Γ is invariant, $\mathcal{H}$ is homomorphic with $\mathfrak{G}$ and the order of $\mathcal{H}$ is the ratio of the order of $\mathfrak{G}$ to that of Γ: $\dim\mathcal{H} = \dim\mathfrak{G}/\dim\Gamma$. Thus, it makes sense to speak of a complementary group with respect to an invariant subgroup, and in this case the right cosets coincide with the left cosets, while in case of a non-invariant subgroup it essentially corresponds to the original group.

The *conjugacy class*, or simply class, of an element of a group is the set of all elements conjugated with the given element, including the element itself:

$$\mathcal{C}(a) = \{g^{-1}ag : g{\in}\mathfrak{G}\} \tag{A.13}$$

Identity forms a class by itself. The other elements form one or more classes that have no elements in common (and do not form groups).

The *composition series* is the sequence of maximum invariant subgroups such that $\mathfrak{G}{\supset}\mathfrak{G}_1 \supset \ldots \supset \mathfrak{G}_{\mu-1} \supset g_1$. Called m_i the orders of the subgroups, the indices of the subgroups are $e_i = m_{i-1}/m_i$ and are called *composition factors* since the order of the group is given by their product: $m = e_1{\cdot}e_2{\cdot}\ldots{\cdot}\,e_\mu$. The composition series allows the structure of a non-simple group to be broken down into successively simpler subgroups, until we arrive at a subgroup $\mathfrak{G}_{\mu-1}$ that is simple and therefore has no invariant subgroup other than the identity.

Given the composition series, an element of a non-simple group can be expressed as the product of an element of each subgroup: $g_i = \gamma_1 \ldots \gamma_\mu$ (the order does not matter because the elements of invariant subgroups commute with each other). The minimum number of elements that multiply with each other to obtain any other element of the group is called the set of group *generators*. A non-simple group is thus obtained by the *semidirect product* of simple groups:

$$\mathfrak{G}{=}\mathfrak{G}_1 \otimes \mathfrak{G}_2 \ldots \otimes \mathfrak{G}_\mu.$$

From the above, it is clear that the study of simple groups provides insight into the behavior of all non-simple groups, that is, all groups that contain invariant subgroups and can therefore be split into a composition series.

As for the symmetries of solids, they are contained in the following classes:

E: identical transformation;

r_n: rotation of angle 360°/*n* around an axis of rotation represented vertically in drawings of solids; if the rotation is 180°, it is also called tilting;

σ_h: reflection with respect to a horizontal plane, called the equator, thus perpendicular to the axis of rotation, if any;

σ_v: reflection with respect to a vertical plane, thus coplanar with respect to the axis of rotation, if any;

I: inversion with respect to the center of inversion, which is the point of intersection between the axis of rotation, if any, and the equator; the inversion can be considered the composition of a tilt and a reflection with respect to the equator.

A.2.2 Character representations and tables

A finite group can be represented by square matrices with complex elements such that each element corresponds to a matrix:

$$A \mapsto \mathbf{R}(A) : A \circ B = C \Rightarrow \mathbf{R}(A)\mathbf{R}(B) = \mathbf{R}(C) \tag{A.14}$$

The set of matrices that correspond to the elements of a group is called the group *representation*. If each matrix corresponds to only one element, the representation is *faithful*; otherwise, it is *unfaithful*. Since matrices can be transformed according to a similarity matrix $\mathbf{T}: \mathbf{R}(A) \mapsto \mathbf{T}^{-1}\mathbf{R}(A)\mathbf{T}$, there are an infinite number of representations equivalent to a given representation.

The *orthogonality lemma* ensures that any representation can be transformed into a representation with unitary matrices, i.e., such that $\mathbf{R}(A)^{-1} = \mathbf{R}(A)^{\dagger}\ \forall A$.

In general, there can be representations of any dimension, including 1—i.e., scalars—which will generally not be a faithful representation, but if it is possible by a transformation to block diagonalize all the matrices of a representation, it is said to be *reducible*, because each block constitutes a separate representation:

$$\mathbf{R}(A) = \begin{bmatrix} \mathbf{R}^{(1)}(A) & \mathbf{0} \\ \mathbf{0} & \mathbf{R}^{(2)}(A) \end{bmatrix}. \tag{A.15}$$

If, on the other hand, a representation cannot be further simplified, it is called *irreducible*. The study of irreducible representations is sufficient to establish the general behavior of other representations. A reducible representation that is composed of two irreducible representations is denoted as $\mathbf{R} = \mathbf{R}^{(1)} \oplus \mathbf{R}^{(2)}$.

The *great orthogonality theorem* [19] ensures that given two irreducible representations $\mathbf{R}^{(\mathrm{i})}, \mathbf{R}^{(\mathrm{j})}$ of dimension l_i, l_j and a group of h elements, the following relation holds:

$$\sum_{A \in \mathfrak{G}} \mathrm{R}^{(\mathrm{i})*}(A)_{\mu\nu} \mathrm{R}^{(\mathrm{j})}(A)_{\alpha\beta} = \frac{h}{l_i} \delta_{ij} \delta_{\mu\alpha} \delta_{\nu\alpha}, \tag{A.16}$$

where the summation is extended to all elements A of the group, represented by the element of the matrix $\mathbf{R}^{(\mathrm{i})}(A)$ in the same position $\mu\nu$ and by the element of the matrix $\mathbf{R}^{(\mathrm{j})}(A)$ in the same position $\alpha\beta$. Therefore, not only all irreducible representations are orthogonal to each other but also are the individual elements of the matrices of different representations.

Given that each irreducible representation has l_i^2 elements that constitute the components of a vector space of dimension h, the *dimensionality theorem* can be proved:

$$\sum_i l_i^2 = h, \tag{A.17}$$

where the summation is extended to all irreducible representations. This places a stringent constraint on the number and size of irreducible representations of a group.

The *multiplication table* is a table that lists all the products between pairs of elements in a group, completely defining the structure of the group. For example, in a group of order 6, we can have the following multiplication table:

∘	E	A	B	C	D	F
E	E	A	B	C	D	F
A	A	E	D	F	B	C
B	B	F	E	D	C	A
C	C	D	F	E	A	B
D	D	C	A	B	F	E
F	F	B	C	A	E	D

The inverse elements are immediately recognized by the fact that the cell corresponding to the intersection of their row and column (or vice versa) shows the identity element *E*; thus, $A^{-1} = A, B^{-1} = B, C^{-1} = C, D^{-1} = F$.

By rearranging the rows in the table so that the inverse elements of those in the corresponding columns appear, we obtain:

∘	E	A	B	C	D	F
E^{-1}	E	A	B	C	D	F
A^{-1}	A	E	D	F	B	C
B^{-1}	B	F	E	D	C	A
C^{-1}	C	D	F	E	A	B
D^{-1}	F	B	C	A	E	D
F^{-1}	D	C	A	B	F	E

This arrangement is particularly convenient because it allows to define the *regular representation* of the group. This representation is of size *h*, and therefore obviously reducible, and consists of matrices whose elements are 1 in correspondence to where the group element considered appears in the multiplication table, and 0 otherwise. For example, in the regular representation, the matrix corresponding to the element *A* is as follows:

$$\mathbf{R}^{\text{reg}}(A) = \begin{bmatrix} 0 & 1 & 0 & 0 & 0 & 0 \\ 1 & 0 & 0 & 0 & 0 & 0 \\ 0 & 0 & 0 & 0 & 0 & 1 \\ 0 & 0 & 0 & 0 & 1 & 0 \\ 0 & 0 & 0 & 1 & 0 & 0 \\ 0 & 0 & 1 & 0 & 0 & 0 \end{bmatrix}$$

The *celebrated theorem* ensures that the regular representation contains all the irreducible representations, each one a number of times equal to its size.

Normally, it is not necessary to explicitly write the matrices of a representation to study the behavior of group elements. Since there are infinite equivalent representations, it is useful to consider the trace, which is an invariant quantity for transformations of matrices:

$$\chi^{(j)}(A) = \text{tr}\mathbf{R}^{(j)}(A) = \sum_{\mu=1}^{l_j} \mathrm{R}^{(j)}(A)_{\mu\mu}. \tag{A.18}$$

Since elements of the same class are connected by a similarity transformation according to (A.13), they have identical trace in a given representation. Said N_k the number of elements of the class C_k, a consequence of the grand orthogonality theorem is that:

$$\sum_k \chi^{(i)*}(C_k)\chi^{(j)}(C_k)N_k = h\delta_{ij}. \tag{A.19}$$

Comparing (A.19) with (A.16) we conclude that the number of classes must be equal to the number of irreducible representations. Furthermore, summing overall irreducible representations yields another orthogonality relation:

$$\sum_i \chi^{(i)*}(C_k)\chi^{(i)}(C_l) = \frac{h}{N_k}\delta_{kl}. \tag{A.20}$$

We can then construct *character tables* that have as rows the irreducible representations and as columns the classes and contain the respective characters.

In the example considered, the classes consist of $C_1 = E$; $C_2 = A, B, C$; $C_3 = D, F$ and there are two irreducible representations of dimension 1: $R^{(1)}, R^{(2)}$ and one of dimensions 2: $R^{(3)}$; therefore, we can construct the following character table:

	C_1	$3C_2$	$2C_3$
$R^{(1)}$	1	1	1
$R^{(2)}$	1	-1	1
$R^{(3)}$	2	0	-1

In the table, the code of the class is preceded by the number of its components, if different from 1. It can be seen that the rows and columns meet the conditions of orthogonality (A.19) and (A.20).

The list of character tables of point symmetry groups is readily available on the Internet. The following site also lists all isomorphisms among the different groups: https://www.staff.ncl.ac.uk/j.p.goss/symmetry/Isomorphism.html

A.2.3 List of point symmetry groups

Listed in the following are all the point symmetry groups that can give rise to a Bravais lattice, that is, a regular structure covering the entire three-dimensional space.

Groups C_n (Table A.1)

- are formed by an axis of rotation of order n; solids can be rotated by an angle of 360° k/n around the axis ($0 \le k \le n-1$);
- consist of n elements.

Groups C_{nh} (Table A.2)

- consist of an axis of rotation of order n and a plane of reflection perpendicular to the axis (the part above the equator of the solid is equal to the part below the equator); solids can be rotated by an angle of 360°·k/n around the axis or reflected with respect to the plane;

Table A.1 Groups C_n

Symbol of Schoenflies	Symmetry operations	Solid	Stereogram
C_1	No symmetry	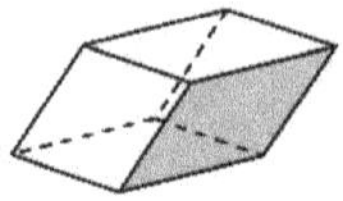	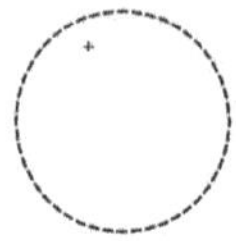
C_2	The solid can rotate 180° around the axis	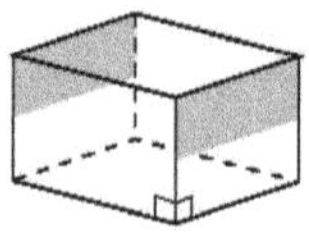	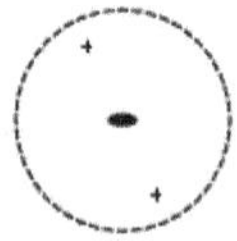
C_3	The solid can rotate 120° and 240° around the axis	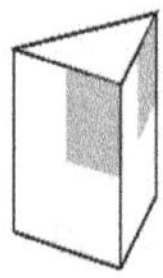	
C_4	The solid can rotate 90°, 180°, and 270° around the axis	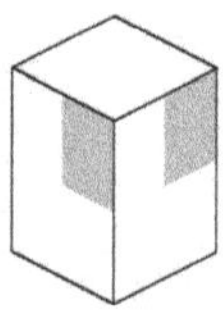	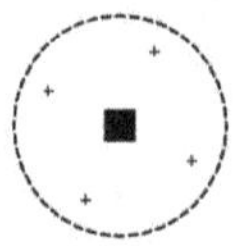
C_6	The solid can rotate by 60°, 120°, 180°, 240°, and 300° around the axis	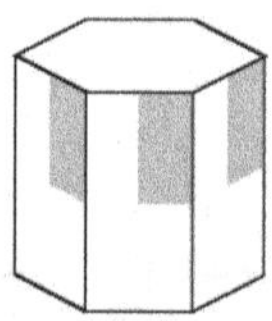	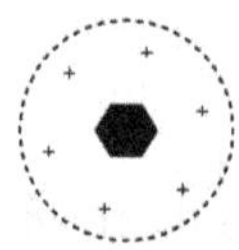

- are obtained from the product of group C_n with group C_s=C_{1h} (formed by E and σ_h); groups C_{2nh} can also be obtained from the product of C_n with C_i (formed by E and I);
- consist of 2*n* elements.

Groups C_{nv} (Table A.3)

- consist of an axis of order *n* and *n* reflection planes[15] passing through the axis; the solids can rotate by an angle of 360°·*k*/*n* around the axis or rotate and reflect with respect to any of the planes because the solid is symmetrical with respect to each of them;

[15]A single plane of reflection is replicated *n* times by the axis of rotation.

Table A.2 Groups C_{nh}

Symbol of Schoenflies	Symmetry operations	Solid	Stereogram
$C_s = C_{1h}$	You can reflect the solid with respect to the plane	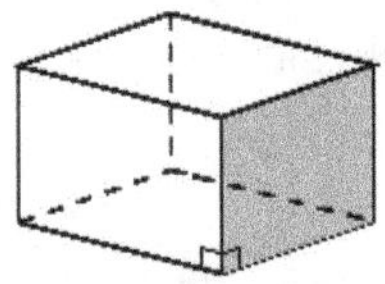	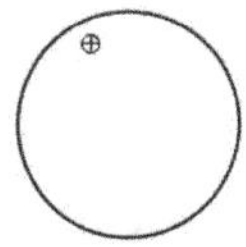
C_{2h}	The solid can rotate 180° around the axis or reflect it with respect to the plane	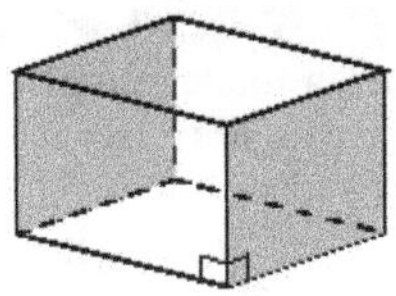	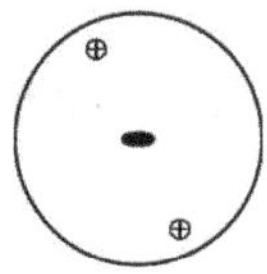
C_{3h}	The solid can rotate 120°, 240° around the axis or reflect it with respect to the plane	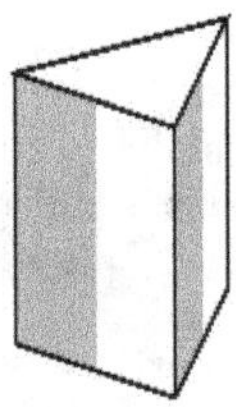	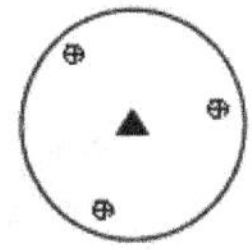
C_{4h}	The solid can rotate 90°, 180°, and 270° around the axis or reflect it with respect to the plane	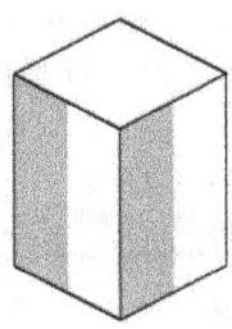	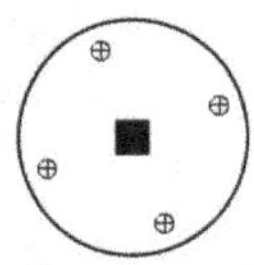
C_{6h}	The solid can rotate by 60°, 120°, 180°, and 240° around the axis or reflect it with respect to the plane	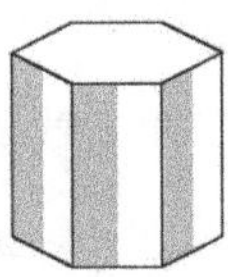	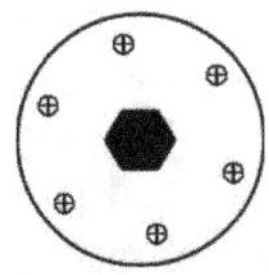

- are obtained from the product of the group C_n with the group C_s (= C_{1h} = C_{1v}), because if there is only one plane of symmetry, it is irrelevant whether it is perpendicular or coplanar to the axis, since there is no axis of symmetry;
- consist of $2n$ elements.

Groups S_{2n} (Table A.4)

are formed by a rotoreflexion (rotation + reflexion) axis of order $2n$; the solids can rotate by an angle of $360°\text{-}k/2n$ around the axis and necessarily reflect k times with respect to the plane of symmetry perpendicular with respect to the axis (so it only flips if k is odd, because an even number of flips equals identity);

Table A.3 Groups C_{nv}

Symbol of Schoenflies	Symmetry operations	Solid	Stereogram
C_{2v}	The solid can rotate 180° around the axis or reflect it with respect to a plane	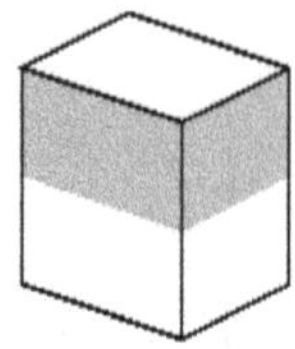	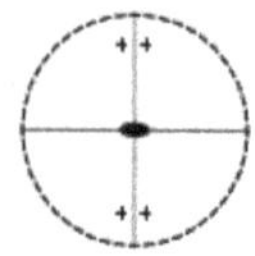
C_{3v}	The solid can rotate 120° and 240° around the axis or reflect it with respect to a plane	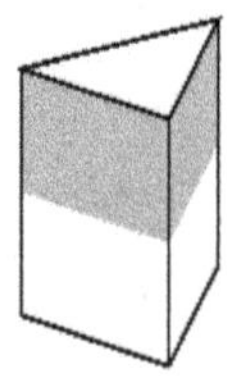	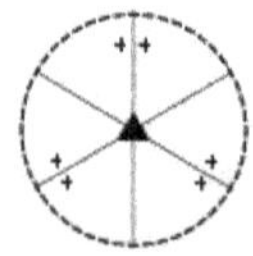
C_{4v}	The solid can rotate 90°, 180°, and 270° around the axis or reflect it with respect to a plane	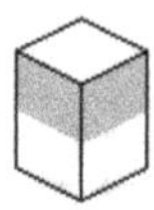	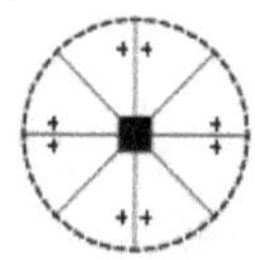
C_{6v}	The solid can rotate by 60°, 120°, 180°, 240°, and 300° around the axis or reflect it with respect to a plane	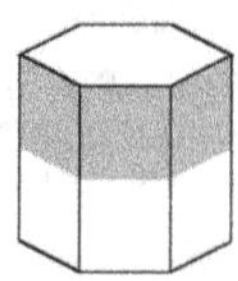	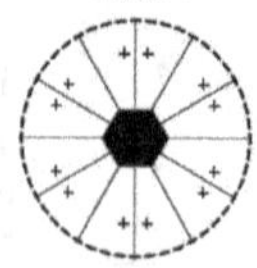

- are obtained from the product of group C_n with group $C_i = S_2$ (formed by E and I);
- consist of 2*n* elements;
- groups S_{2n+1} would be equivalent to $C_{2n+1\ h}$ groups and therefore are not considered;

  Groups D_n (Table A.5)

are formed by an axis of rotation of order *n* and *n* axes of rotation of order 2 (*n*/2 if *n* is even >2) in the plane perpendicular to the first axis[16]; the solid can rotate 360°·*k*/*n* with respect to the axis of order *n* or flip it (= rotate it 180°) with respect to any of the axes of order 2 because the solid is symmetrical with respect to each of them;

- consist of 2*n* elements.

[16] A single axis of order 2 perpendicular to the axis of order *n* is replicated *n* times by the latter.

Table A.4 S Groups$_{2n}$

Symbol of Schoenflies	Symmetry operations	Solid	Stereogram
$C_i = S_2$	The solid can be inverted with respect to its center	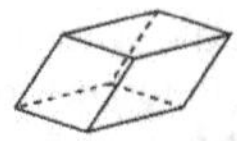	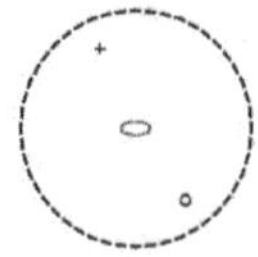
S_4	The solid can rotate 90° and 270° and then reflect it or rotate it 180°	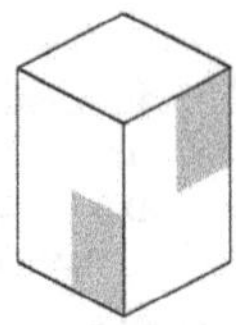	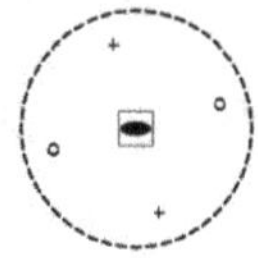
S_6	The solid can rotate by 60°, 180°, and 300° and then reflect it or rotate it by 120° and 240°	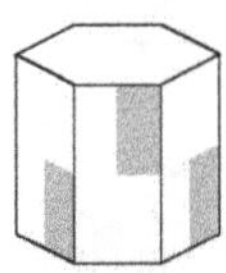	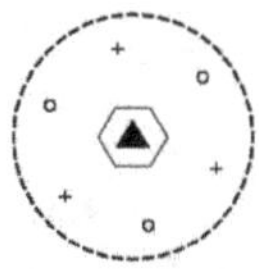

Groups D_{nh} (Table A.6)

are formed by an axis of rotation of order *n* and *n* axes of rotation of order 2 (*n*/2 if *n* is even >2) in the plane perpendicular to the first axis, which is also a plane of reflection (equator); one can rotate the solid 360°·*k/n with* respect to the axis of order *n or* tilt it with respect to one of the axes of order 2 or reflect it with respect to the plane;

- are obtained from the product of group D_n with group $C_{1h}=C_s$;
- consist of 4*n* elements.

Table A.5 Groups D_n

Symbol of Schoenflies	Symmetry operations	Solid	Stereogram
D_2	The solid can be flipped with respect to each of the three axes of order 2	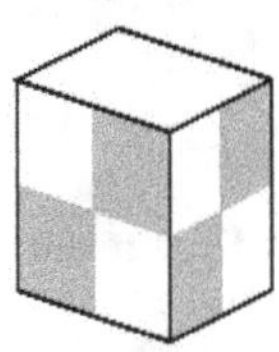	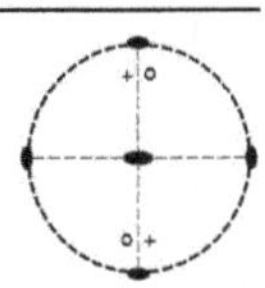
D_3	The solid can rotate 120° and 240° with respect to the axis of order 3 or tilt it with respect to each of the three axes of order 2	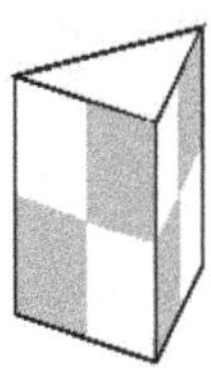	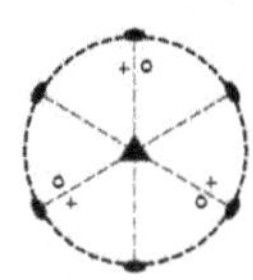

(Continues)

Table A.5 (*Continued*)

Symbol of Schoenflies	Symmetry operations	Solid	Stereogram
D_4	The solid can rotate 90°, 180°, and 270° with respect to the order 4 axis or tilt it with respect to each of the two order 2 axes (the other two axes are identical)	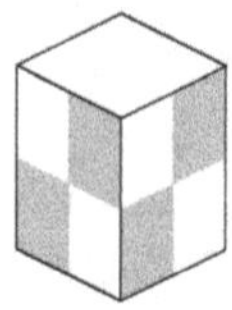	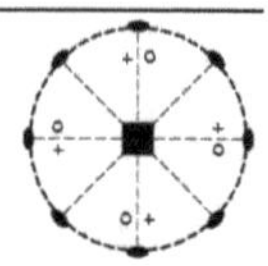
D_6	The solid can rotate by 60°, 120°, 180°, 240°, and 300° with respect to the order 6 axis or tilt it with respect to each of the three order 2 axes (the other three axes are identical)	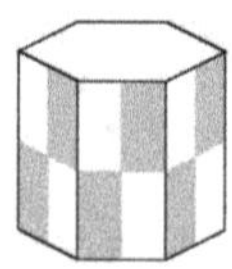	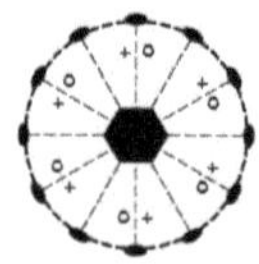

Groups D_{nd} (Table A.7)

are formed by an axis of rotation of order *n*, *n* axes of rotation of order 2 in the plane perpendicular to the first axis, and *n* vertical planes of reflection bisecting the axes of order 2; one can rotate the solid 360°·*k*/*n* with respect to the axis of order *n* *or* tilt it with respect to one of the axes of order 2 or reflect it with respect to any of the planes because the solid is symmetrical with respect to each of them;

Table A.6 Groups D_{nh}

Symbol of Schoenflies	Symmetry operations	Solid	Stereogram
D_{2h}	The solid can be flipped with respect to each of the three axes of order 2 and reflect it with respect to the equator	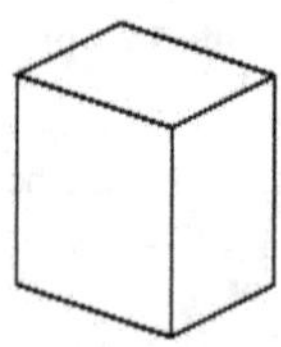	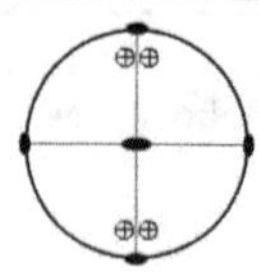
D_{3h}	The solid can rotate 120° and 240° with respect to the order 3 axis or tilt it with respect to each of the three order 2 axes and reflect it with respect to the equator	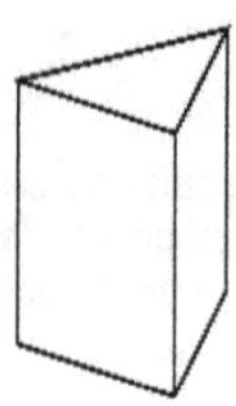	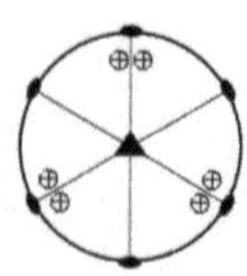

(*Continues*)

Table A.6 (*Continued*)

Symbol of Schoenflies	Symmetry operations	Solid	Stereogram
D_{4h}	The solid can rotate 90°, 180°, and 270° with respect to the order 4 axis or tilt it with respect to each of the two order 2 axes and reflect it with respect to the equator	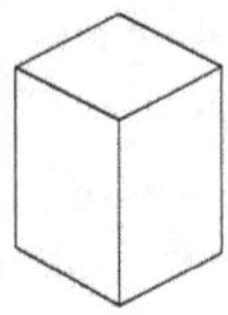	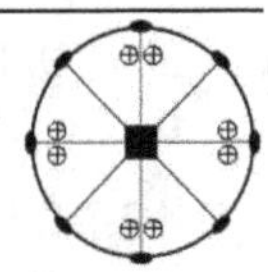
D_{6h}	The solid can rotate by 60°, 120°, 180°, 240°, and 300° with respect to the order 6 axis or tilt it with respect to each of the three order 2 axes and reflect it with respect to the equator	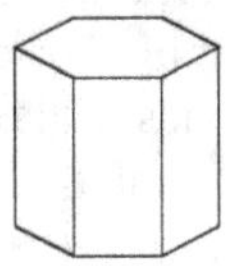	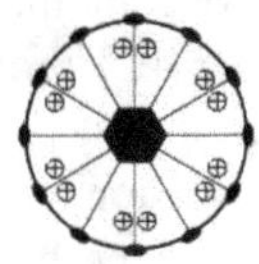

- are obtained from the product of group D_n with group $C_{1v} = C_s$;
- consist of 4*n* elements.

Group of the tetrahedron T

Table A.7 Groups D_{nd}

Symbol of Schoenflies	Symmetry operations	Solid	Stereogram
D_{2d}	The solid can be flipped with respect to each of the three axes of order 2 or reflect it with respect to a plane	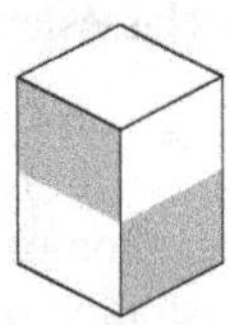	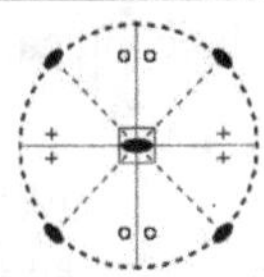
D_{3d}	The solid can rotate 120° and 240° with respect to the order 3 axis or tilt it with respect to each of the three order 2 axes or reflect it with respect to a plane	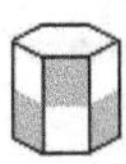	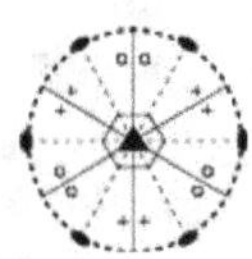

- consists of all possible rotations that leave a tetrahedron unchanged;

is formed by all 90° rotations of even order along the diagonals of the faces of the cube[17];

[17]The correspondence between tetrahedron and cube is obtained by inscribing a tetrahedron in the cube with vertices in the nonadjacent vertices of the cube.

- therefore corresponds to the alternate group of permutations on four elements and thus has 4!/2 = 12 elements;

It includes the following conjugacy classes:

- E: Identity;
- $3C_2$: 180° rotation along the three Cartesian axes passing through the center of the cube faces;

$4C_3$: 120° clockwise rotation along one of the four diagonals of the cube that corresponds to a 90° rotation along one Cartesian axis followed by a 90° rotation along another axis, leading one face to coincide with one of the four adjacent faces;

$4C_3^2$: rotations of 120° counterclockwise along one of the four diagonals of the cube that corresponds to the same operations described above, but in the opposite direction leading to one face coinciding with one of the four adjacent faces, rotated, however, by 180°.

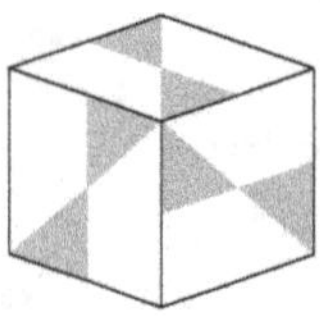

Group of the tetrahedron with inversion T_h

- is obtained from the product of the T group with the group C_i (formed by E and I);
- therefore, has $2\times 12 = 24$ elements.

The conjugacy classes are as follows:

- the ones of group T;
- the inversion with respect to center I;
- $4S_6$: an operation of $4C_3$ followed by I;
- $4S_6^5$: an operation of $4C_3^2$ followed by I;
- $3S_4$: an operation of $3C_2$ followed by C_i.

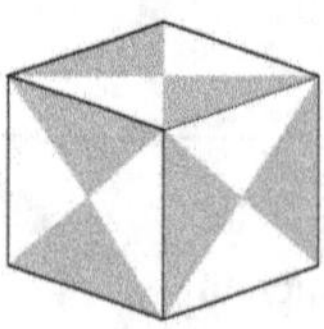

Group of the tetrahedron with tilting T_d

- is obtained from the product of group T with group C_s (formed by E and σ_h)[18];
- therefore, has $2\times 12 = 24$ elements.

[18]The vertical plane passes along one of the diagonals of the faces bordering the dark figure in the group T drawing.

The conjugacy classes are as follows:

- E: Identity;
- $3C_2$: 180° rotation along one of the three Cartesian axes passing through the center of the cube faces;
- $8C_3$: 90° rotation along two axes clockwise or counterclockwise (the σ element$_h$ unifies the 4C classes$_3$ and $4C_3^2$ of the T group);
- $6C_{3v}$: inversion with respect to one of the six diagonals of three non-opposite faces;
- $6S_4$: 90° rotation along an axis clockwise or counterclockwise followed by tilting with respect to a plane coplanar to the axis.

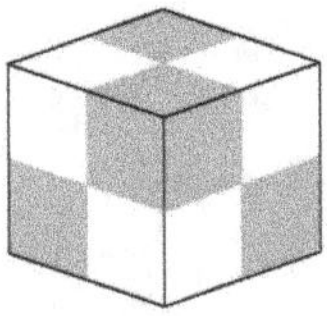

Octahedron group O

- is the symmetry group of the octahedron[19] without reflections;
- is formed by all 90° rotations along the diagonals of the faces of the cube[20];
- therefore, is isomorphic to the group of permutations on four objects, so it has 4! = 24 elements.

The conjugacy classes are as follows:

- E: Identity;
- $8C_3$: 120° rotation according to one of the four diagonals of the cube clockwise or counterclockwise;
- $6C_4$: 90° rotation along one of the three Cartesian axes clockwise or counterclockwise;
- $3C_2$: 180° rotation along one of the three Cartesian axes;
- $6C_2^2$: 180° rotation according to one of the six diagonals passing from the centerline of one edge to the centerline of the opposite edge.

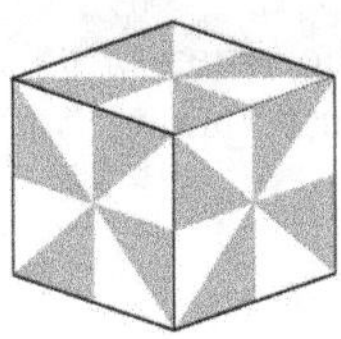

Group of the octahedron with inversion O_h

- is the symmetry group of the octahedron with reflections;

[19]The octahedron and the cube are dual (i.e., you can match the vertices of one with the faces of the other), so they have the same symmetries.

[20]It is obtained from the product of group T with group C_2 of the diagonals not highlighted in the figure of group T.

- is obtained from the product of O with C_i (formed by E and I);
- therefore, it has $24 \times 2 = 48$ elements.

The conjugacy classes are as follows:

- the ones of group T;
- the inversion with respect to center I;
- $8S_6$: an operation of $8C_3$ followed by I;
- $6S_4$: an operation of $6C_4$ followed by I;
- $3C_{2h}$: a $3C_2$ operation followed by I;
- $6C_{2v}$: a $6C_2^2$ operation followed by I.

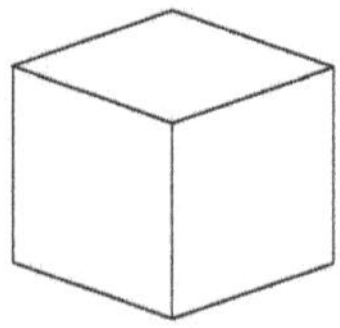

A.3 Crystal symmetry groups

A.3.1 Bravais reticle

The crystal symmetry groups are the *space groups*, which are the transformations of points in space $\boldsymbol{x} \mapsto \boldsymbol{y}$ of the type:

$$\boldsymbol{y} = \boldsymbol{M}\boldsymbol{x} + \boldsymbol{d} \tag{A.21}$$

that leave the structure of the crystal unchanged.

The crystal is formed by a regular arrangement of atoms. We consider a regular arrangement of nodes, called a *Bravais lattice*, which are obtained by translations of the type:

$$\boldsymbol{t} = n_1\boldsymbol{a}_1 + n_2\boldsymbol{a}_2 + n_3\boldsymbol{a}_3, \tag{A.22}$$

with n_1, n_2, n_3 integers and $\boldsymbol{a}_1, \boldsymbol{a}_1$, $\boldsymbol{a}_3$ noncoplanar vectors (*primitive vectors*).

We define a *base*, consisting of a geometric arrangement of atoms, one of which is conventionally placed in the origin and any other atoms (that may be identical or different from the atom in the origin) arranged at defined distance and orientation with respect to the atom in the origin.

Placing a base at each node of the Bravais lattice, with the origin of the base corresponding to a node of the lattice (although different arrangements do not change the overall symmetry), results in the crystal lattice:

$$\text{crystal} = \text{Bravais lattice} \otimes \text{base} \tag{A.23}$$

The base must contain the minimum set of atoms that, when replicated, create the entire lattice. The *primitive cell* is the portion of space that contains a base and that, replicated on all the nodes of the lattice, fills the whole space. There are several ways to choose the primitive cell; in the Wigner–Seitz method, the base stands in the center of the primitive cell.

To classify possible Bravais lattices, it is more convenient to consider not the primitive cell, but the *conventional cell*, which is a solid with greater point symmetries than the primitive cell. The conventional cell consists of one or more primitive cells translated according to primitive vectors. The conventional cell may also contain portions of primitive cells, but the composition of all the portions must result in an integer number of primitive cells.

Conventional cells are obtained by composing the seven *lattice systems* (Table A.8) with the four *lattice centrations*. The lattice systems are as follows:

- Triclinic;
- Monoclinic;
- Orthorhombic;
- Tetragonal;
- Rhombohedral;
- Hexagonal;
- Cubic.

They define the *symmetry classes* of the Bravais lattice, i.e., the rotation, reflection, or inversion symmetries of the cell, which correspond to the *point groups* of symmetry of the corresponding solid:

Of course, the cubic system also has all the symmetries of the lower systems (except the hexagonal), etc. The hexagonal system also has the symmetries of the rhombohedral system (see below).

In total, there are 32 classes of point symmetry including C_1, that is, no point symmetry.

The set of symmetry classes of the same lattice system constitutes a *crystalline system*. Each lattice system corresponds to a crystalline system, except for the crystalline system called *trigonal*, which corresponds to the symmetries of the rhombohedral lattice system, also attributed to the hexagonal lattice system. In fact, it is also possible to obtain the hexagonal lattice system by replicating a rhombohedral cell built on the diagonal of the rhombic-based right prism that forms the basis of the hexagonal system and, conversely, to obtain the rhombohedral lattice system by replicating the rhombic-based right prism on the diagonal of the rhombohedron (see figure).

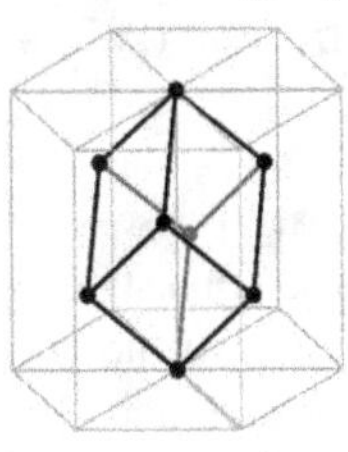

Base centrations in the conventional cell can be:

- primitive (P): no other points than the vertices of the cell;
- body-centered (I): the vertices and a point at the center of the cell;
- face-centered (F): the vertices and a point at the center of each face;
- with one face centered (C): the vertices and a point at the center of the two faces in one direction.

However, not all crystal-centrations combinations result in different lattice types, as some of them are equivalent. For example, a monoclinic I lattice is equivalent to a monoclinic C lattice by changing the choice of primitive vectors.

Table A.8 Lattice systems

Reticular system	**Solid**	**Sides**	**Corners**	**Classes of symmetry**
Triclinic	Parallelepiped	$a \neq b \neq c$	$\alpha \neq \beta \neq \gamma$	C_1, C_i
Monoclinic	Oblique prism with a rectangular base, in which one side face is perpendicular to the base	$a \neq b \neq c$	$\alpha = \gamma = 90°$, $\beta \neq 90°$	C, C_{s2}, C_{2h}
Orthorhombic	Straight prism with rectangular base	$a \neq b \neq c$	$\alpha = \beta = \gamma = 90°$	C, D_{2v2}, D_{2h}
Tetragonal	Right prism with a square base	$a = b \neq c$	$\alpha = \beta = \gamma = 90°$	S_4, D_{2d}, C_4, C_{4h}, C_{4v}, D_4, D_{4h}
Rhombohedral	Equal-sided rhombohedron	$a = b = c$	$\alpha = \beta = \gamma \neq 90°$	C_3, S_6, C_{3v}, D_3, D_{3d}
Hexagonal	Right prism with a rhombic base, consisting of two equilateral triangles	$a = b\ (\neq c)$	$\alpha = \beta = 90°$, $\gamma = 120°$	C_{3h}, D_{3h}, C, C_{66h}, C_{6v}, D_6, D_{6h}
Cubic	Cube	$a = b = c$	$\alpha = \beta = \gamma = 90°$	T, T_h, T_d, O, O_h

The centering of the base in the conventional cell defines whether the cell is primitive or not: it must be counted how many bases are contained in the conventional cell, considering that the bases at the edges of the cell shared with other cells must be counted fractionally (e.g., if the base in a vertex borders eight cells, it is counted as one eighth of base).

In total, the 14 Bravais lattices are represented in Table A.9.

The Bravais lattice, in addition to all the point symmetries of the related solid, also possesses the translation symmetries given by the transformation (A.21).

A.3.2 Space groups

It must be considered that the crystal lattice does not necessarily possess all the point symmetries of the relevant lattice system, since if the basis consists of more than one atom this may not remain unchanged by a transformation of the lattice's point symmetry group. Conversely, the composition of a point symmetry with a translation that is not part of the translational symmetry group of the Bravais lattice may lead, for a complex basis, to obtaining a new symmetry.

The group of all point symmetries of a crystal lattice R is denoted by H(R) and is called the *substationary group* (it is a subgroup of O(3), the group of three-dimensional rotations). The group of all translations that leave the lattice unchanged (which are of the type (A.21)) denoted by L(R) is called the *translational group*. The group of all point transformations of a crystal lattice R, i.e., point transformations that possibly followed by a translation (not necessarily of the type (A.21)) constitute a symmetry of the crystal, is denoted by S(R), and is called the

Table A.9 Bravais lattices

	P	**I**	**F**	**C**
Triclinic	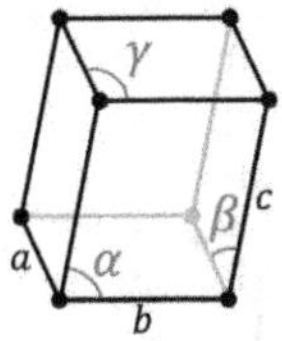			
Monoclinic	$\beta \neq 90°$ $a \neq c$ 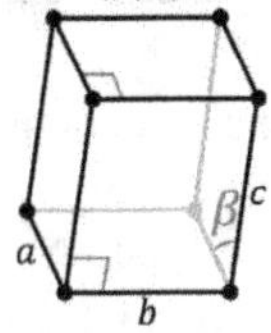			$\beta \neq 90°$ $a \neq c$ 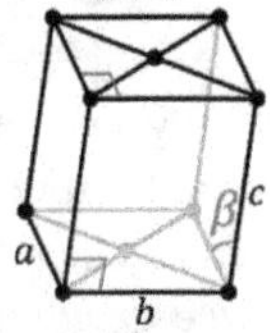
Orthorhombic	$a \neq b \neq c$ 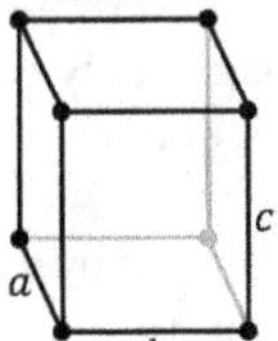	$a \neq b \neq c$ 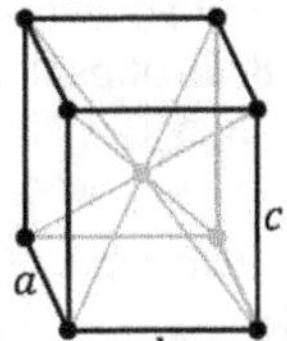	$a \neq b \neq c$ 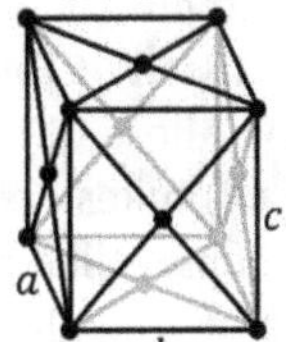	$a \neq b \neq c$ 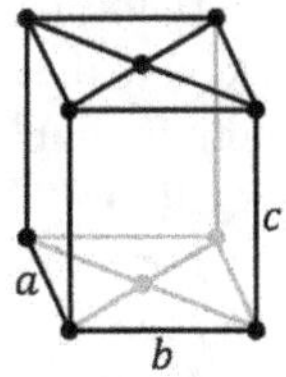
Tetragonal	$a \neq c$ 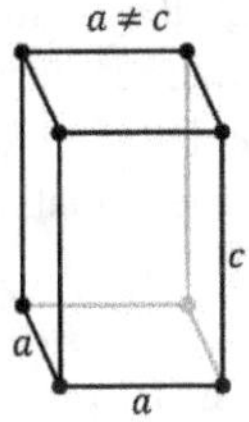	$a \neq c$ 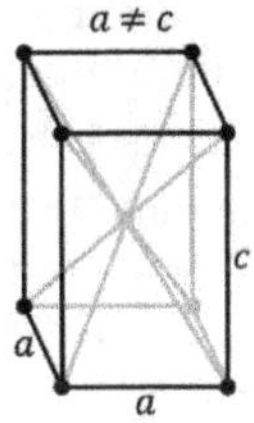		
Trigonal	Rhombohedral P = hexagonal R $\alpha \neq 90°$ 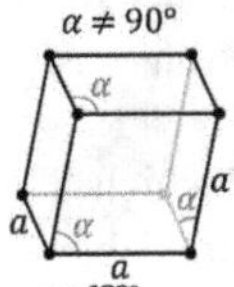 $\gamma = 120°$ 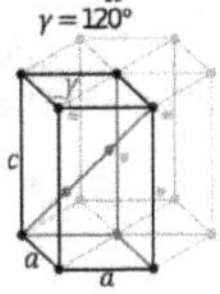		Hexagonal P = rhombohedral D $\gamma = 120°$ 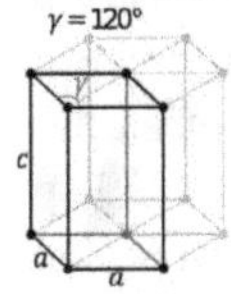 $\alpha \neq 90°$ 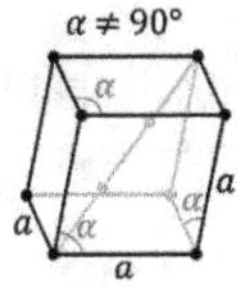	
Cubic	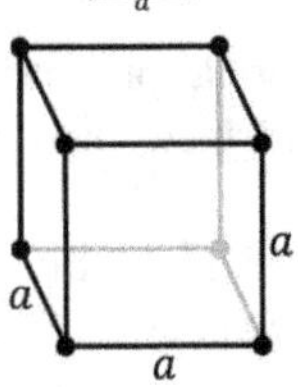	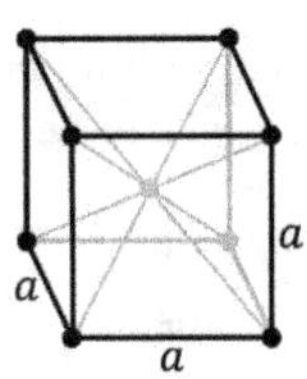	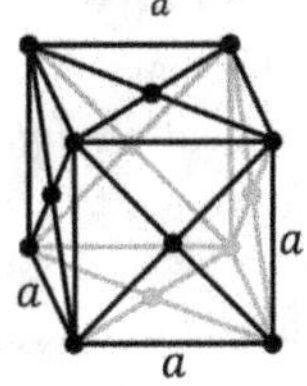	

Table A.10 List of spatial groups

Crystal system	No. of space groups	No. of symmorphic groups	No. of non-symmorphic groups
Triclinic	2	2	0
Monoclinic	13	6	7
Orthorhombic	59	13	46
Tetragonal	68	16	52
Trigonal	25	13	12
Hexagonal	27	8	19
Cubic	36	15	21

point group (it can be shown to be a group). This group also contains elements that are not lattice symmetries per se, because they require translation to restore symmetry. The group of all crystal lattice symmetries of the type (A.21) is called the *space group* and is denoted by $G(R)$; clearly $H(R) = G(R) \cap O(3)$.

If the base is simple (only one atom), there is the correspondence $H(R) = S(R)$ = to the classes listed in Table A.8. Groups that contain only point symmetries and pure translations (those of the relevant Bravais lattice) are called *symmorphic*, and $G(R) = H(R)\ L(R)$ holds.

For the other groups, called *non-symmorphic*, $G(R)$ is given by the union of $H(R)$ $L(R)$ and $T(R)$, consisting of all non-simple transformations (point transformations + translations); in this case $H(R) = S(R) \cap G(R)$ holds, i.e., $H(R)$ is the subset of $S(R)$ formed by the point transformations that constitute symmetries per se of the lattice (i.e., without the need to perform translations to leave the lattice unchanged).

The composition of all the point symmetries with the translation symmetries that leave the crystal lattice unchanged leads to the definition of the spatial symmetry groups, which have the form (2.51). If $\mathbf{d} = \mathbf{0}$, the transformation is a point symmetry $H(R)$; if $\mathbf{M} = \mathbf{1}$, the transformation is a pure translation $L(R)$; in the other cases, it is a composition of the two actions $T(R)$ (screw translation = rotation + translation along the axis of rotation, glide reflection = translation + reflection in a plane parallel to the direction of translation, rotoinversion = rotation + inversion with respect to a point on the axis of rotation).

In total, there are 230 spatial symmetry groups of which 73 are symmorphic and 157 are non-symmorphic, broken down as shown in Table A.10 (from Table 1.7 in [10]).

A complete list of all spatial groups with representations of their symmetries is available at: http://img.chem.ucl.ac.uk/sgp/mainmenu.htm.

A.3.3 Magnetic groups

If we consider that the base is endowed with an additional property, on which an inversion operation A called *antisymmetry* can be performed, then it is possible to extend the spatial groups.

The property can be understood in a physical sense as the time-averaged current density, and thus the magnetic moment associated with the basis, and the antisymmetry corresponds to the reversal of time (which changes the direction of the velocity, and consequently of the current density and magnetic moment). In a more abstract sense, lattice points can be endowed with a discrete characteristic that can assume only three states (called positive, negative, and neutral) or colors (white, black, and gray).

Point groups, also considering antisymmetry, are divided into three types:

- the ordinary 32 "white" groups without the antisymmetry operation (to this type belong diamagnetic, paramagnetic, and ferromagnetic materials, whose bases have a determined magnetic moment direction, which is therefore not invariant by antisymmetry);
- the product of the ordinary 32 groups with the group of the antisymmetry operation $\{E, A\}$, which constitutes another 32 "gray" groups each with twice as many elements as the ordinary ones (to this type belong the nonmagnetic materials, whose bases have no magnetic moment and are therefore invariant by antisymmetry);
- a total of 58 "white-black" groups called *magnetic* in which the antisymmetry operation enters only in combination with other elements of the ordinary group. To this type belong ferrimagnetic and antiferromagnetic materials, in which the magnetic moments of some of the atoms of the base have alternate direction with respect to those of the corresponding atoms of neighboring bases, but diamagnetic, paramagnetic, and ferromagnetic elements can also belong to it if the composition of the point symmetry and antisymmetry operation restores the original arrangement of magnetic moments.

The construction of the groups of the first two types is straightforward. In contrast, to construct the groups of the third type, one must consider that the elements of that group are divided into two sets:

- the elements g_1 that do not contain the antisymmetry operator, which evidently constitute a subgroup that is denoted by Γ;
- the Ag_2 elements that contain the antisymmetry operator, which do not constitute a subgroup but a class that is denoted by C.

Since the antisymmetry operator commutes with any element of the ordinary point group and is idempotent (i.e., $A^2 = E$), the subgroup Γ must be invariant, since $(Ag_2)^{-1} g_1 A g_2 = g_2^{-1} g_1 g_2 \in \Gamma$. Therefore, C constitutes a coset of Γ with respect to an idempotent symmetry element that we denote by g', viz.:

$$G = \Gamma \otimes C = \Gamma \otimes Ag'\Gamma. \tag{A.24}$$

It then turns out that the subgroup Γ is of index 2 with respect to the group G. Consequently, in order to construct magnetic groups, all invariant subgroups with index 2 of the point symmetry groups must be identified. Since a group can also contain multiple invariant subgroups of index 2, magnetic groups are more numerous than ordinary groups. Their construction is described in [2] Section 38.

Associated with these three types of point groups are the corresponding spatial groups:

- the ordinary 230 space groups;
- the 230 space groups completed with the antisymmetry operation;
- the additional 1191 magnetic space groups, in which the antisymmetry operation enters in combination with the composition of point translations and symmetries, and operations without antisymmetry constitute a subgroup of index 2, which in turn are subdivided into:
- a total of 674 spatial groups in which the element g' generating the class with antisymmetry is a point operation;
- a total of 517 spatial groups in which the element g' that generates the class with antisymmetry is a translation.

Thus diamagnetic, paramagnetic, and ferromagnetic materials can only belong to type IIIa, while ferrimagnetic and antiferromagnetic materials belong to type IIIb. The magnetic cell of type IIIb groups has twice the volume of the ordinary space group cell because it is stretched by the translation that generates the magnetic group.

A.3.4 Miller's indices

Miller indices allow a compact expression of directions and planes in the space occupied by a lattice system. The index $[h\,k\,l]$, where h, k, l are integer numbers (fractional numbers are multiplied by a common factor and negative numbers are over-lined, e.g., $\overline{1} = -1$), represents a direction, while the index $(h\,k\,l)$ represents the plane intersecting the axes arranged according to the primitive vectors at the points $(h, 0, 0), (0, k, 0), (0, 0, l)$; if the plane does not intersect an axis, its index is set to 0. Only in the cubic system, the plane is perpendicular to the direction with the same indices, viz., $[h\,k\,l] = (h\,k\,l)$ only for the cube (Figure A.3).

The symbol $\langle h\,k\,l\rangle$ denotes the equivalent directions in a crystal, i.e., all the directions that can be transformed into each other by crystal symmetry operations. For example, in a body-centered cubic lattice with a simple base (such as iron-α), the

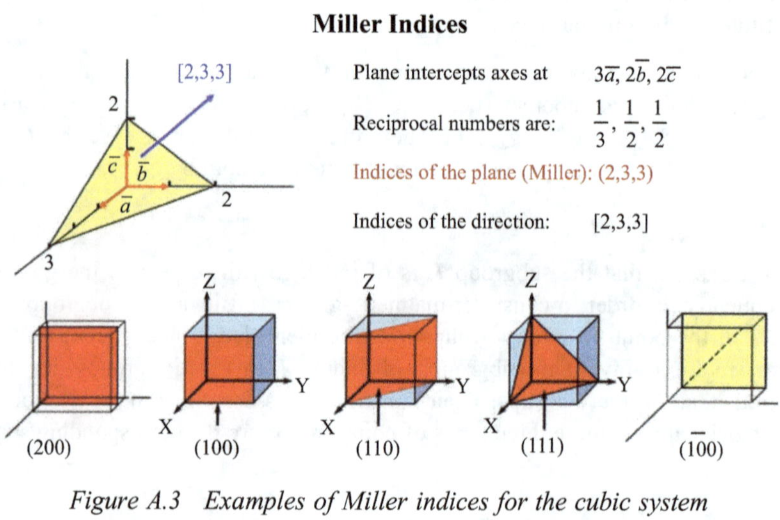

Figure A.3 Examples of Miller indices for the cubic system

symbol $\langle 1\ 0\ 0\rangle$ represents the directions $[1\ 0\ 0]$, $[\bar{1}\ 0\ 0]$, $[0\ 1\ 0]$, $[0\ \bar{1}\ 0]$, $[0\ 0\ 1]$, $[0\ 0\ \bar{1}]$. The index of any equivalent direction can be used to define the family of directions; for example, the symbol $\langle 0\ 0\ 1\rangle$ has the same meaning $\langle 1\ 0\ 0\rangle$ for the *bcc* lattice with simple base.

The plans constructed according to Miller's indices in the cubic system lattices can be seen interactively at: http://calistry.org/calculate/latticePlanesMillerIndices.

A.4 Analytical study of magnetic properties

A.4.1 Anisotropy

The *thermodynamic potential* Φ is defined as the energy per unit volume of a magnetic body, including both thermodynamic (temperature, entropy, etc.) and magnetic quantities where $\boldsymbol{M}$ and $\boldsymbol{B}$ are considered independent variables. Starting from (2.6), we can write:

$$\frac{\partial W}{\partial V} = \Phi = \int_C \boldsymbol{H} \cdot d\boldsymbol{B}, \tag{A.25}$$

from which:

$$\frac{\partial \Phi}{\partial \boldsymbol{B}} = \boldsymbol{H}. \tag{A.26}$$

To find the thermodynamic potential from (A.26), we must add an integration constant equal to the value of the potential in the absence of induction:

$$\Phi(\boldsymbol{M}, \boldsymbol{B}) = \Phi_0(\boldsymbol{M}) + \boldsymbol{H} \cdot \boldsymbol{B}. \tag{A.27}$$

In anisotropic bodies such as crystalline ferromagnets, the product $\boldsymbol{H} \cdot \boldsymbol{B}$ generates anisotropy terms, which account for the dependence of the magnetic energy on the direction of magnetization. By denoting by $\widehat{\boldsymbol{M}} = \boldsymbol{M}/M$ the magnetization direction, we can write the anisotropy terms in first approximation as a quadratic form[21]:

$$\Phi_{\text{an}} = \sum_{i,k} K_{ik} \widehat{M}_i \widehat{M}_k, \tag{A.28}$$

where K_{ik} is a coefficient with the dimensions J/m^3, which may depend on the magnetization intensity M. From matrix theory, we can diagonalize the quadratic form along the principal axes, which are denoted as x, y, z:

$$\Phi_{\text{an}} = K_x \widehat{M}_x^2 + K_y \widehat{M}_y^2 + K_z \widehat{M}_z^2. \tag{A.29}$$

Considering further that $\widehat{M}_x^2 + \widehat{M}_y^2 + \widehat{M}_z^2 = 1$ and that a nondirectional term in the anisotropy potential is irrelevant, we can rewrite the previous expression as

$$\Phi_{\text{an}} = (K_x - K_z)\widehat{M}_x^2 + (K_y - K_z)\widehat{M}_y^2. \tag{A.30}$$

[21]Since the magnetization energy is a scalar, thus independent of the direction from the magnetization, which instead changes direction by time reversal, there can be no odd-degree terms in the series development of the anisotropy energy.

Thus, there are only two independent components of anisotropy. If the coefficients differ from each other (this happens in monoclinic, triclinic, and orthorhombic systems), the crystal is called *biaxial*. If, on the other hand, $K_x = K_y$, the crystal is called *uniaxial* and we can write:

$$\Phi_{an} = K\sin^2\theta, \tag{A.31}$$

where θ is the colatitude in spherical coordinates (angle between $\boldsymbol{M}$ and z, which constitutes the easy axis).

Continuing the development to fourth-degree terms, the peculiar uniaxial reticular system is highlighted:

- in the crystals of the tetragonal system, terms such as $\left(\widehat{M}_x^2 + \widehat{M}_y^2\right)^2$ and $\widehat{M}_x^2\widehat{M}_y^2$;
- in the crystals of the rhombohedral system, terms such as $\left(\widehat{M}_x^2 + \widehat{M}_y^2\right)^2 = \sin^4\theta$ and $\cos\theta\sin^3\theta\cos 3\phi$;
- only terms of the type can appear in the crystals of the hexagonal system $\left(\widehat{M}_x^2 + \widehat{M}_y^2\right)^2$;
- only terms of the type $-\left(\widehat{M}_x^4 + \widehat{M}_y^4 + \widehat{M}_z^4\right)$ o $\widehat{M}_x^2\widehat{M}_y^2 + \widehat{M}_x^2\widehat{M}_z^2 + \widehat{M}_y^2\widehat{M}_z^2$ (the two expressions differ only by a constant).

Also considering sixth-degree terms, the anisotropy energy of a crystal in the cubic system has expression [2] (40,8):

$$\Phi_{\mathrm{an}} = K_1\left(\widehat{M}_x^2\widehat{M}_y^2 + \widehat{M}_x^2\widehat{M}_z^2 + \widehat{M}_y^2\widehat{M}_z^2\right) + K_2\widehat{M}_x^2\widehat{M}_y^2\widehat{M}_z^2. \tag{A.32}$$

The constant K_1 determines the direction of the easy axis in the cube: If it is positive, the easy axis is along the edges of the cube; if it is negative, the easy axis is along the spatial diagonal of the cube.

For greater precision, one would need to consider how magnetization changes the crystal structure of the solid. In particular, a cubic crystal magnetized along the spatial diagonal of the cube becomes slightly rhombohedral. If magnetized along an edge, it becomes slightly tetragonal.

A.4.2 Hysteresis

The thermodynamic potential (A.27) considering the second-degree anisotropy energy in a uniaxial crystal (A.31) is written[22]:

$$\tilde{\Phi}(\boldsymbol{M},\boldsymbol{H}) = \Phi_0(\boldsymbol{M}) - \mu_0\boldsymbol{H}\cdot(\boldsymbol{H}+\boldsymbol{M}) + \Phi_{an}. \tag{A.33}$$

Writing the anisotropy constant as [2] (41,1):

$$K = \frac{1}{2}\beta\mu_0 M^2, \tag{A.34}$$

[22]The potential $\tilde{\Phi}(\boldsymbol{M},\boldsymbol{H})$ is the Legendre transform of the potential $\Phi(\boldsymbol{M},\boldsymbol{B})$, in which the variable $\boldsymbol{B}$ is replaced by the variable $\boldsymbol{H}$. This is why the second addend changes sign.

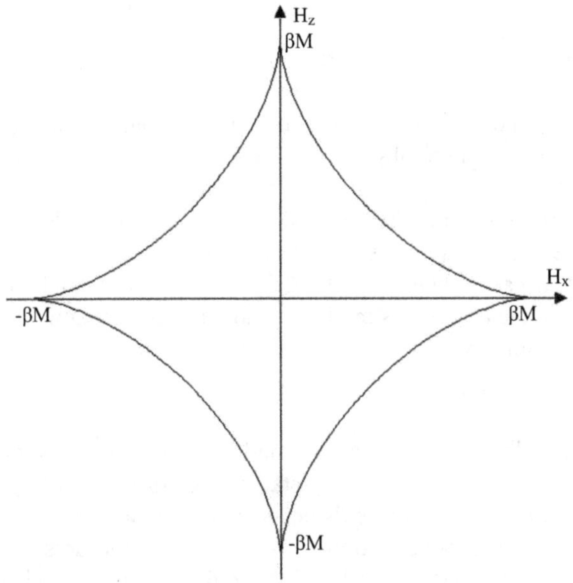

Figure A.4 Limits of the area where hysteresis is possible

where β is a dimensionless constant, and placing the plane in which the vector lies $\boldsymbol{M}$ on the x–z axes, we obtain the thermodynamic potential in the form:

$$\tilde{\Phi}(\boldsymbol{M},\boldsymbol{H}) = \Phi_0(\boldsymbol{M}) - \mu_0 M(H_x \sin\theta + H_z \cos\theta) - \mu_0 H^2 + \frac{1}{2}\beta\mu_0 M^2 \sin^2\theta. \tag{A.35}$$

From the equilibrium condition $\frac{\partial\tilde{\Phi}}{\partial\theta} = 0$, the following is obtained:

$$\beta M \sin\theta \cos\theta = H_x \cos\theta - H_z \sin\theta. \tag{A.36}$$

This can be transformed into a fourth-degree equation in $\sin\theta$ by rearranging the terms and raising it to the square. This equation can have four real solutions[23]; since these solutions correspond to the extremes of the thermodynamic potential, in the first case, it has two maxima and two minima (thus two equilibrium points), and in the second, case it has one maximum and one minimum (thus one equilibrium point). Situations in which there are two equilibrium points make hysteresis possible. The value of the magnetizing force for which the different situations merge is when a point of maximum and a point of minimum become a single horizontal inflection point, in which $\frac{\partial^2\tilde{\Phi}}{\partial\theta^2} = 0$; from which the following

[23]The relation $M = \chi H$ ensures that the real solutions are in absolute value less than unity, thus admissible.

expression is found:

$$H_x = \sin^3\theta\beta M, H_z = -\cos^3\theta\beta M \Rightarrow H_x^{2/3} + H_z^{2/3} = \beta M^{2/3} \tag{A.37}$$

which is represented graphically (in the approximation of constant M) in Figure A.4. The area bounded by the curve (A.37) represents the area where hysteresis is possible.

However, measurement interpolation curves and finite element models are used to calculate hysteresis cycles in real materials [20].

In the special case where the magnetizing force is parallel to the easy axis ($H_x = H, H_z = 0$), the (A.36) is simplified; under the assumption that $H < \beta M$, the potential has a minimum for

$$M \sin\theta_m = M_x = H/\beta, \tag{A.38}$$

to which two possible solutions correspond θ_m and $\pi - \theta_m$, symmetrical with respect to the x-axis, which are equally stable since the value of $\tilde{\boldsymbol{\Phi}}$ is identical in the two cases. This leads to the existence of the contact magnetic domains, in which the direction of magnetization M is opposite in the presence of the same magnetization force H, so the overall induction B is lower and so is the associated energy.

A.4.3 Magnetostriction

The change in the size of a ferromagnetic body subjected to a magnetizing force, i.e., magnetostriction, is due to a dependence between the stress tensor of the solid σ_{ik} and the magnetization, which is represented by a fourth-order tensor a_{iklm}, symmetrical with respect to the index pairs ik and lm:[24]

$$\boldsymbol{\Phi}_{\text{ms}} = -a_{iklm}\sigma_{ik}\widehat{M}_l\widehat{M}_m. \tag{A.39}$$

In cubic symmetry crystals, this potential takes on the expression [2] (42, 2):

$$\begin{aligned}\boldsymbol{\Phi}_{\text{ms}} = &-a_1\left(\sigma_{xx}\widehat{M}_x^2 + \sigma_{yy}\widehat{M}_y^2 + \sigma_{zz}\widehat{M}_z^2\right)\\ &- 2a_2\left(\sigma_{xy}\widehat{M}_x\widehat{M}_y + \sigma_{xz}\widehat{M}_x\widehat{M}_z + \sigma_{yz}\widehat{M}_y\widehat{M}_z\right).\end{aligned} \tag{A.40}$$

The strain tensor is obtained by deriving $\boldsymbol{\Phi}$ with respect to the components of the stress tensor:

$$u_{ik} = \frac{\partial\boldsymbol{\Phi}}{\partial\sigma_{ik}}. \tag{A.41}$$

[24]The expression (2.39) is written with a minus sign for convenience, since magnetostriction is normally a compression effect.

In the presence of elastic energy, one must also add its contribution to the potential, which for cubic symmetric crystals is written:

$$\Phi_{\mathrm{el}} = \frac{m_1}{2}\left(\sigma_{xx}^2 + \sigma_{yy}^2 + \sigma_{zz}^2\right) + \frac{m_2}{2}\left(\sigma_{xx} + \sigma_{yy} + \sigma_{zz}\right)^2 + m_3\left(\sigma_{xy}^2 + \sigma_{xz}^2 + \sigma_{yz}^2\right), \tag{A.42}$$

where m_1, m_2, m_3 are positive coefficients of elasticity.

Overall, the following formulas result:

$$u_{xx} = (m_1 + m_2)\sigma_{xx} + m_2\left(\sigma_{yy} + \sigma_{zz}\right) + a_1\widehat{M}_x^2, u_{xy} = m_3\sigma_{xy} + a_2\widehat{M}_x M_y\,, \tag{A.43}$$

and similarly for the other components.

Some formulas have been developed to estimate the noise produced by measuring the magnetostriction of magnetic steel sheets [21]; however, the acoustic behavior of the complete core does not appear to be easy to extrapolate by measurements on a sample of the sheets; better results are obtained from measurements on sample cores reproduced at a smaller scale.

References

[1] EU Regulation No. 548/2014: *Implementing Directive 2009/125/EC of the European Parliament and of the Council with Regard to Small, Medium and Large Power Transformers*, OJ L 152, 22/5/2014, as amended.

[2] Lev D. Landau, and Evgenij M. Lifšitz: *Course in Theoretical Physics - Part 8: Electrodynamics of Continuous Media*, Editori Riuniti, 1986.

[3] IEC 60404-2: *Magnetic Materials – Part 2: Methods of Measurement of the Magnetic Properties of Electrical Steel Strip and Sheet by Means of an Epstein Frame*, ed. 3.1, 2008.

[4] IEC 60404-3: *Magnetic Materials – Part 3: Methods of Measurement of the Magnetic Properties of Electrical Steel Strip and Sheet by Means of a Single Sheet Tester,* ed. 2.2, 2010.

[5] M.S. Muhit: *Magnetic and electrical characterization of materials for electrical machines*, Master thesis at the Royal Institute of Technology, Stockholm, August 2011.

[6] Robert M. Del Vecchio, Bertrand Poulin, Pierre T. Feghali, Dilipkumar M. Shah, and Rajendra Ahuja: *Transformer Design Principles – With Applications to Core-Form Power Transformers*, CRC Press, 2002.

[7] Charles Kittel: *Introduction to Solid State Physics*, Wiley, 7th ed., 1996.

[8] A. Shadowitz: *The electromagnetic field, Dover*, 1988.

[9] John P. Lowe, and Kirk A. Anderson: *Quantum Chemistry*, 3rd ed., 2006.

[10] C.J. Bradley, and A.P. Cracknell: *The Mathematical Theory of Symmetry in Solids*, Clarendon, 1972.

[11] Walter Nicodemi: *Metallurgia*, Zanichelli editore, 2nd ed. 2007.

[12] Anthony Moses, Philip Anderson, Keith Jenkins, and Hugh Stanbury: *Electrical Steels - Volume 1: Fundamentals and Basic Concepts*, IET, 2019.

[13] A.J. Moses: *Electrical steels: Past, present and future developments*, IEE Proceedings A, vol. 137, no. 5, September 1990.

[14] I. Petryshynets, V. Puchý, F. Kovác, and M. Šebek: *Effect of laser scribing on soft magnetic properties of conventional grain-oriented silicon steel*, Proceedings of the 16th Czech and Slovak Conference on Magnetism, Košice, Slovakia, June 13–17, 2016.

[15] Ludger Lahn, Chaoyong Wang, Andreas Allwardt, Thierry Belgrand, and Jonathan Blaszkowski: *Improved transformer noise behavior by optimized laser domain refinement at ThyssenKrupp Electrical Steel*, IEEE Transactions on Magnetics, vol. 48, no. 4, April 2012.

[16] IEC 60404-8-7: *Magnetic Materials – Part 8-7: Specifications for Individual Materials – Cold-Rolled Grain-Oriented Electrical Steel Strip and Sheet Delivered in the Fully-Processed State,* ed. 3.0, 2008.

[17] IEC 60404-8-7: *Magnetic Materials – Part 8-7: Specifications for Individual Materials – Cold-Rolled Grain-Oriented Electrical Steel Strip and Sheet Delivered in the Fully-Processed State*, ed. 5.0, 2020.

[18] European Commission: *Working Document on the Review of Commission Regulation (EU) No. 548/2014*, 05/10/2017.

[19] Michael Tinkham: *Group Theory and Quantum Mechanics*, McGraw-Hill, 1964.

[20] A.P.S. Baghel, and S.V. Kulkarni: *Modeling of magnetic characteristics including hysteresis effects for transformers*, Cigré Transformer Colloquium, Split, October 15–17, 2014.

[21] Belgrand Thierry, Lemaître Régis, and Lahn Ludger: *How grain oriented electrical steel may lower transformer noise*, Cigré Transformer Colloquium, Split, October 15–17, 2014.

[22] M. Alderson *et al.*: *Power Transformers – Vol. 2: Expertise*, Alstom, 2011.

[23] Christian Freitag: *Magnetic properties of electrical steel, power transformer core losses and core design concepts*, Thesis at the Karlsruher Institutes für Technologie, February 2017.

[24] William D. Callister: *Materials Science and Engineering. An Introduction*, Edises, 2007.

[25] Linus Pauling, and E. Bright Wilson: *Introduction to Quantum Mechanics – With Application to Chemistry*, McGraw-Hill, 1935.

[26] Rüdiger Koshorst: *How grain oriented electrical steel can help on continuous improvement on transformer efficiency*, ThyssenKrupp Electrical Steel Presentation at Coiltech, September 27–28, 2017, Pordenone.

[27] Alex Hubert, and Rudolf Schaefer: *Magnetic Domains*, 3rd ed., 2009.

[28] Arnold Magnetic Technologies: *Soft Magnetic Applications Guide*, 2015.

[29] Franca Albertini: *Magnetic materials for energy*, Presentation at the Italian School of Magnetism, Milan, April 18–22, 2016.

[30] Paola Tiberto: *Electric motors*, Presentation at the Italian School of Magnetism, Milan, April 18–22, 2016.

[31] M. Solzi: *Basic magnetic properties*, Presentation at the Italian School of Magnetism, Rome, February 3–7, 2020.
[32] Stefano Cicalé: *Improvement of electrical steels products and related production technologies*, RINA Consulting Presentation at CWIEME 2017, Berlin.

Chapter 3

Conductors for power transformer windings

3.1 Generalities

Windings influence all the constructional and functional characteristics of each type of transformer. This chapter describes the state of the art of conductor technology adopted for power transformer windings and, in particular, copper conductors, to provide indications for the accurate drafting of technical specifications in order to require the correct characteristics and achieve the best performance of power transformers, with high product quality.

3.2 Electrical conductors

3.2.1 Electronic bands

Electrons in the bound state of the isolated atomic nucleus arrange themselves on *orbitals* of quantized energy. When different atoms bond in a complex structure, such as molecules or crystals, the orbitals are determined not only by interaction with the single nucleus but are influenced by all adjacent nuclei. The resulting structure of orbitals is arranged in virtually continuous energy bands, called *electronic bands* (Figure 3.1).

Electrons, according to the Pauli exclusion principle, dispose themselves onto orbitals of increasing energy until they reach the *Fermi energy*.[1]

The electrical behavior of materials depends on the arrangement of the partially occupied band (called the *conduction band*) with respect to the Fermi energy. Metals have the conduction band with many available energy levels and high density of states[2] (Figure 3.2).

[1]According to the Fermi–Dirac statistic, the Fermi energy is the upper energy limit of electrons bound at a temperature of 0K. For nonzero temperatures, electrons can have higher energies, and the Fermi energy represents the level at which the probability density of finding an electron is 0.5.

[2]The density of state represents the number of electrons that can be present simultaneously in an energy level and is represented by the horizontal bandwidth in Figure 3.2.

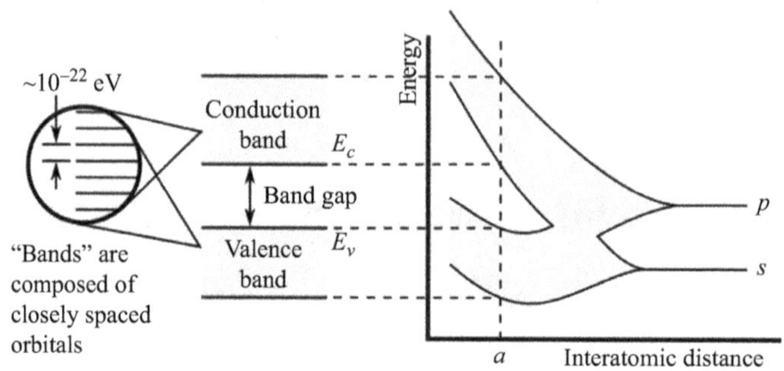

Figure 3.1 Structure of electronic bands

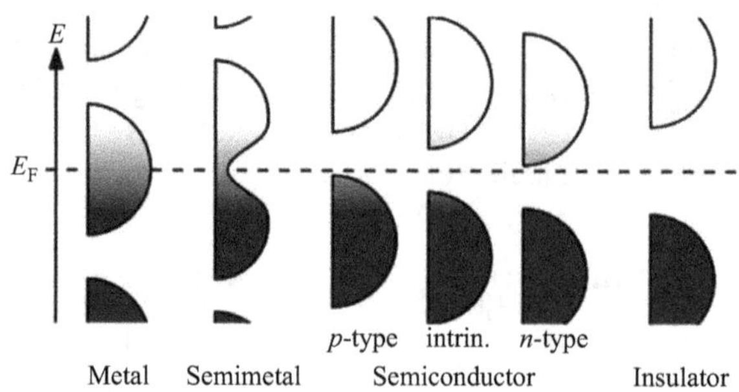

Figure 3.2 Arrangement of electrons in energy levels (black occupied, white unoccupied)

3.2.2 Electrical conductivity

The occupation of energy levels in a material can be represented in the *momentum space* $\boldsymbol{p}$, where the axes are given by the wave vector:

$$\boldsymbol{k} = \frac{\boldsymbol{p}}{\hbar}. \tag{3.1}$$

The *Fermi sphere* represents the radius of the vector $\boldsymbol{k}$ to which the Fermi level corresponds. The application of an electric field $\boldsymbol{E}$, and thus of a force $\boldsymbol{F} = -\mathrm{e}\boldsymbol{E}$, on each electron causes a displacement of the sphere, within the limit of the conduction band (Figure 3.3):

$$\delta k = -\frac{\mathrm{e}\boldsymbol{E}t}{\hbar}, \tag{3.2}$$

where t is the field application time.

Since the sphere is no longer centered in the origin of the momentum space, there is a net momentum and thus electron motion in the material. However, imperfections in the atomic structure of materials restrain the motion of electrons,

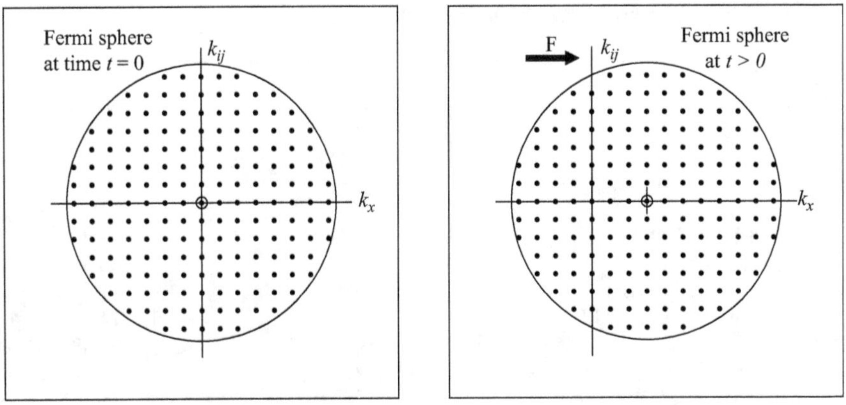

Figure 3.3 Displacement of the Fermi sphere following the application of an electric field

and a *collision time* τ can be defined, after which an electron accelerated by the electric field terminates its motion in the crystal lattice (or other structure of the material).

The momentum reached by the electron before collision is $\boldsymbol{p} = m_e\boldsymbol{v} = -e\boldsymbol{E}\tau$. Consequently, the current density that is established is given by ([1], ch. 6, eq. (42))

$$\boldsymbol{J} = n(-e)\boldsymbol{v} = \frac{ne^2\tau\boldsymbol{E}}{m_e}, \tag{3.3}$$

where

- n is the volume density of electrons in the conduction band;
- $\boldsymbol{v}$ is speed;
- m_e is the mass of the electron.

Equation (3.3) constitutes *Ohm's law* at the microscopic level; the magnitude

$$\sigma = \frac{ne^2\tau}{m_e}, \tag{3.4}$$

is called *electrical conductivity*, measured in siemens per meter (S/m), and represents the ability of materials to conduct electric current. Its reciprocal $\rho = \sigma^{-1}$ is called *electrical resistivity*, measured in ohms meter (Ω m).

3.3 Metals for electrotechnical use

3.3.1 Comparison of aluminum and copper

The metals, pure or alloyed, used for electrotechnical applications are aluminum and copper. Table 3.1 compares the main characteristics of the two metals.

Because of its lower cost, aluminum is commonly used in small medium voltage/low voltage (MV/LV) distribution transformers. For larger power transformers,

Table 3.1 Physical properties of aluminum and copper [2]

Properties	Aluminum (Al)	Copper (Cu)
Atomic number	13	29
Density (g/cm^3)	2.6989	8.93
Vickers hardness (kgf/mm^2)	15	50
Breaking load (MPa)	47	210
Elastic limit load (MPa)	11	33.3
Elongation at break (%)	–	60
Modulus of elasticity (GPa)	68	110
Compressibility modulus (GPa)	76	140
Poisson's ratio	0.36	0.343
Shear modulus (GPa)	25	46
Electrical resistivity (Ω m)	$2.7{\cdot}10^{-8}$	$1.7{\cdot}10^{-8}$
Relative magnetic susceptibility	$6.00{\cdot}10^{-7}$	$-8.00{\cdot}10^{-8}$
Coefficient of thermal elongation (Å/m K)	24.0	16.4
Specific heat (J/g K)	0.9	0.385
Thermal conductivity (W/m K)	210	385
Melting temperature (°C)	660.37	1083.6
Boiling temperature (°C)	2519	2562

above 4 MVA, the use of aluminum, due to its higher resistivity, implies higher operating losses and bulkier machines; therefore, copper windings are preferred.

In addition, as shown in Table 3.1, pure aluminum has poor mechanical properties, which are indispensable in high-power transformers to withstand electrodynamic stresses. As a result, it must be used alloyed with other metals with a further reduction in conductivity. For example, an alloy containing 0.4% magnesium, 0.1% titanium, 0.02% manganese or chromium causes a 10% reduction in conductivity.

In the following, it is assumed the use of copper and its alloys, the only metals used in the construction of the windings of large power transformers installed in the transmission grids.

3.3.2 Copper mining and production

Copper ores are found in the Earth's crust in various forms and concentrations. In the outermost 10 km of the Earth's crust, there are about 33 g of copper per ton of rock. However, in some locations, the concentration is much higher, and copper mining becomes economically viable. In particular, this is currently the case when the copper content is more than 0.2% by weight, about 2 kg/ton of rock.

Copper can be found in the form of sulfides, oxides, carbonates, silicates, or other compounds; more rarely, it is found in its native state. The most abundant ores, covering 80% of the world's known deposits, are chalcopyrite and bornite, and both contain copper and iron sulfides.

The processes for processing sulfur-containing copper oxides are as follows (Figure 3.4).

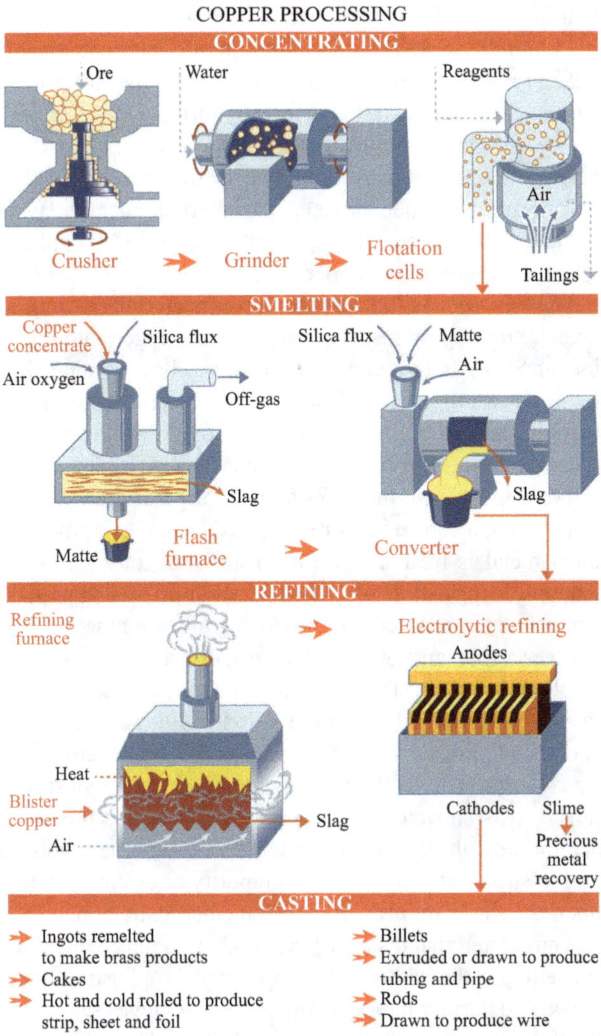

Figure 3.4 Diagram of the copper processing

Copper mining: Copper ores are extracted from both open pit and tunnel mines, with a 90% prevalence of the former. Mining methods are similar to those used for other ores.

Crushing and grinding (crushing): The extracted ore contains a significant number of aggregates (*gangue*) and must be reduced by means of powerful mills to a powder of appropriate grain size to optimize further processing.

Flotation: This operation allows separation of the copper-containing fractions from the aggregates. The powder, emulsified with surfactant liquids, is fed into large tanks from where the foamy surface layer containing the most copper-rich particles is removed. In fact, by selective surface tension, the sulfides become covered with minute surfactant bubbles that prevent contact with water and are drawn upward by air that is bubbled into the tank. The concentration of copper, obtainable in this way, can vary from 30% to 50% depending on the content of the starting ore.

Concentrating: The sludge is mechanically dried before being fed into the roasting oven since a high amount of water would produce a waste of heat energy for its evaporation. This results in a compound mainly based on sulfide minerals.

Smelting: Concentration then proceeds by thermal means. By heating in air between 500 °C and 700 °C, the sulfur content is reduced by exploiting its higher chemical affinity for oxygen than copper.

Melting and *conversion*: In the furnace, the concentrate (*matte*) is melted in an environment made oxidizing by insufflation of air or oxygen into the bath, resulting in the formation of SO_2 gas that separates from the liquid metal. In addition, the addition of silicon allows the removal of the iron present, which forms *slag*, composed mainly of silicates, which floats and is removed by skimming.

Thermal refining: This treatment is carried out for the purpose of further reducing the concentration of impurities in the copper, and while it is practically a continuation of the previous operation, it can be implemented in different equipment and at different times. The molten metal is treated again by insufflation of air or oxygen and subsequent scorification of impurities. At this stage, the bath is also partially reoxidized. Next, the oxygen content is reduced by introducing a green pine log into the furnace, which releases water vapor and other reducing gases as it burns. This operation is called "*pinning*." It makes it possible to obtain anodes that will then undergo electrolytic refining; however, if thermal refining is extremely thorough, it is possible to use the copper thus obtained directly for the production of certain semifinished products.

Electrolytic refining: This is the process by which the highest levels of copper purity are achieved. It is carried out using the soluble anode process in tanks containing a copper sulfate solution and in which anodes alternate with cathodes. The latter consist of thin sheets of copper with high purity of electrolytic origin. Between anodes and cathodes, which are immersed in the electrically conducting solution, an electric current is circulated that dissolves the anode into ions, which migrate through the solution and settle on the cathode. By appropriately adjusting the parameters of potential difference and current intensity, the process achieves selective deposition of copper against other metals that constitute impurities. The metals less noble than copper remain in solution, and the more noble ones precipitate in the sludge, from which they are periodically separated with obvious economic advantages. The process stops as the anodes are exhausted. The cathodes (Figure 3.5) with a purity of 99.99% are extracted and sent for further processing or trade. Today about 90% of the copper extracted from mines is refined by this process.

Remelting and casting: Cathodes cannot be machined by plastic deformation directly because they are made of inconsistent material and must therefore undergo remelting, which is a delicate operation, since misdirection of the furnace would lead to pollution of the metal. On the contrary, at this stage, it is still possible to intervene on the bath, changing its composition, to obtain special properties. The liquid metal is periodically extracted from the furnace and stored in special ladles or other furnaces to await the subsequent casting operations. The latter is done

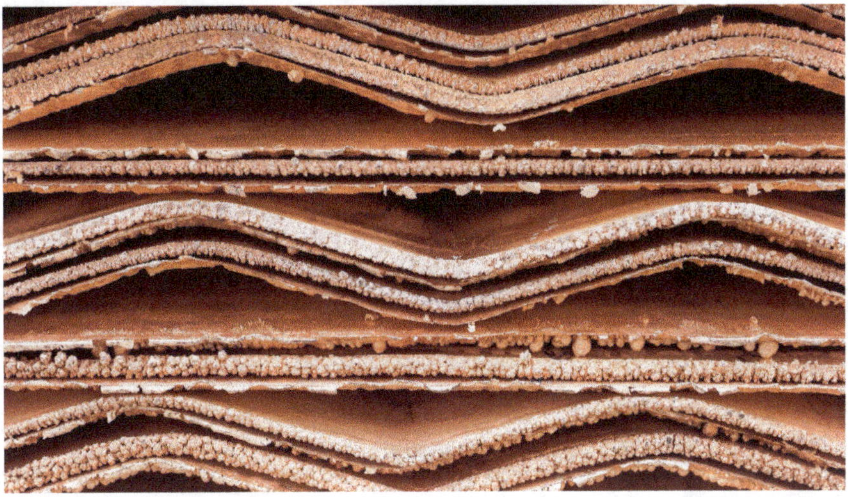

Figure 3.5 Semifinished copper cathodes

Figure 3.6 Semifinished copper wire rods

today mainly by continuous casting processes for wire rod, and semicontinuous for billets and plates. The shape of the solid metal is optimized according to the semifinished product to be produced: wire rod to obtain wire (Figure 3.6), billets for pipes and bars, and plates for sheets and strips.

Leaching: Leaching is an alternative mining technique, which involves spraying ore heaps, collected on impermeable soils, or large ponds, with a water-diluted solution of sulfuric acid. The acid percolates and is enriched with copper sulfate, which is sent to electrolysis cells. These are quite similar to conventional cells, except that the anode is made of insoluble metal alloys, the copper being made available directly from the solution. The process is known as direct electrodeposition (SX–EW: *solvent extraction/electrowinning*). The advantages are as follows: less energy consumed compared to traditional processes, no gaseous emissions, low capital investment, and the ability to operate cost-effectively even on a small scale. Minerals with a copper content of 0.1% can be processed, and for this reason, it is gaining in importance. It was estimated that in 2011, leaching and direct electrodeposition accounted for 17% of refined copper production.

Recycling: Another important source of copper is scrap recycling, described as secondary copper production. In 2011, it covered 41% of all refined copper production. Recovery and recycling of end-of-life products is an example of the sustainability of copper.

3.3.3 Characteristics of copper for electrical applications

Copper alloys for electrotechnical uses must contain a content of other metals, excluding silver, not exceeding 0.10% in order not to reduce conductivity.

The various alloys differ from each other in the composition of the share of the remaining 0.10%, which is also responsible for significant variations in technological characteristics. From the perspective of quantities produced and consumed worldwide, the most important types of copper are shown as follows (Table 3.2):

- copper Cu-ETP (*electrolytic tough pitch*) is an alloy obtained by electrolytic refining and treatment with pine log (as described above). It is characterized by the presence of oxygen and the absence of phosphorus.
- copper Cu-DHP (*deoxidized high residual phosphorus*) is a totally oxygen-free alloy in which a relatively high phosphorus content of 0.015%–0.04% is maintained to ensure deoxidation.
- copper Cu-PHC, Cu-HCP (*oxygen-free low-phosphorus* content) is a totally oxygen-free alloy with a low-phosphorus content of 0.002%–0.007%.

The presence of phosphorus makes it possible to eliminate brittleness phenomena in a reducing environment, improve cold plastic deformability, and, most importantly, increase brazing aptitude. The presence of oxygen reduces the strength of the joints. Therefore, copper of laminates intended to be used for roofing and tinwork, and also for the production of pipes for fluid distribution systems, is the Cu-DHP type.

Phosphorus, as opposed to oxygen, greatly reduces electrical conductivity. Therefore, the copper alloy suitable for electrical and electrotechnical applications is the Cu-ETP type.

Table 3.2 Composition of copper alloys

Copper type	Cu (+Ag) (%)	O (%)	P (%)
Cu-DHP	99.90 min.	–	0.015–0.040
Cu-ETP	99.90 min.	0.015–0.040	–
Cu-PHC, Cu-HCP	99.95 min.	–	0.002–0.007

Table 3.3 Nominal resistivity of unprocessed copper alloys

Copper type	Maximum resistivity (μΩ m)	Ratio to IACS res. (%)
Cu-HCP	0.01754	98.3
Cu-PHC, Cu-OF, Cu-ETP	0.01724	100
CuAg0.04, CuAg0.07, CuAg0.10	0.01724	100
Cu-OFE, Cu-OF1, Cu-ETP1	0.01707	101

Other technological properties, such as melting point or thermal conductivity, are not significantly affected.

The presence of oxygen can be compensated by the addition of silver according to three contents: CuAg0.04–CuAg0.07–CuAg0.10.

To obtain a copper with very high conductivity, even higher than that of common Cu-ETP, the oxygen content in it must be minimized. The result is an alloy called Cu-OFHC (*oxygen-free high-conductivity copper*), with an oxygen content ranging from 0.001% to 0.003%, and a maximum content of other impurities not exceeding 0.03% in total.

In the Cu-OFHC category, two alloys are particularly important:

- Cu-OF, with 99.95% copper;
- Cu-OFE (*oxygen-free electronic*), with 99.99% copper, no Ag, and oxygen under 0.0005%.

Alloys in the Cu-OFCH category are used where maximum conductivity combined with maximum purity is needed, such as vacuum capacitors, thermionic valves, particle accelerators,[3] and waveguides. Because of their high cost, they are not normally used for power transformers.

The intermediate products of the process, i.e., wire rods to be retrained, are specified according to UNI EN 1977 [3], which prescribes the maximum resistivities given in Table 3.3, taking as reference the IACS resistivity.

Conductors obtained by rethreading and possible subsequent quenching are specified according to UNI EN 13601 [4]. Their maximum resistivity values depend on the physical state of the material, i.e., the hardness possibly required (Table 3.4).

The degree of tensile strength depends on the work hardening process: Grades with higher hardness are also more prone to hardening loss when annealed, and this fact must be taken into account in the thermal and mechanical design of the transformer (Figure 3.7).

To achieve high and stable hardness at high temperatures, a silver content is added (Figure 3.8), at an additional cost and increased processing difficulty. This is employed only when necessary, such as for high expected electrodynamic stresses, or where enhanced safety rules are required (such as in nuclear power plants or cavern plants).

[3]For example, in the Large Hadron Collider at CERN in Geneva, the super-magnet windings are constructed of Cu–OFE and niobium–tantalum fibers contained in a high-conductivity copper matrix.

Table 3.4 Nominal resistivity of machined copper alloys

Physical state	Vickers hardness (kgf/mm^2)	Maximum resistivity (μΩ m)	Ratio to IACS res. (%)
D	–	0.01786	96.6
H035	35–65	0.01724	100.0
H065	70–95	0.01754	98.3
H085	90–115	0.01786	96.6

3.4 Conductors for windings

3.4.1 Types of conductors

The conductors used in transformers can be cold workable *wires* or *rods* and *profiles* (Figure 3.9). The former are used to make the windings and connections to the bushings and on-load tap changer. The latter are used for high-current connections, for example, arc furnace outlets, and are ordered already machined to final dimensions. In the following, we will discuss only winding wires, which have the greatest diversity and complexity in construction.

The production of winding conductors takes place in factories specializing in the production of these products for electrical applications (e.g., Essex, Asta, Erikoğlu).

The starting semifinished product is the wire rod, which is drawn to give the wires the appropriate shape. Then the wires are covered with the appropriate insulating vanishes and thermosetting resins and then (where applicable) assembled into a cable, which is finally covered with the appropriate tape of insulating material.

The simplest conductor is the enameled round wire (Figure 3.10).

The processing is done by a machine called *drawing die* (Figure 3.11), which produces by drawing the conductor, which is then covered with a varnish that provides protection from oxidation and some dielectric insulation.

These conductors are only suitable for windings with low currents and low voltages; therefore, they are used in small LV transformers.

For higher currents, enameled copper strands are used, which can be single, double, or triple (rarely quadruple) stranded, shown in cross section in Figure 3.12. These strands are typically covered with an insulating tape.

Flat conductors reach about ~150 mm^2 in useful cross section. Since the design current density is typically 2–4 A/mm^2 ([6], § 4.2 p. 124), for currents above 300 A, multiple conductors are used in parallel. However, to ensure that the different conductors are subject to the same induction, *transpositions* must be made along the winding (Figure 3.13), so that on average each conductor is invested by the same flux. Otherwise, this would create a potential difference and thus current circulation between conductors in parallel.

For currents above 200 A, the number of transpositions of conductors in parallel would greatly complicate the construction of the winding. Therefore, in these cases of very high currents, the *continuously transposed cable* (CTC) is used that is a conductor

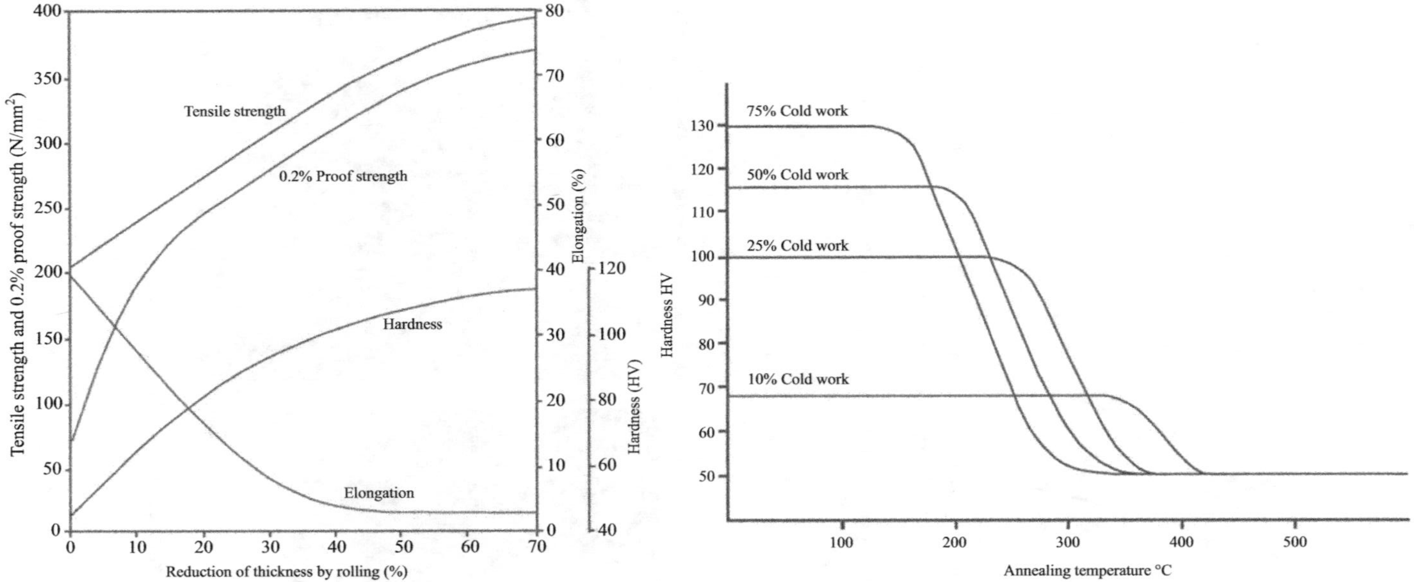

Figure 3.7 Left, relationship between quenching and mechanical strength; right, relationship between tempering temperature and hardness loss for various quenches

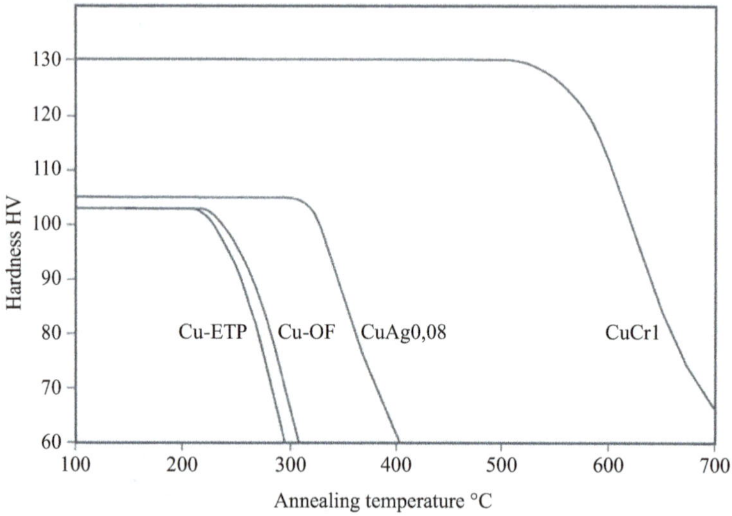

Figure 3.8 Relationship between tempering temperature and hardness loss for various alloys

Figure 3.9 Examples of rigid copper profiles and connectors

Figure 3.10 Coils of enameled wire

Figure 3.11 Drawing die for the production of wire with circular cross section

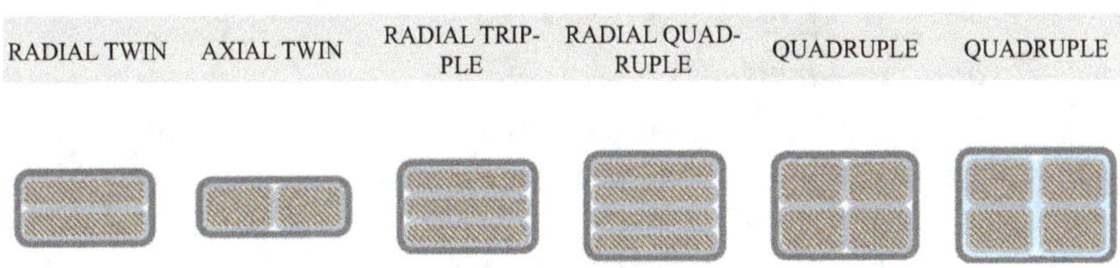

Figure 3.12 Types of multistranded conductors

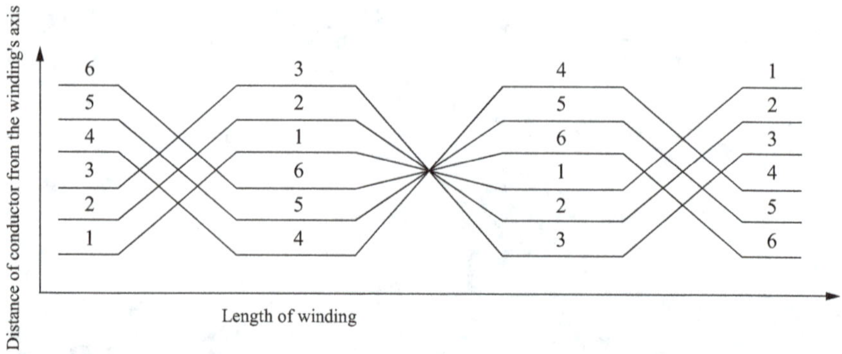

Figure 3.13 Diagram of transposition of six conductors in parallel

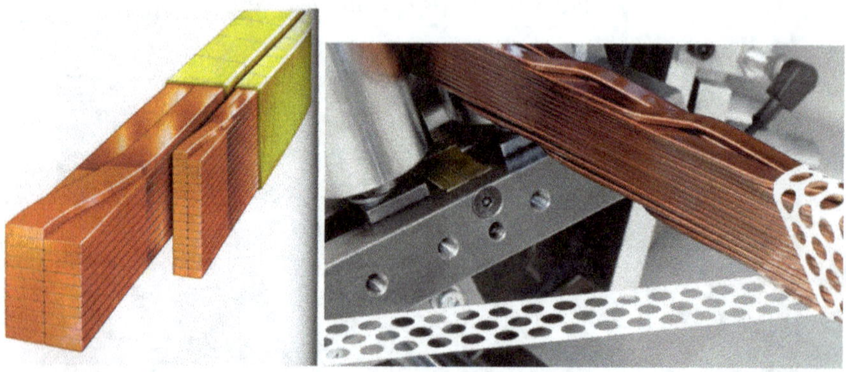

Figure 3.14 Drawing and photograph of a continuously transposed cable

in which several strands (even dozens of them) are already transposed at fixed intervals within the bundle, which results in two adjacent stacks of strands (Figure 3.14).

Manufacturing with transpositions is done by moving the upper strand from the right stack to the left stack, and simultaneously moving the lower strand from the left stack to the right stack (Figure 3.15).

This operation is carried out by automatically controlled machines (Figure 3.16).

3.4.2 Insulating varnishes

The most widely used varnish, guaranteed up to the use temperature of 120 °C, is *polyvinyl acetal* (PVA). It is a polymer of vinyl (Figure 3.17), given by combining one or more aldehydes with polyvinyl alcohol $[CH_2CH(OH)]_n$, obtained by hydrolyzing polyvinyl acetate $(C_4H_6O_2)_n$, best known as a constituent of vinyl glue (Figure 3.18).

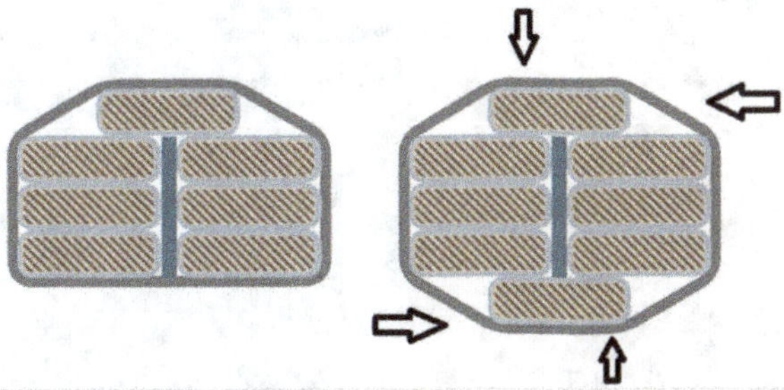

Figure 3.15 Diagram of a CTC transposition

Figure 3.16 Transposing machine and detail of the entry of wires into the machine

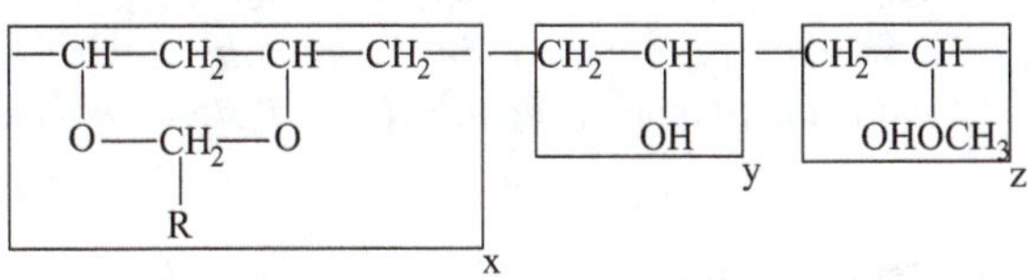

Figure 3.17 Generic formula of a polyvinyl acetal, with acetal (x), hydroxyl (y), and acetyl (z) groups [7]

For higher temperatures, up to 180 °C, *polyesterimide resin* (PEI), obtained by the reaction of an aromatic diamide with trimellitic acid and trimellitic anhydride, is used (Figure 3.19), to form a dicarboxylic acid monomer of diimide (DIDA), which is then polymerized. The strength is given by the presence of the nitrogen atoms in the structures called *imides*.

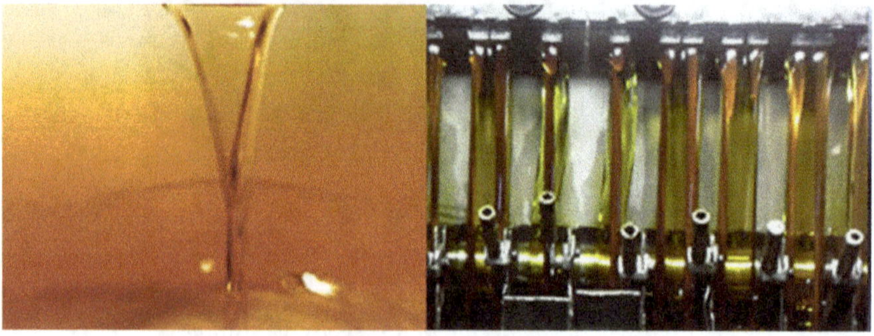

Figure 3.18 Appearance of a liquid state insulating varnish (left) and conductor coating process (right)

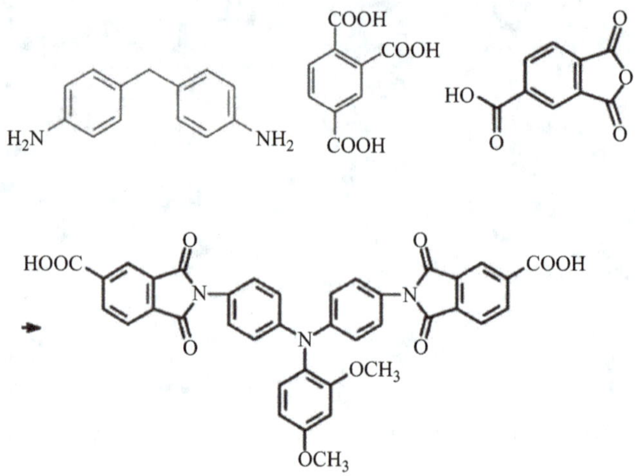

Figure 3.19 Top, from left: a molecule of 4,4′-methylenedianiline (MDA), the most widely used aromatic diamide, a molecule of trimellitic acid, and a molecule of trimellitic anhydride. Bottom: a monomer of DIDA

At even higher temperatures, up to 220 °C, *polyamide imide resin* (PAI) is used, obtained by a process similar to that used to produce PEI (Figure 3.20). This compound, after application, is also very resistant to solvents and oxidizing agents, making it an excellent insulator for the bare conductor. However, the cost of this varnish is much higher than that of the more common PVA varnish.

There are also other insulating varnishes, such as *polyurethane* varnishes with thermal class 155 °C and polyester varnishes with thermal class 180 °C, which are directly solderable, i.e., the varnish does not have to be hot-removed before making joints between strands, but they are used on dry-type transformers because of their poor resistance to insulating oil.

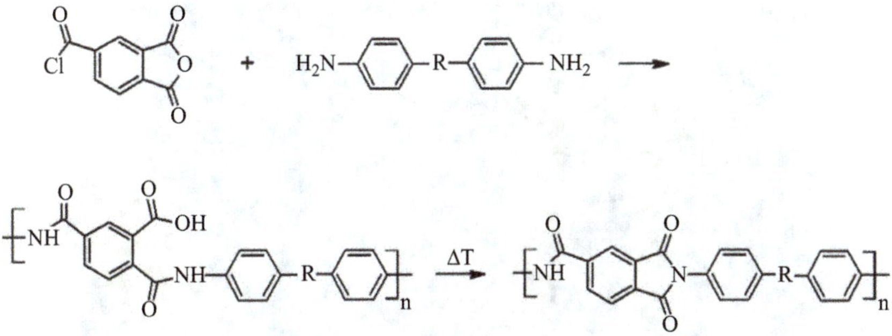

Figure 3.20 Synthesis reaction of polyamide imide

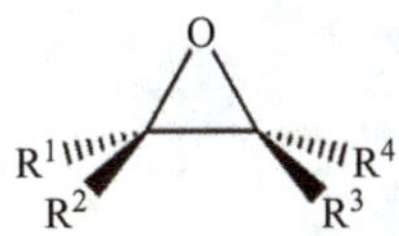

Figure 3.21 Generic epoxy group

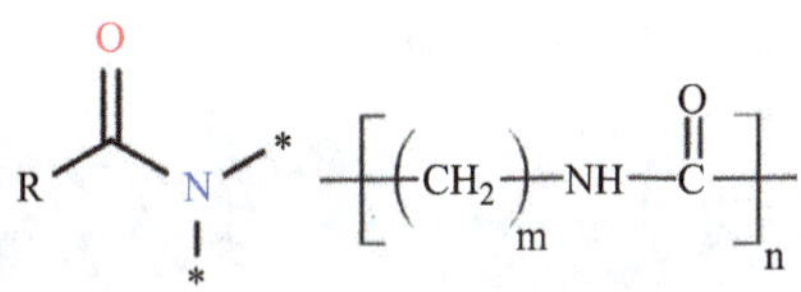

Figure 3.22 An amide functional group (left) and an aliphatic polyamide monomer (right)

3.4.3 Thermosetting resins

Transposed cables that need high mechanical strength are also coated with a *thermosetting resin* (or *bondable resin*), which cures and cements after the winding is baked in an oven, making the cable structurally a single rigid body.

The resin normally used for temperatures up to 120 °C–140 °C is *epoxy resin*, a mixture of a prepolymer, i.e., an oligomer capable of forming further polymerization, with epoxy groups (Figure 3.21) and a diluent, which solidifies at high temperature.

For higher temperatures, up to 180 °C, *aliphatic polyamide resins* are used, formed from chains containing the amide group (Figure 3.22). These are thermoplastics, which are also the basis of commercial products such as nylon.

3.4.4 Insulating tapes

Conductors for medium and high voltages are typically covered with a continuous tape, wound helically along the length of the conductor (Figure 3.23).

Figure 3.23 Operation of automatic taping of a conductor

Figure 3.24 Different types of tapes

The tapes can be made of a continuous band or strips that only partially cover the conductor (corded cable) or perforated with holes (perforated tape) or with a taping pitch greater than the width of the band (diamond taping), to allow more coolant circulation (Figure 3.24).

The most commonly used material for taping is *kraft paper*, a paper of forestal origin generally derived from Scandinavian spruce, which is highly refined but not bleached ([8] § 2.1), so as to obtain a cellulose concentration of 76%–86%, within a semi-cellulose matrix of 10%–15% and with a lignin residue of 4%–8%.

Cellulose is a polymer of glucose (Figure 3.25) with a degree of polymerization (DP), i.e., the number of component glucose monomers, at the origin of ~1200

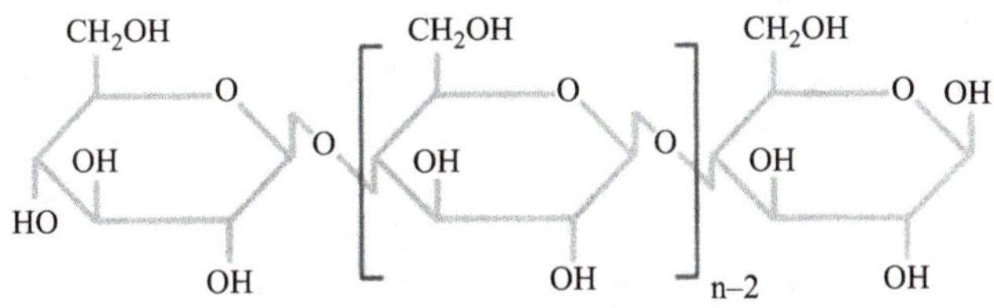

Figure 3.25 Chemical structure of cellulose polymer

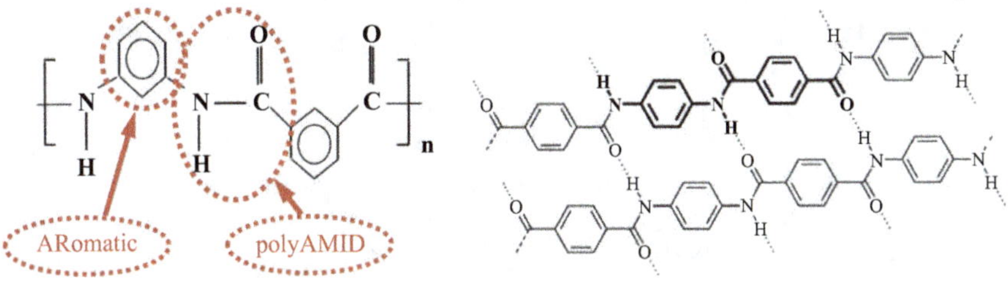

Figure 3.26 Aramid monomer (left) and polymer structure of Kevlar (right)

([9] § 2.2). These very long polymers form the fibrils that give the high mechanical strength to kraft paper, which is necessary to withstand the operations of taping the conductor and bending the winding coils.

Kraft paper is guaranteed up to thermal class 105 °C; beyond this temperature rapid degradation of cellulose molecules begins by pyrolysis and, in the presence of oxygen and oil acidity, by oxidation and hydrolysis.

For applications at higher temperatures, up to 120 °C, or to obtain a paper with a reduced degradation rate, chemically treated paper is used to remove compounds most prone to hydrolysis, e.g., polyacrylamide is used, and a treatment with reagents that replace oxygen atoms with nitrogen atoms, reducing the oxidizability of the paper. This chemically treated paper is called *thermally upgraded paper* (TUK), normally green in color so to distinguish it from the beige kraft paper.

For higher temperatures, up to 220 °C, the most popular material is *aramid* (an acronym given by the haplology of the words *aromatic polyamide*), a synthetic fiber developed by DuPont and marketed as Nomex® 410 for electrical applications (white in color), and as Kevlar® for the textile industry. Aramid is a polymer, the monomer of which is composed of an aromatic part and a polyamide (Figure 3.26).

The aging curve, shown in Figure 3.27, compares the expected service life of three materials as a function of temperature. According to Arrhenius' law, service life decreases exponentially as a function of operating temperature.

Recently, DuPont has introduced other varieties of Nomex tape, with special characteristics. Nomex 926 has greater oil impregnability and is therefore more suitable for use in transformers filled with insulating liquid than Nomex 306, which was developed for dry-type transformers. Nomex 818 has superior dielectric

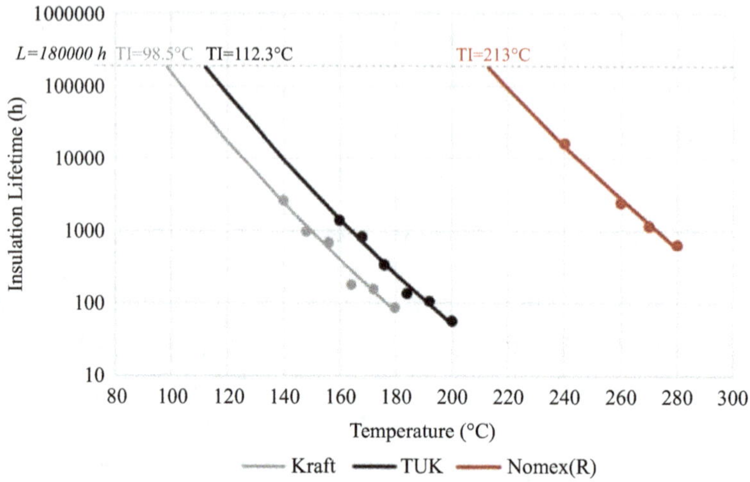

Figure 3.27 Thermal aging curve

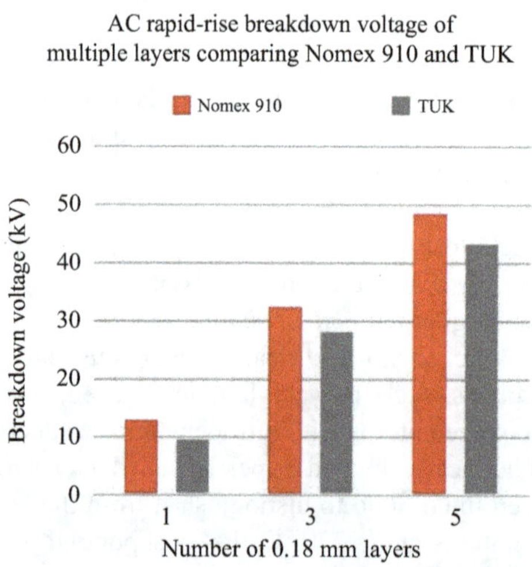

Figure 3.28 Comparison of dielectric characteristics of Nomex 910 and TUK

characteristics to Nomex 410, with a continuous voltage withstand of 3.2 kV/mm instead of 1.6 kV/mm due to mica content. Nomex 910 is a hybrid cellulose and aramid tape with intermediate characteristics between TUK paper and Nomex 410 (Figure 3.28).

DuPont claims a thermal class of 130 °C for Nomex 910 paper when used with mineral oil and 140 °C when used with certain esters. Therefore, this product is particularly attractive for use in transformers insulated with natural ester.

3M has developed another synthetic fiber, called *3M Liquid-Filled Transformer Insulation*, with thermal class 155 °C and reduced hygroscopicity. However, the poor tensile strength, at 35 N/cm compared to 141 N/cm of Nomex for 0.13 mm thick tape, prevents its use for individually taping conductors. This product remains attractive for uses in distribution transformers like taping complete windings.

3.5 Windings

3.5.1 *Types of winding*

A *coil* is a solenoid formed by several turns surrounding the magnetic core. The coil is the basic component of the *windings* of the transformer. These are a set of coils connected so that the desired circuit is made, leaving the availability of two or more terminals[4] for external connections. Transformers with core-type active parts are more common in transmission grids, and therefore, windings described below are referred to that type of construction.

The simplest coiling is the *crossover coil*, without transpositions (Figure 3.29). This is used with conductors of circular cross section for low voltages and currents (maximum ~20 A).

For voltages typically up to 6.6 kV, the *spiral coil*, also called *cylindrical coil*, is used, consisting of taped conductors *flat wound* or, more rarely, of *edge wound*. Coiling takes place on the bakelite cylinder (or other similar material) that separates the column from the winding, since the spiral coil has no independent structural resistance (Figure 3.30).

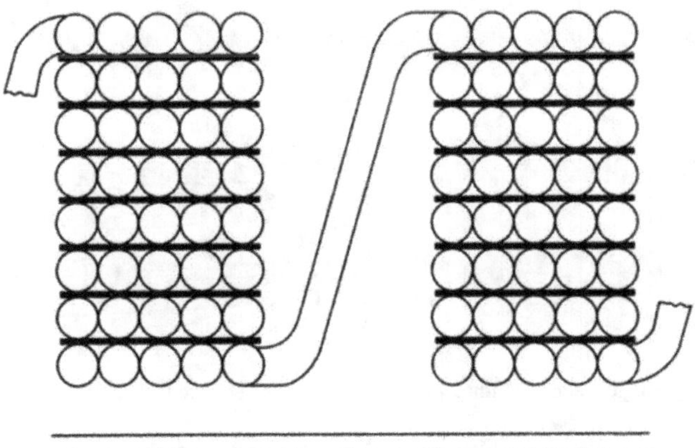

Figure 3.29 Coil with adjacent sections

[4]Intermediate terminals are called *taps* and are typically used for connection to the transformation ratio's changeover switch, namely the on-load or off-circuit tap changer, but taps can also be used to make particular connections in special transformers, as in the case of phase shifting transformers.

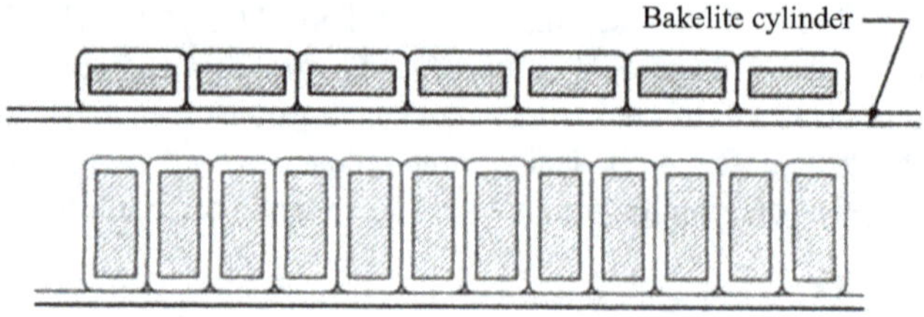

Figure 3.30 Spiral coiling with flat wound (top) and edge wound (bottom)

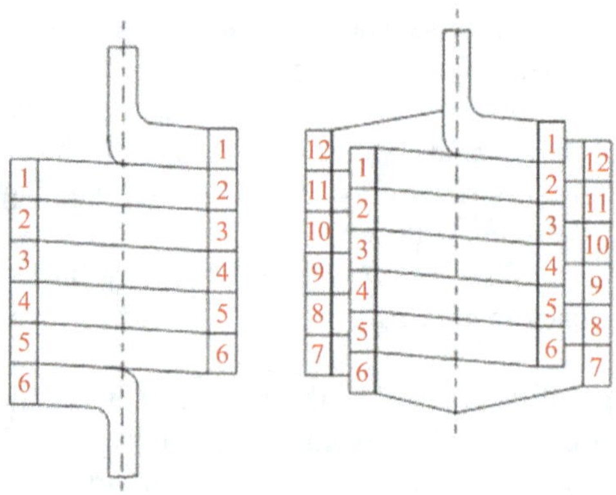

Figure 3.31 Single- and double-wound spiral coil

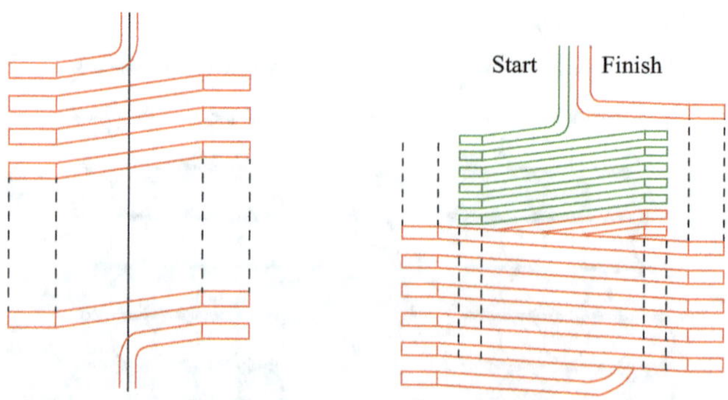

Figure 3.32 Single- and double-wound helical coils

If the number of coils requires it, the spiral coil can be made with several successive passes, forming concentric layers (Figure 3.31).

For higher voltages up to 33 kV, spiral coiling is normally not sufficient to provide the required insulation between turns, so *helical coiling*, consisting of one or more conductors in parallel (Figure 3.32), where each coil is separated from the adjacent one by a channel formed by spacers made of insulating material (Figure 3.33), so as to obtain a higher dielectric withstand voltage and allow the passage of insulating liquid for cooling.

For voltages up to 132 kV, the number of turns and the insulation between them and between the layers would cause excessive size and cost with the helical coil, so *continuous disc coil* is used, winding the turns concentrically a certain number of turns from inside to outside to form a disc, bounded by insulating spacers, and then from outside to inside in the next disc (Figure 3.34).

This solution has its limitation of use at 132 kV because the adjacent arrangement of successive coils within the same disc does not provide sufficient stray capacitance distribution to allow the appropriate withstand to lightning

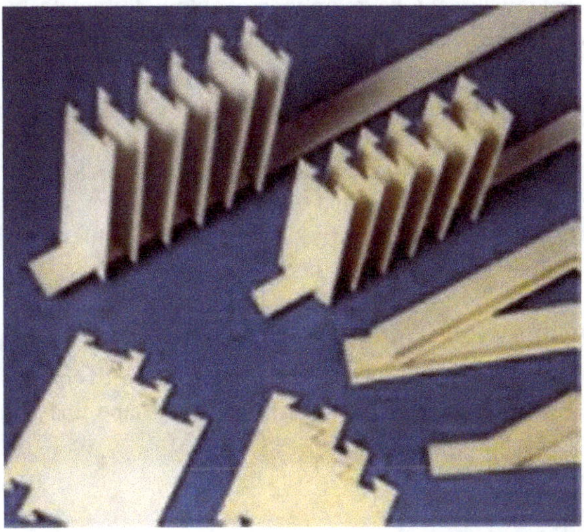

Figure 3.33 Various types of "dovetail" spacers and their use

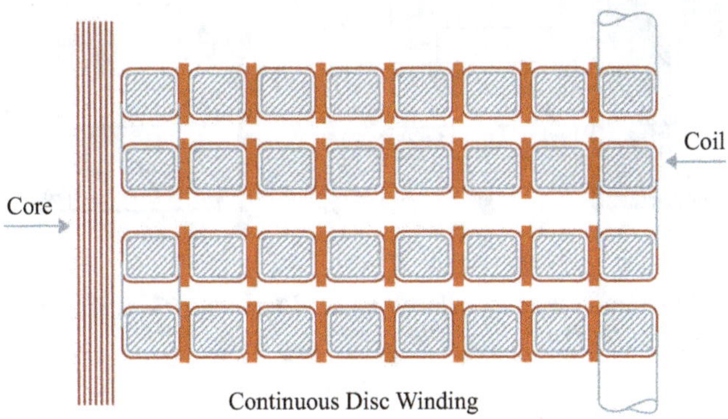

Figure 3.34 Continuous disc coil

impulse.[5] Therefore, since the 1960s, a different system of coiling called *interleaved disc coil* was introduced,[6] in which the coils are not each connected to the adjacent one, but to the next one, with appropriate transpositions (complex to achieve), in order to have an optimal distribution of stray capacitances (Figure 3.35).

In special cases, where special withstand capabilities to impulse are required (e.g., for truncated waveforms), the *shielded disc coil*, in which conductors, not connected to the winding circuit but held at line or ground potential, are inserted, with the sole function of shielding between turns and between discs (Figure 3.36). Clearly, this solution involves additional material expenditure and thus higher costs.

The coils of the regulating windings (connected to the tap changer) are spiral or helical, since they do not have only two output terminals but also several intermediate taps, they are not wound with the turns of the successive taps progressively on the same axis, but with the turns alternating between the different taps (Figure 3.37). This prevents the axial portion related to the unplugged taps from being completely devoid of induced current, which would create imbalances between the magnetomotive forces. This clearly has the disadvantage that the upper terminal of each tap must be connected to the lower terminal of the next one (and vice versa), with a considerable employ of connections to the tap changer, so that the set of connections is called a *harp* in the jargon.

3.5.2 Arrangement of windings

Individual coils are electrically connected to each other with conductors consisting of copper wire braids insulated with several layers of tape, the number of which

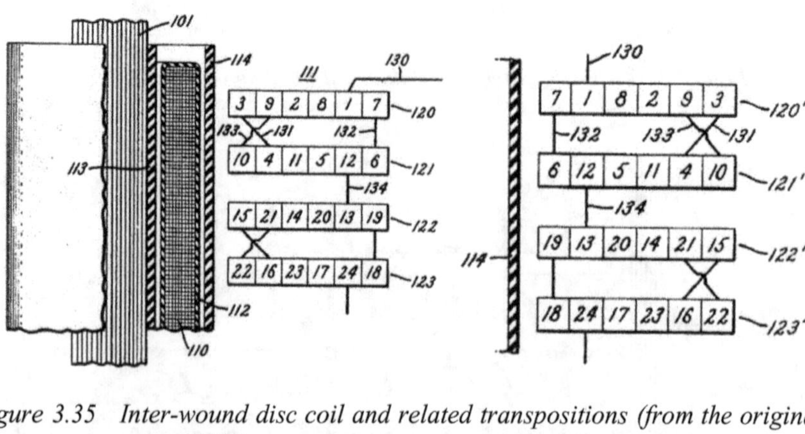

Figure 3.35 Inter-wound disc coil and related transpositions (from the original patent [11])

[6]The original term *interlaced disc* is now obsolete.

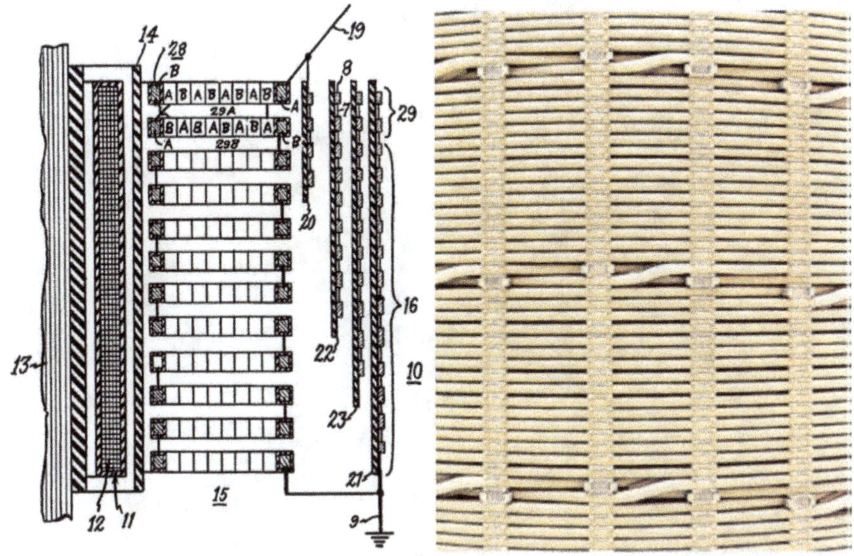

Figure 3.36 *Left, inter-wound and shielded disc coil (from the original patent [12]); right, disc coil in which transpositions between conductors can be seen*

depends on the degree of insulation of the connection. The coils are keyed one inside the other, with the interposition of insulating screens made of pressboard, that is, a pressed cellulosic material, or by other solid insulation with different thermal classes, where necessary, to form the winding, that is, the set of turns with a given rated voltage constituting one of the transformer circuits.

The windings are pressed so to give them compactness and the size necessary by the design requirements, treated in an oven to remove paper's moisture and bake the thermosetting resins, and finally keyed onto the magnetic core column.

Normally, the windings are keyed radially: from the innermost lower voltage winding to the outermost higher voltage winding (Figure 3.38), to achieve more efficient and less bulky sizing of the winding relative to the core.

However, since the leakage inductance depends on the mutual size and position of the windings, the necessity to meet special design requirements for this parameter may lead to different arrangements.

The same happens in presence of the regulation windings because the change in the short circuit voltage is a function of the number of regulation turns inserted, and is influenced by the type of coil arrangement (Figure 3.39).

For HV autotransformers, which are the most common equipment in electrical transmission networks, the windings are normally made up as follows, in order:

- regulation winding, spiral coil with taps;
- common winding, continuous disc coil with transposed cable;
- series winding, interleaved disc coil with triple stranded conductor or similar.

Figure 3.37 Regulating winding, with the coils of the same tap colored identically

The terminals of the star-connected windings are normally arranged with the line terminal on the top of the winding, for easier connection to the bushings on the tank's cover, and the neutral terminal on the bottom the winding, where the star-center connection is brought back to the cover, so that the connection to the neutral bushing can be made.

The line terminal of the series winding is sometimes arranged at mid-height (*central terminal*) to allow for greater isolation between the connection and the tank's cover (Figure 3.40); in this case, the series winding is made from two coils in parallel, with neutral terminals at the upper and lower ends.

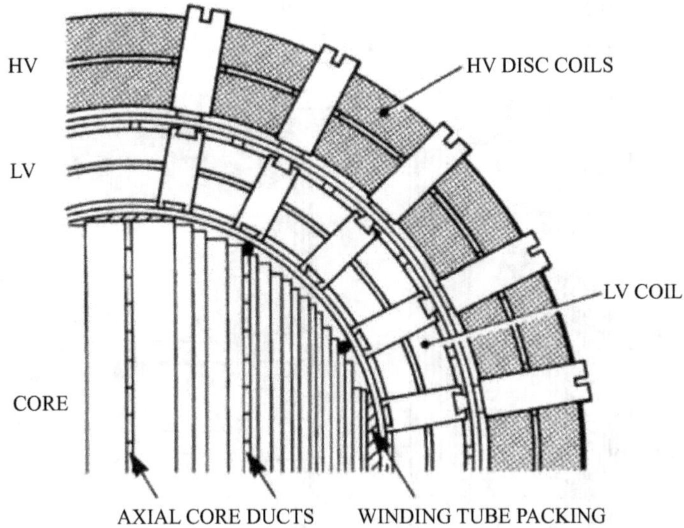

Figure 3.38 Arrangement of windings with respect to the core column

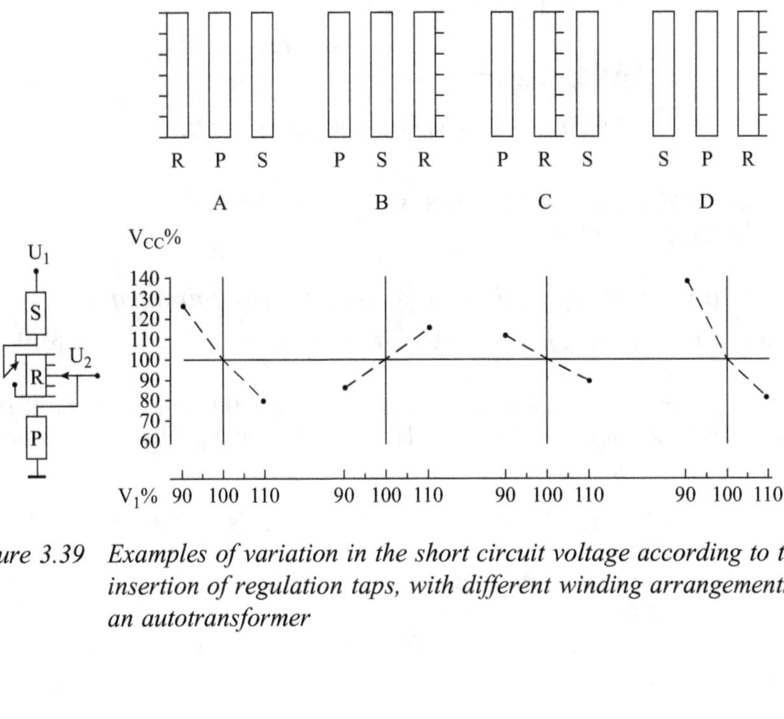

Figure 3.39 Examples of variation in the short circuit voltage according to the insertion of regulation taps, with different winding arrangements in an autotransformer

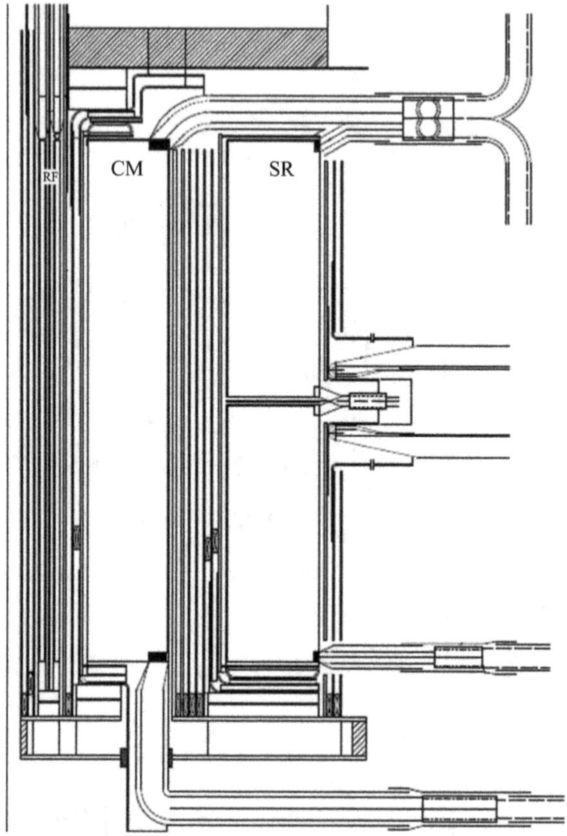

Figure 3.40 Winding with central outlet

3.6 Regulatory requirements and technological developments

3.6.1 Applicable regulations and suggested requirements

Winding conductors are specified in the standards issued by the IEC, in particular by its TC 55 named "winding wires."

Requirements for conductors are described in the IEC 60317 series of standards, titled "Specifications for particular types of winding wires." Of particular interest are the parts:

- 0–2 "General requirements—Enamelled rectangular copper wire," dealing with the enameled copper wires;
- 18 "Polyvinyl acetal enamelled rectangular copper wire, class 120," dealing with the PVA-enameled wires;
- 27 "Paper tape covered rectangular copper wire," dealing with the wires wrapped in paper.

The test procedures are specified in the IEC 60851 series of standards, titled "Methods of test for winding wires" (referenced to in the abovementioned standards), which set out the levels of testing and acceptance.

Solid insulators are specified in the standards managed by TC 15 named "Solid electrical insulating materials."

The requirements for paper are contained in the IEC 60554 series of standards, titled "Specification for cellulosic papers for electrical purposes." In particular, kraft and TUK papers are described in Parts 3–5 "Special papers" and testing in Part 2.

Taping paper must meet at least the requirements for type 5A2. For special uses, such as high-tension gradients, a paper with a higher specific density of Type 5A4, called *supercalendered*, may be specified. The standard does not distinguish between kraft and TUK paper, and there is no aging test; therefore, this requirement must be specified explicitly. In addition, the standard does not specify a minimum DP. Therefore, to ensure the quality of the material, a minimum DP value of 1000 should be specified, to be measured after oven treatment of a sample of taped conductor.

Unfortunately, there are no specific standards for cemented transposed cables. However, the industry has developed test methods that have become de facto standards, particularly the bending test called "p2k," which measures the mechanical strain/deformation curve of the transposed cable after cementing. This test should also be mentioned in the technical specifications of winding wires.

Normally, conductors enameled with PVA varnish and the transposed cable with epoxy resin are suitable for the temperature requirements of autotransformers in electric transmission grids.

Recently, vegetable oil (natural ester) is gaining wider application as an insulating fluid, with thermal performance superior to mineral oil. Planning and operating requirements for transmission networks do not provide for the continuous use of autotransformers with load factors close to unity, let alone under continuous overload, but the possibility of overloading isolated transformers with vegetable oil can be considered.

The thermal aspects of transformer operation are complex topics that will be described in a later chapter. In summary, the standards related to it include the IEC 60076-7 "Loading guide for mineral-oil-immersed power transformers" and IEC 60076-14 "Liquid-immersed power transformers using high-temperature insulation materials." In particular, the latter provides special requirements for being able to declare a transformer thermal class higher than the standard 105 °C. Another chapter describes the caveat of using TUK paper and vegetable oil, which per se does not allow the declaration of a class higher than 120 °C and a resulting permanent overload.

It follows that in order to pursue the increase of the thermal operating limits of new transformers (which would be cost-effective if the rated power were reduced and compensated by an increase of the overload capability), also solid insulation with high-temperature materials should be properly specified.

Tape with hybrid paper-aramid insulation is a compromise solution for overtemperatures above the current but still contained within 130 °C, but both the insulating varnish and thermosetting resin would have to be changed. The solid insulation in contact with the conductors would have to be replaced, the pressboard being guaranteed for class 105 °C, while the rest of the solid insulation not in contact with the hot-spot could remain unchanged. This technological solution, which requires careful consideration by the manufacturer and careful design review, is referred to in IEC 60076-14 as *mixed hybrid insulation*. At this time, it is reiterated that there are no technical considerations and cost-benefit analyses that support the pursuit of this solution.

3.6.2 Technological innovations

Enameled and taped copper winding conductor is a product with an established technology; however, research and development seek to achieve optimized products with ever-increasing performance.

The main topics of innovation are as follows:

- the process of quenching copper to increase its mechanical properties;
- the production of transposed cable with smaller strands (Figure 3.41), to reduce eddy current losses and conductor size;
- insulating varnishes and bonding resins with higher thermal performance, or with a better trade-off between cost and performance;
- synthetic insulating tapes with intermediate thermal class between paper and aramid [13], and competitive cost.

The previous areas of innovation are combined with increasingly accurate quality controls to detect impurities in the alloy and any inclusions, even microscopic ones, with the aim of preventing latent defects that can evolve into reduced performance and in-service failures (Figure 3.42).

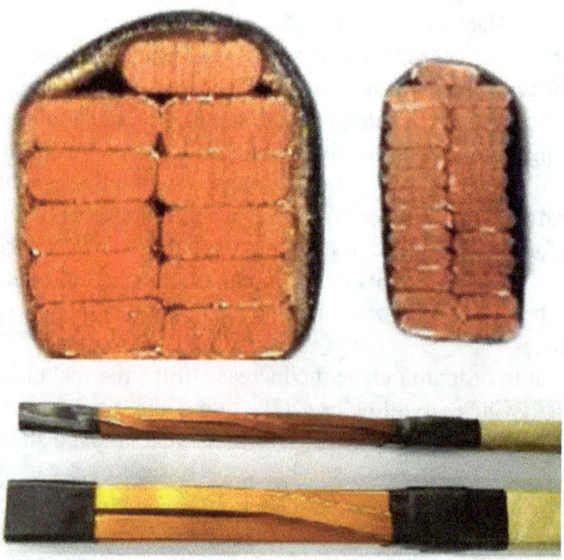

Figure 3.41 Comparison of normal transport cable and one with reduced flatlets

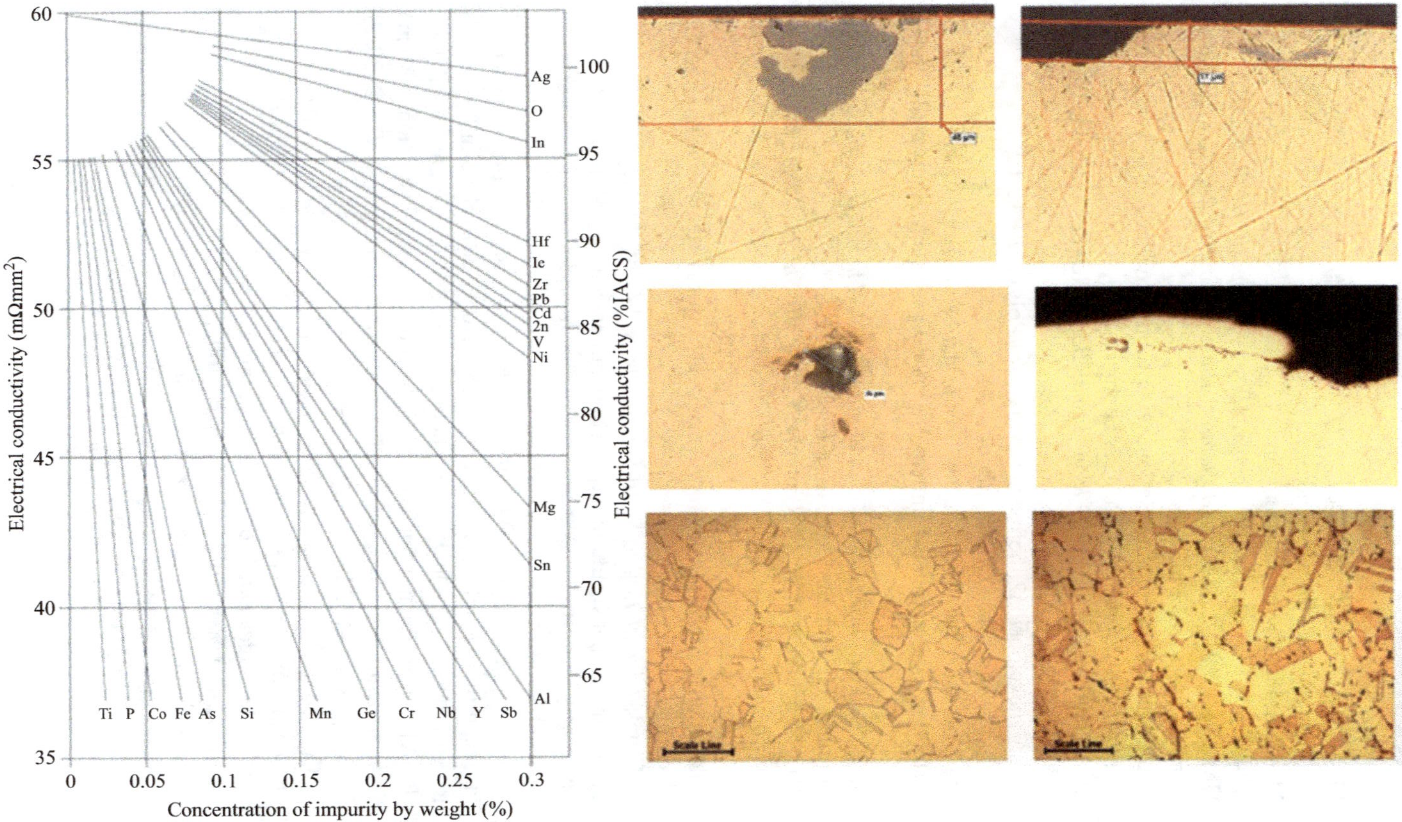

Figure 3.42 *Left: effect of impurities on copper conductivity; right: examples of impurities and (below) hydrogen embrittlement seen under a microscope*

References

[1] Electrical4U: *Different Types of Transformer Winding*, https://www.electrical4u.com/transformer-winding/

[2] MatWeb: http://www.matweb.com/

[3] UNI EN 1977: *Copper and Copper Alloys – Copper Wire Rod*, 2013.

[4] UNI EN 13601: *Copper and Copper Alloys – Copper Rods and Wires for General Electrical Purposes*, 2013.

[5] Copper Development Association: *High Conductivity Copper for Electrical Engineering*, Publication No. 122, February 2016.

[6] M. J. Heathcote: *The J & P Transformer Book*, Newnes, 12th ed. 1998.

[7] The Great Soviet Encyclopedia: *Polyvinyl Acetal*, 3rd ed. 1970–1979, https://encyclopedia2.thefreedictionary.com/Polyvinyl+Acetal

[8] Nynas: *Transformer Oil Handbook*, 2004.

[9] Cigré brochure No. 393: *Thermal Performance of Transformers*, 2009.

[10] Bharat Heavy Electricals Limited: *Transformers*, Tata McGraw-Hill, 2nd ed. 1998.

[11] G. E. Sauber: *Interlaced Disc Winding with Improved Impulse Voltage Gradient*, U.S. Patent 3528046, Filed Nov. 22, 1966.

[12] R. C. Degeneff, and W. N. Kennedy: *EsTernally Shielded Disc Windings for Transformers*, U.S. Patent 4318066, Filed Mar. 2, 1982.

[13] De Angeli: *CTC with Smaller Strips*, brochure, 2016.

[14] Wikipedia: *Electronic Band Structure*, https://en.wikipedia.org/wiki/Electronic_band_structure

[15] C. Kittel: *Introduction to Solid State Physics*, Wiley, 7th ed., 1996.

[16] European Copper Institute: https://copperalliance.it

[17] Polymer Properties Database: *Polyamideimide (PAI)*, https://polymerdatabase.com/polymer%20classes/Polyamideimide%20type.html

[18] Radosław Szewczyk (DuPont): *High temperature solid insulation*, Presentation at Doble Conference, 26–28 October 2016, Dublin.

[19] R. Szewczyk, J. C. Duart, H. Wilhelm, and P. Fernandes: Diagnostic tools (DGA) for resilient transformers with aramid-based insulation systems, *27th International Conference on Electricity Distribution (CIRED 2023)*, 2023, pp. 2828–2832.

[20] F. Coppadoro: *Variation of short circuit voltage in autotransformers for interconnection on different regulating sockets as a function of patterns and winding arrangement*, L'elettrotecnica, vol. LX no. 3, March 1973.

[21] ABB: *Transformer Windings*, brochure, 2008.

[22] DuPont: *Nomex® 910 Engineered Cellulose*, data sheet, 2020.

[23] J. P. Dubey: *Role of copper in continuously transposed conductors*, Transformers Magazine, vol. 4, no. 4, October 2017.

Chapter 4
Insulating materials and dielectric design of transformers

4.1 Generalities

The dielectric design of transformers is of paramount importance in ensuring their operational reliability.

Its correctness involves both the choice of insulating materials and the design calculation for their optimal use, which are necessary conditions for complying with dielectric withstand requirements under all testing and operating conditions of the transformer.

To appreciate the peculiarities of the technology associated with insulating materials, the available materials and design criteria currently used in transformer construction will be described below.

4.2 Insulating materials

4.2.1 Solid insulation

Solid insulating materials are used for taping the winding conductors,[1] for the construction of insulating barriers, and for the supporting frame of the leads (connections to the bushings) and connections among windings and with the on-load tap changer (Figure 4.1).

Cellulosic materials are normally used to produce insulating tapes (kraft and TUK paper), and for barriers and mechanical supports of rigid or semirigid material of varying thickness, called *pressboards*, obtained by pressing or bonding several layers of paper.

Kraft paper is made from high-purity unbleached wood because bleaching agents reduce dielectric strength; it is typically derived from northern spruce. In general, wood is made of three main components (Figure 4.2):

- cellulose, about 40% by mass. It is a polysaccharide formed by a linear chain of a high number of glucose monomers (7000–15,000);

[1]No description is given here of the insulating varnishes that individually coat the strands because this topic is described in Section 3.4.2.

Figure 4.1 Active part with taped connections, insulating cylinders on windings and supporting frame

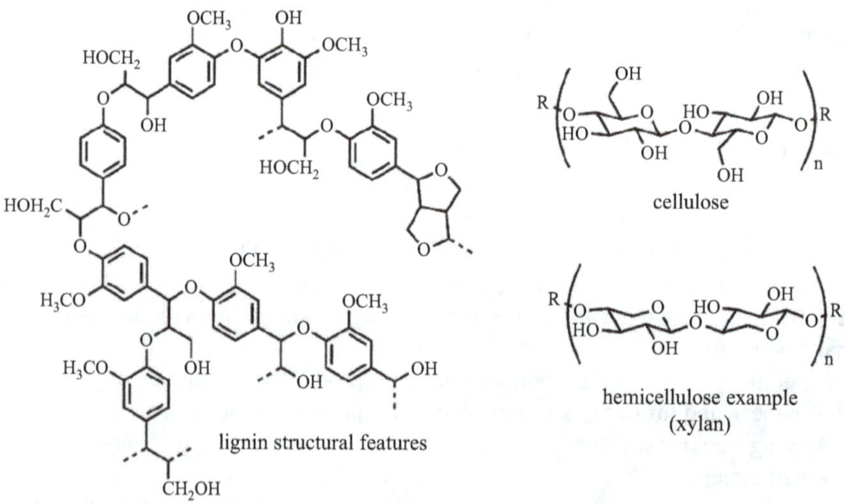

Figure 4.2 Chemical structure of the main components of wood

- semi-cellulose, about 30%. It is a matrix of polysaccharides that is not perfectly linear, consisting of fewer glucose monomers than cellulose (500–3000);
- lignin, about 27%. It is a complex polymer that imparts structural strength to wood.

The component that provides the tensile strength and flexibility needed for kraft paper is the cellulose. Therefore, wood is processed to obtain a compound with a high cellulose content (Figure 4.3).

(a) (b)

Figure 4.3 Left, schematic of the pulpwood and papermaking semifinished products process; right, a pulp making machine for Kraft paper

Initially, the wood is cut, chopped, and reduced to chips by means of abrasive grinding wheels. Then, the chips are processed into *paper pulp* by special machines, which through a mechanical and chemical process, employing sulfates, produce a semifinished product with a high cellulose content (~80%). During processing, the degree of polymerization (DP) of cellulose is reduced to about 1200 glucose monomers, sometimes up 2000 for the best grades.

Semi-cellulose, with a DP after processing of about 200, helps to impart firmness to the pulp but adversely affects tensile strength and has a higher moisture absorption capacity than cellulose; this makes the drying process of insulation materials difficult. Therefore, a good pulp should have a semi-cellulose concentration limited to 15%. Caustic soda (NaOH) is employed to dissolve lignin.

The resulting pulp is sent to specialized plants that subject the pulp-water mixture to mechanical processes through multiple passes in different machines (Figure 4.4), including centrifuges for slag removal and rollers for compression.

This produces a kraft paper layer with the cellulose molecules aligned in the direction of the lamination length and connected to each other by intermolecular hydrogen bonds favored by the aqueous solution, in order to obtain a material with high tensile strength (Figure 4.5).

These layers are then subjected to further processing, varying according to the desired end product:

- to obtain semirigid and rigid structures, the layers are pressed together with regulated pressures to achieve the required density and thus stiffness, or they are dried directly by rolling without pressing to obtain a very flexible *calendered board*, or several layers are bonded together by casein glue, or polyester resin for higher mechanical strength applications, to form rigid laminated structures;

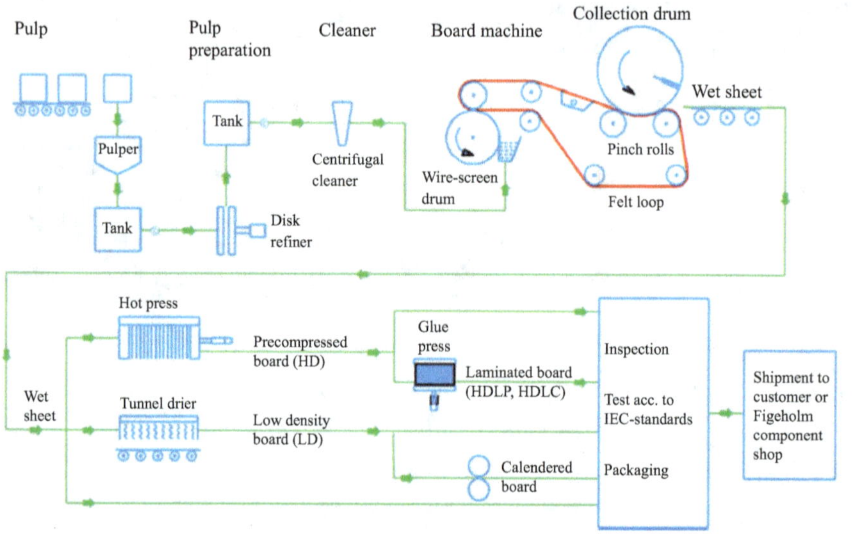

Figure 4.4 Schematic diagram of the process of producing pressboard from paper pulp

- to obtain flexible sheets, the layers are passed through rollers, which can give a smooth finish (*flat paper*), suitable for taping winding conductor paper, or through calendering rollers (Figure 4.6), can also give a crepe finish (*crepe paper*), suitable for taping irregular surfaces given its greater extensibility.

Pressboard for special structures (rings, tubes, etc.) is placed in hot presses that give the product its final shape before gluing (Figure 4.7).

The processing of flexible pressboard is difficult and can lead to a sub-optimal result due to shrinkage of the material because of the oven treatment; therefore, for the parts in contact with the conductors, it is preferable to use custom-made rigid structures, which, although at a higher cost, provide superior performance.

Laminated pressboard consists of several layers. In some cases, an aluminum foil is interposed between them to better distribute the electrical potential (Figure 4.8). This expedient is especially required for high-voltage connections to bushings, where the electric field gradient is high and must be properly controlled to avoid dielectric breakdown.

Kraft paper and paperboard have thermal class 105, i.e., they are guaranteed up to a continuous operating temperature of 105°C. In cases where it is necessary to achieve higher thermal performance, the paper can be subjected to chemical treatment before the formation of the insulating structures, to achieve greater resistance to thermal degradation. This is the case of *thermally upgraded kraft* or TUK, which can reach thermal class 120. For higher thermal classes, synthetic compounds are used.

Figure 4.9 represents the typical insulating elements of a medium- or high-voltage winding, made with paper for conductor insulation and pressboard for the rest.

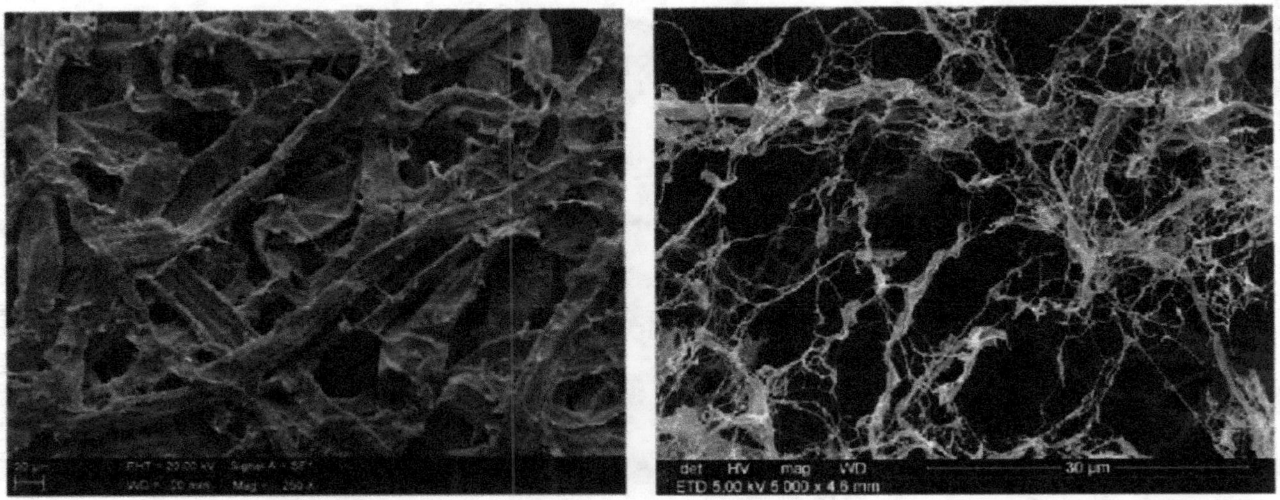

Figure 4.5 *Microscopic structure of paper pulp before treatment (left) and micro-fibrillated cellulose after treatment (right)*

Figure 4.6 Left, crepe paper; right, calender roller

Figure 4.7 Different types of pressboard insulating structures

Figure 4.8 Sheet of paper-aluminum

For more complex structural parts such as the bolts for fixing the support frame, it is not possible to use pressboard, which cannot be precisely molded, so laminated wood or synthetic materials must be used, the latter typically based on phenolic resins, the most common of which is known as *Bakelite* (Figure 4.10).

Bakelite is obtained through the reaction between formaldehyde and phenol, so that a cross-linked, thermosetting, and very strong polymer structure is obtained (Figure 4.11).

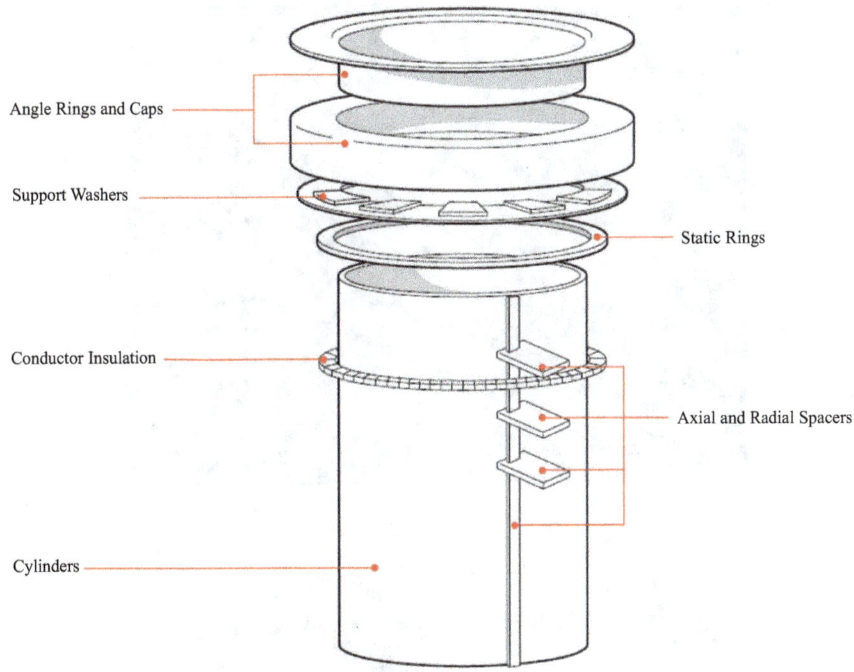

Figure 4.9 Main insulating elements in a winding

Possibly, fiberglass can be used to improve the strength of the synthetic material. For applications with high mechanical strength, epoxy resin is preferred.

Cellulose-based materials have good dielectric strength, low dielectric losses, and excellent impregnability to insulating liquids; however, they absorb moisture, i.e., they are hygroscopic, and are prone to thermal degradation by pyrolysis and chemical degradation by oxidation in the presence of oxygen and by hydrolysis in the presence of water (Figure 4.12).

Of the three mentioned degradation mechanisms, hydrolysis is the most dangerous under normal conditions, because among its degradation products, there is water and thus the mechanism is self-accelerating.

The overall effect of degradation is a reduction in the DP, which mainly causes structural brittleness of the paper (Figure 4.13). Instead, the reduction in dielectric strength occurs due to the very presence of moisture (Figure 4.21).

Recently, Weidmann introduced a new paper grade, called *INSULutions®* DPE [1], for which it claims thermal class 130 with mineral oil and 140 with ester. In addition, this product has a higher rate of insulating liquid impregnation and drying. Although there is no in-service data on this product, it may be of interest to use with natural ester to increase transformer performance. In fact, this material has the advantage over aramid of being completely of natural origin.

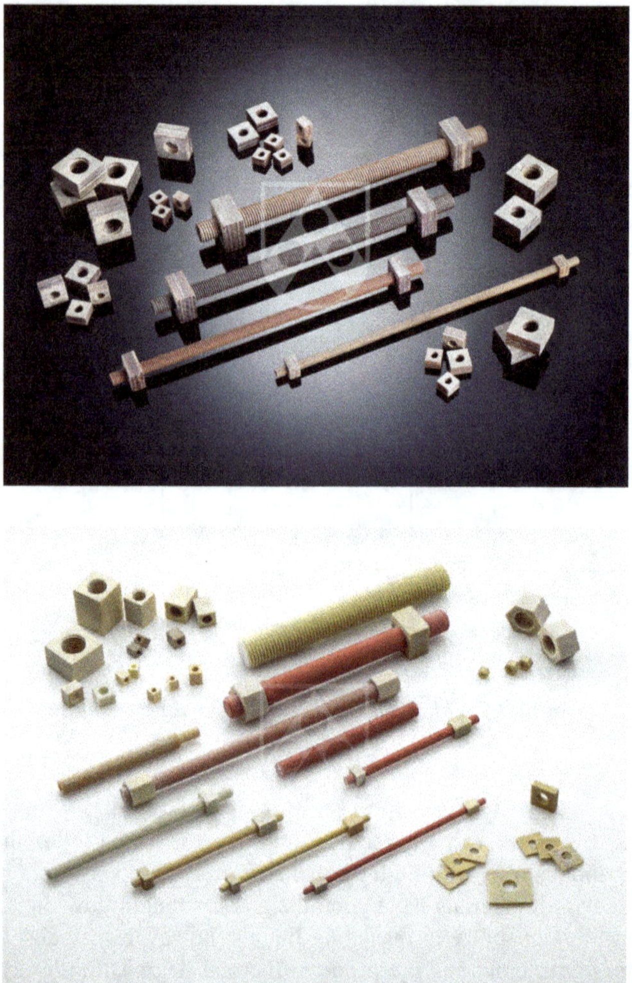

Figure 4.10 Bolts made of wood (top) and synthetic material (bottom)

The choice of insulation material is critical to ensure the reliability of the transformer. Poorly processed insulating components can release particles that pollute the insulating liquid (Figure 4.14). In addition, pressboard elements that are too flexible, although easy to work with, can easily deform because of mechanical stresses or the drying process, which causes non-negligible shrinkage of the pressboard dimensions (Figure 4.15). This fact must be considered both when pressing the windings, which must be repeated after the oven treatment, and when calculating the stresses on the materials in service, because the presence of moisture leads to an increase in size with increased pressures on the winding compression elements.

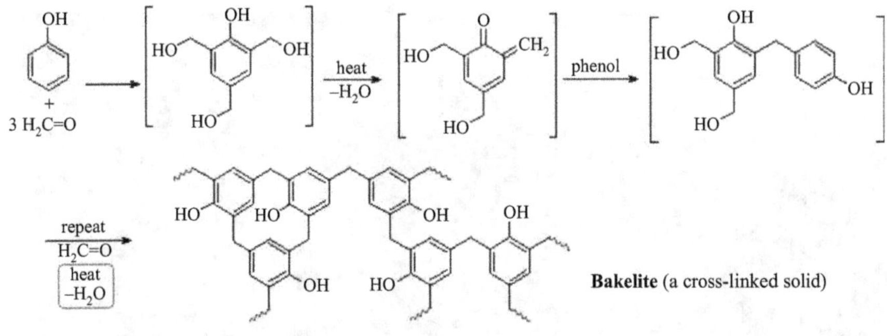

Figure 4.11 Bakelite preparation reactions

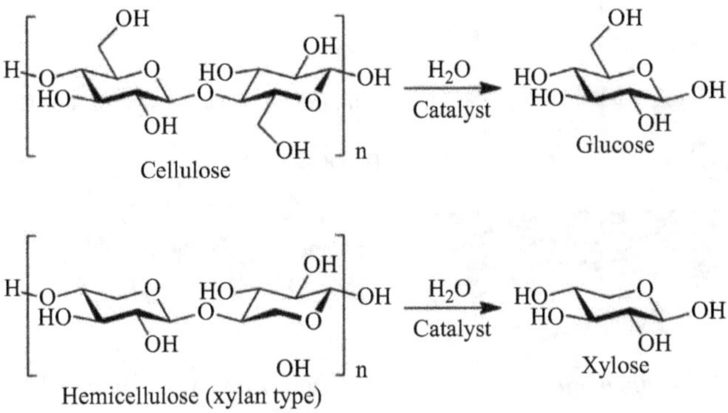

Figure 4.12 Hydrolysis of cellulose and semi-cellulose

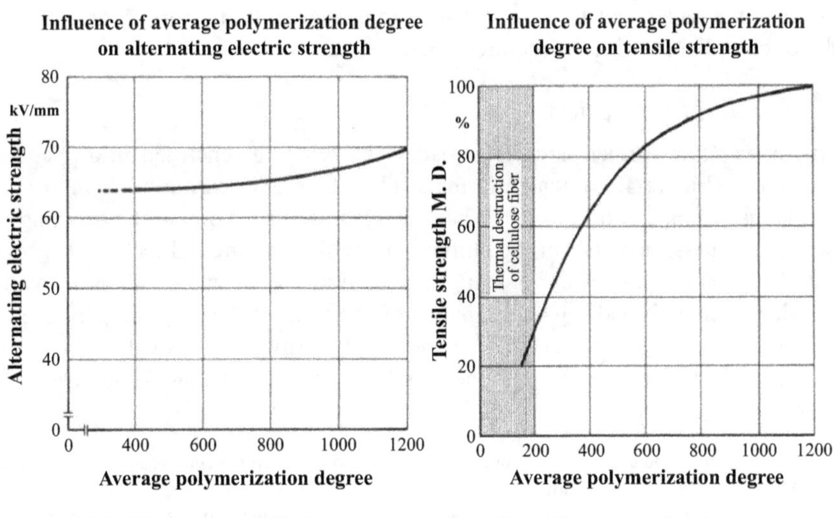

Figure 4.13 Variation of dielectric strength and tensile strength of paper as a function of DP

Figure 4.14 Comparison of spacers with processing defects and without defects

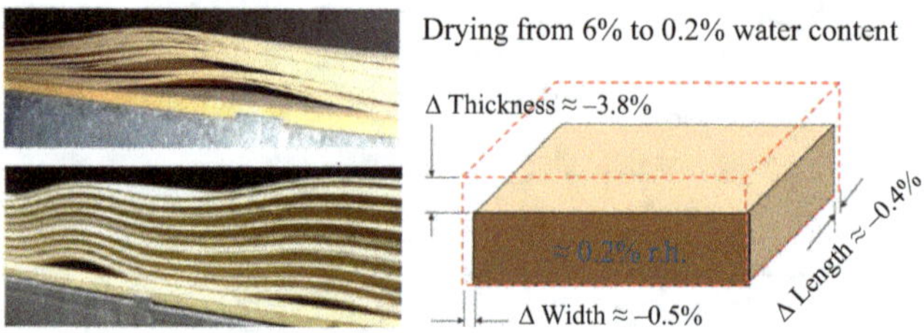

Figure 4.15 Deformation of layers in pressboard and shrinkage following drying

Given the previous issues, when selecting and drafting the insulation material specification, it is important to require compliance with the IEC 60554 series standards for paper and the IEC 60641 series standards for paperboard, as this ensures compliance with the required characteristics.

4.2.2 Liquid insulation

Small power transformers are often made with windings enclosed in a phenolic resin matrix (Figure 4.16), which forms a solid block after curing in oven.[2]

The use of only solid insulators has the advantage of avoiding the handling of insulating liquids, resulting in the elimination of fire and ground leakage hazards, as well as substantial material savings. In addition, the transformer has greater mechanical strength and a greater chance of overloading. Finally, the absence of liquid eliminates problems related to its degradation (moisture, oxidation, etc.). On the other hand, the inclusion of the windings in a solid matrix prevents any

[2]There is no clear boundary of convenience between dry and oil-filled realization; indicatively, up to 3150 kVA transformers are predominantly dry-type, although oil-filled realizations are possible for low powers if space is scarce. A specific application is that of pole-mounted transformers in the public distribution network.

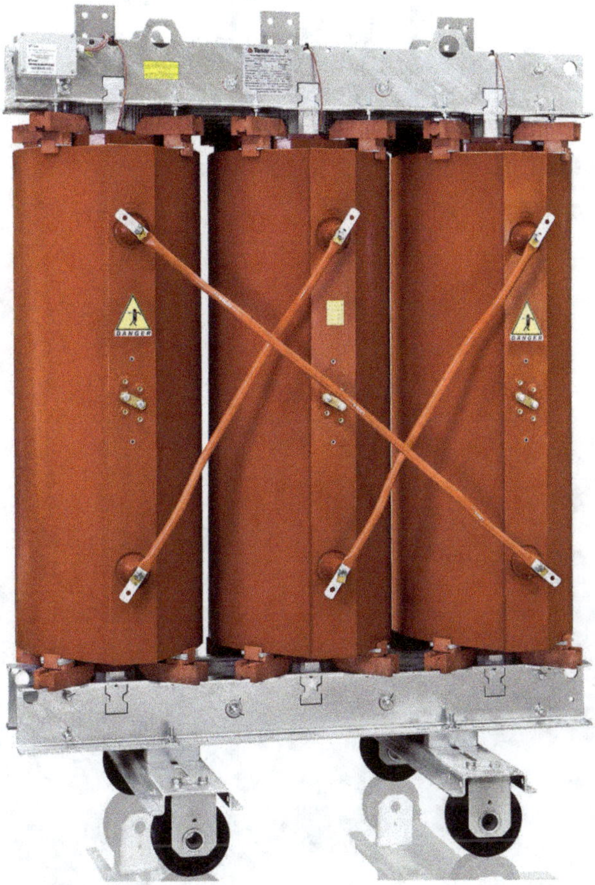

Figure 4.16 Three-dimensional representation of the active part of a dry-type transformer

inspection and repair work and complicates disposal operations because of the difficulty of separating the copper from the solidified resin, and also the design requires larger dimensions because the insulation between the active part and the tank is made of air, which has lower dielectric strength and lower heat dissipation capacity than an insulating liquid.

Medium and large power transformers consist of an active part immersed in an insulating liquid inside a metal case.

The insulating liquid performs the following main functions:

- increases the dielectric strength between parts at different potentials;
- conveys the heat produced by thermal dissipation from conductors and core to the cooling system;
- lubricates the impellers of the cooling system's pumps, where present.

Figure 4.17 Top: a small-power transformer with corrugated tank; bottom: medium-power transformers with conservator and radiators

Since liquids are incompressible, to compensate for volume variations at different operating temperatures, an expansion volume of the insulating liquid must be provided. For small power transformers, the elastic flexibility of the steel tank is sufficient to compensate for volume variations of the small amount of liquid present, while in medium and large transformers, an expansion vessel of adequate volume, called *conservator*, must be installed above the cover (Figure 4.17).

The insulating liquid must be kept separate from the air in the conservator to avoid contamination. This is done by inserting an expansion rubber bag inside the conservator (Figure 4.18), which inflates and deflates according to the volume of liquid in the conservator compartment not occupied by the bag itself.[3] In addition,

[3]The expansion bag is also called *oil-air separation membrane*.

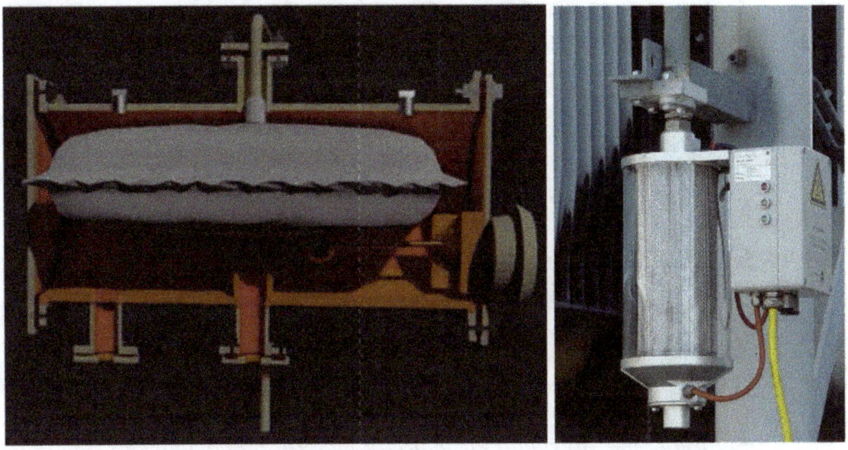

Figure 4.18 Left, schematic of a conservatory with expansion bag; right, a self-restoring desiccator

to prevent moisture from entering the bag, a *desiccator* with hygroscopic salts is installed on the end of the air supply pipe. Modern dryers are equipped with a heating coil that evaporates the moisture in the salts during the liquid expansion stages, so that the moist air is expelled outside, thus restoring the moisture-absorption capacibility of the salts without the need to replace them.

Small power transformers do not need to dispose of large power losses; therefore, cooling is achieved by radiation of the tank surface, called *corrugated* due to the presence of lamellae that increase its heat exchange surface area. In contrast, for medium and high power transformers, it is necessary to install a cooling system, consisting of radiators or heat exchangers.

The insulating fluids used in transformers are mineral oil (basically, a highly refined petroleum product similar to diesel), natural or synthetic ester, and silicone oil. Silicone oil is used only in special high-temperature applications, where space does not permit the installation of an efficient cooling system (this is the case in railway traction). The characteristics and use of esters are covered in Chapter 6; therefore, in the following, we will refer only to mineral oil.

Mineral oil is obtained by refining from petroleum[4] and consists of low molecular mass hydrocarbon molecules, which are liquids at room temperature, mainly of three types (Figure 4.19):

- paraffinic molecules, with linear or branched structure;
- naphthenic molecules, with a circular structure (cycloalkane), to which one or more linear structures, including branched structures, are attached;

[4]The Shell company also uses natural gas to produce, by a process called "gas to liquid" (GTL), a synthetic oil with high purity properties, high flash point and low volatility, at a cost clearly higher than oil obtained by petroleum refining.

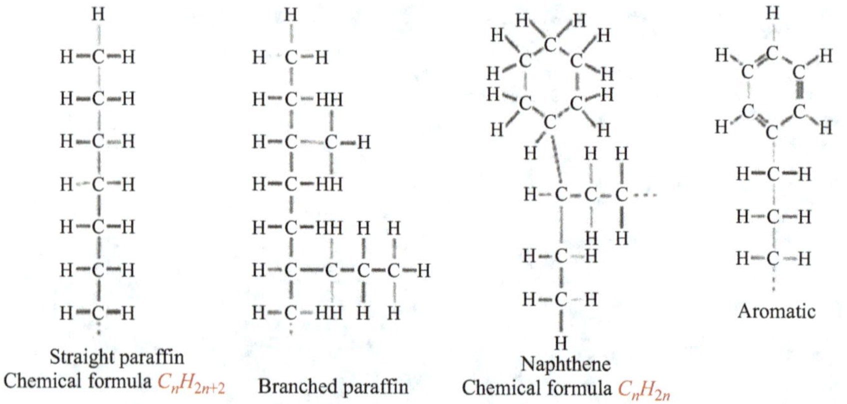

Figure 4.19 Molecular structure of mineral oil components

- aromatic molecules, in which there is a circular structure with double bonds, or several such structures (polyaromatic), and possibly one or more linear structures, including branched structures.

Paraffinic molecules have greater resistance to oxidation, but when they oxidize, they form acids that remain dissolved in the oil; they possess not only low volatility and higher flash point, but also higher viscosity and high pour point.

Naphthenic molecules have lower resistance to oxidation, and when they oxidize, they form *sludge*, semisolid compounds that tend to settle to the bottom of the case, but can also clog the circulation channels in the windings. Naphthenic molecules possess not only higher volatility and lower flash point but also lower viscosity and low pour point.

Aromatic molecules, especially polyaromatic ones, have high resistance to oxidation, and when they oxidize, they form phenols that have free radical-inhibiting abilities and thus mitigate the oxidative process. In addition, they possess good absorption capacity of the gases present in the oil. However, they have lower dielectric strength than other types of molecules.

There are also other types of molecules containing, in addition to hydrogen and carbon, other elements such as nitrogen, sulfur, and oxygen. In general, nitrogen and oxygen can promote the oxidation process and reduce the dielectric strength of the oil, while sulfur acts as an inhibitor for oxidation but can corrode copper and silver, the latter being used for the tap changer's contacts.

The characteristics required in a new mineral oil for use as an insulating fluid in a transformer, and how to measure them, are given in IEC 60296.

The main characteristics that identify a type of mineral oil are as follows:

- the *viscosity*, which represents the resistance to flow, and the *pour point*, which is the minimum temperature at which oil can flow as a liquid;
- the *breakdown voltage*, which is the voltage at which arcing occurs between two normalized electrodes immersed in oil 2.5 mm apart;

- the *density*, which is lower for mineral oil than for water (≤0.895 g/mL);
- the *dielectric dissipation factor*, which is the tangent of the dielectric loss angle of oil subjected to an electric field;
- the *acidity*, measured by the neutralization number, which is the concentration of basic reagent (KOH) that makes the pH of the oil neutral;
- the *flash point*, which is the temperature at which the oil catches fire when subjected to flame;
- the *oxidation stability*;
- the presence of stabilizing additives, particularly antioxidants.

In general, a naphthenic oil has a lower viscosity and thus is more suitable for use in high-power transformers where continuous oil circulation is required even at low temperatures. However, a paraffinic component may be useful to ensure a flash point above regulatory requirements (≥135°C).

The presence of impurities in oil should be avoided at all costs because even small amounts of them reduce its performance and can cause damage to active parts. In particular, the presence of water reduces the dielectric withstand of the oil and accelerates its degradation through hydrolysis. Therefore, it is necessary to treat the oil before filling the transformer and whenever the result of periodic oil analyses requires it. Special treatment machines allow filtering, degassing, and dehumidification of the oil (Figure 4.20).

Oil analysis provides an indication of the presence of dissolved gases that may come from oil degradation, caused by hot-spots or partial/total discharges in the active part; these are combustible gases such as hydrogen, methane, ethane,

Figure 4.20 Oil treatment machine

ethylene, acetylene. Additionally, gases from contamination with air (oxygen, nitrogen) or from the degradation of solid insulation (carbon monoxide and carbon dioxide) can be present in oil. IEC 60599 provides guidance for the interpretation of dissolved gas concentrations in oil.

Finally, chemical and physical analysis for determination of acidity, interfacial tension, dissipation factor, presence of sludge or cellulose decomposition products (furans) can give important clues as to whether to treat the oil or replace it. Sludge, which is impossible to remove, can clog circulation channels between turns; therefore, it is important to perform periodic oil analysis to avoid transformer damage. IEC 60422 provides guidelines in this regard.

4.2.3 *Effect of moisture on dielectric strength*

Moisture is an important accelerating factor in the degradation of solid insulation. The load guide ([2] Tab. A.2), following laboratory research, predicts that the nominal life of kraft paper will decrease from 11.7 years with moisture content[5] 0.5%, to 3.2 years with moisture content 1.5%, and to 1.1 years with moisture content 3.5%, assuming the absence of oxygen. In the presence of oxygen, the standard provides a nominal service life of 3.4 years already with moisture content 0.5%.

In addition, moisture in the oil and paper greatly reduces the dielectric strength (Figure 4.21).

The entry of air and moisture into the oil can occur due to rupture of the expansion bag in the conservator and due to saturation of salts in the desiccator. The presence of oxygen accelerates the oil degradation process, which can also be detected by the dark hue that oxidized oil assumes (Figure 4.22).

Moisture is mainly contained in paper and pressboard, due to their high hygroscopicity; especially in crepe paper and low-density pressboard, where water molecules are more likely to bind to the –OH groups of cellulose (Figure 4.23, top). The process of moisture absorption and release exhibits hysteresis and is highly temperature dependent: as the temperature increases, the impregnating capacity of the paper is reduced (Figure 4.23, bottom, and Figure 4.24, top).

The distribution of moisture between solid and liquid insulation is not uniform, as it tends to equilibrium between relative humidities[6] in different materials, which in turn depends on temperature, which is neither uniform nor constant in the transformer. To assess the approximate moisture distribution, isotherm curves have been developed (Figure 4.24, bottom), based on the studies by T.V. Oommen and others, which allow to derive the estimated paper humidity, given that of the oil and the steady-state temperature.

If the temperature increases, for example due to increased current circulating in the windings, moisture tends to migrate into the oil, conversely if the temperature

[5]Although not explicitly stated in the mentioned standard, the content seems to be expressed as the proportion of mass of water contained in the solid insulation to the mass of the latter.

[6]Relative humidity is the ratio of absolute humidity to saturation humidity.

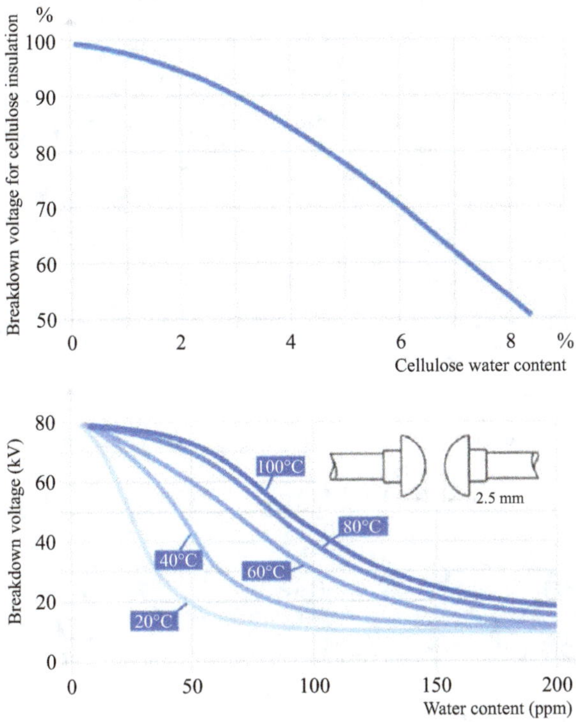

Figure 4.21 Trend of dielectric strength of paper and oil as a function of humidity

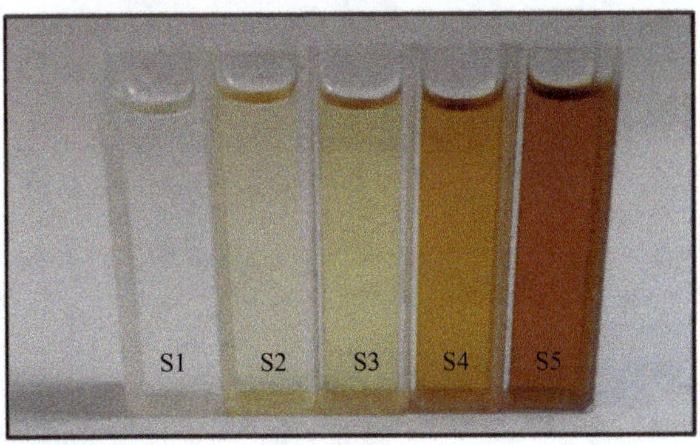

Figure 4.22 Oil samples at different stages of oxidation

decreases, moisture tends to migrate into the paper (Figure 4.25). However, for practical purposes, the total moisture in the paper can be considered almost constant, given the total water content, since 99% of the moisture is contained in the solid insulation.

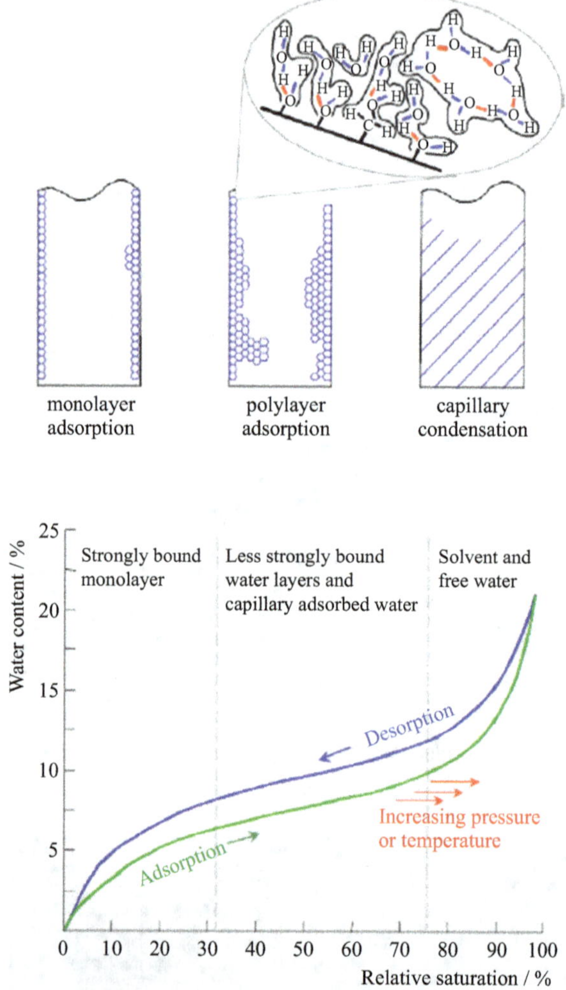

Figure 4.23 Top: water absorption process in cellulose fibrils; bottom: moisture absorption and release curves in paper

In the presence of excessive moisture, or sudden increase in load, it is possible for the water content in the liquid insulation to exceed its moisture absorption capacity (Figure 4.26). In this case, water is present in the liquid or gaseous phase in the form of water vapor bubbles; these are separate from the oil and, being in circulation, can cause a discharge between the coils due to failure of the insulation.

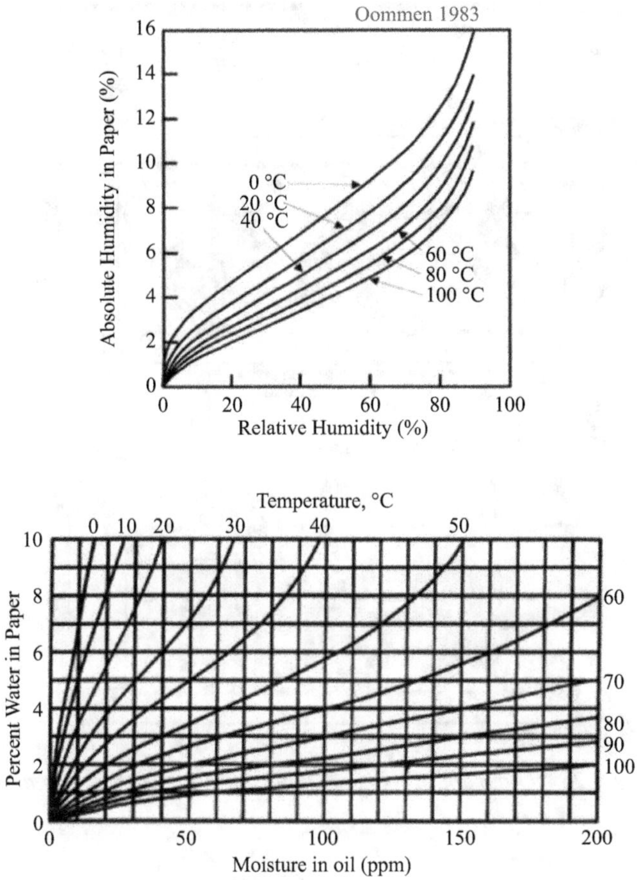

Figure 4.24 Top: paper absorption curves at different temperatures; bottom: moisture balance curves in paper and oil

4.3 Dielectric design

4.3.1 General criteria

Transformer dielectric design consists of verifying that the arrangement of insulating materials ensures dielectric withstand under all expected operating conditions.

The IEC 60076-3 standard [3] concerning dielectric tests of power transformers distinguishes two types of electrical insulation voltages:

- industrial frequency voltages (applied voltage, induced voltage);
- impulsive voltages (switching impulse, lightning impulse).

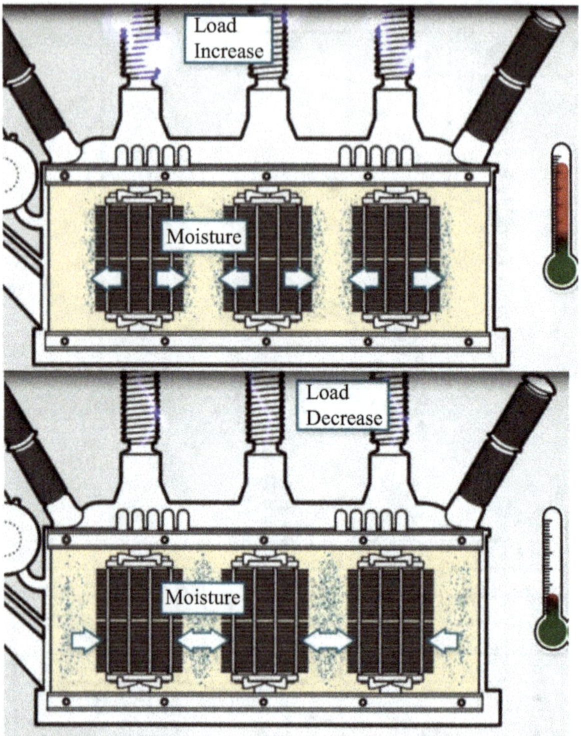

Figure 4.25 Diagram of moisture migration between paper and oil

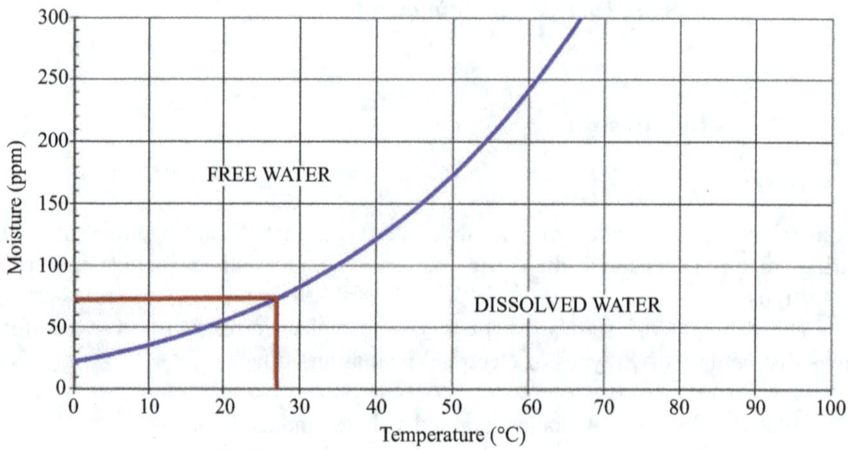

Figure 4.26 Oil saturation diagram: for concentrations above the blue curve at a given temperature, water is present in a separate phase from the oil

The normalized values from which to choose insulation voltages are given in the standard itself, depending on the maximum winding voltage considered.[7]

Dielectric withstand has different behavior depending on the voltage waveform; therefore, withstand to industrial frequency, voltages must be checked separately from withstand to impulse voltages.

4.3.2 Design for voltages with industrial frequency

In designing for voltages with industrial frequency, the behavior of the insulation and transformer can be considered a steady state. Therefore, the voltages are distributed in the windings according to the turns' ratio and according to the capacitive behavior of the insulating materials. In fact, insulating structures subjected to a potential difference between two terminals exhibit behavior similar to that of an electrostatic capacitor. If the structure consists of several flat layers of different materials that are not too thick (Figure 4.27, left), the peak electric field in each layer i can be calculated as follows:

$$E_i = \frac{V}{\varepsilon_i\left(\frac{d_1}{\varepsilon_1} + \frac{d_2}{\varepsilon_2} + \frac{d_3}{\varepsilon_3}\right)}, \tag{4.1}$$

where:

V is the applied voltage (peak value),
d_i is the thickness of the layer,
ε_i is the relative dielectric permittivity of an insulating layer.

In the case of a conductor of circular cross section wrapped with several layers of different insulating materials (Figure 4.27, right), the electric field depends on the distance r from the center according to the formula:

$$E_i(r) = \frac{V}{r\varepsilon_i\left(\frac{1}{\varepsilon_1}\ln\frac{r_2}{r_1} + \frac{1}{\varepsilon_2}\ln\frac{r_3}{r_2} + \frac{1}{\varepsilon_3}\ln\frac{r_4}{r_3}\right)}. \tag{4.2}$$

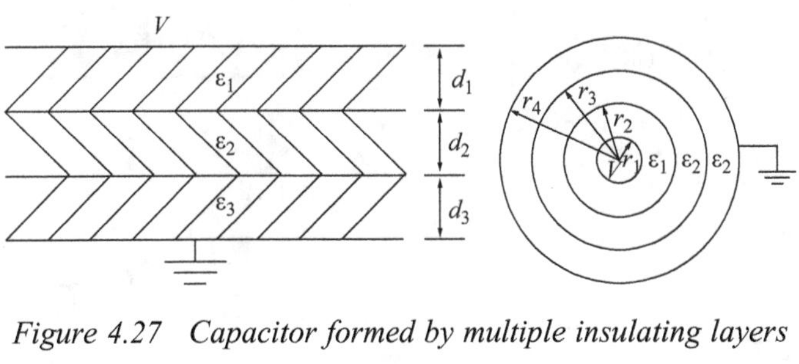

Figure 4.27 Capacitor formed by multiple insulating layers

[7]Note that the maximum voltage itself is not a dielectric sizing parameter but determines the actual insulation voltages. However, it remains a magnetic sizing parameter, as the core must avoid saturation up to the prescribed maximum operating voltage.

For more complicated geometric configurations, it is necessary to resort to calculation by numerical methods. In all cases where the electric field is not uniform but depends on position, one must calculate the field at the most unfavorable position. It is useful to define a *disuniformity factor* given by:

$$\eta = \frac{E_{\max}}{V/d}, \tag{4.3}$$

where d is the total distance between the two electrodes at potential difference V.

It results $\eta \geq 1$, with equality only in the uniform flat-layer configuration.

As a rule, the smaller the radius of curvature of the insulating surface, the greater the electric field gradient at that point. Therefore, edges and sharp live parts should be avoided.

Since the relative dielectric permittivity of paper is equal to 4.2 and that of mineral oil 2.2, it follows that the electric field is greater in oil than in paper (Figure 4.28). It is observed that a partial discharge in oil causes only gas formation without appreciable reduction in the overall dielectric strength, while partial discharges in paper irreparably damage the dielectric withstand capability of paper. Therefore, dielectric design must mainly consider the limits of paper withstand capability, which are provided by the manufacturer, usually through curves depending on the thickness of the paper/paperboard and the oil channel.

The above formulas assume constant permittivity within the materials, a fact normally verified for thin materials a few millimeters thick. For greater thicknesses, phenomena of nonuniform charge distribution in the dielectric intervene,

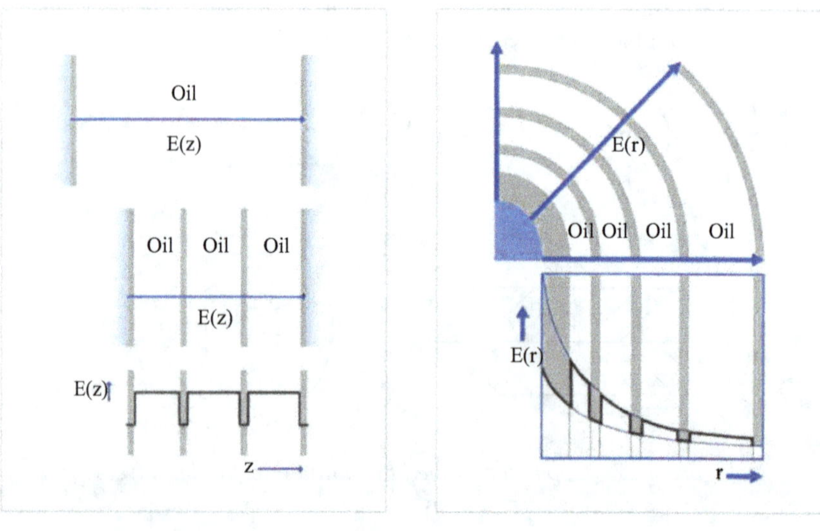

Figure 4.28 Qualitative trend of peak electric field in the presence of mixed paper-oil insulation

Figure 4.29 Winding with multiple insulating barriers between different coils highlighted

and the real maximum electric field is greater than the calculated one. Therefore, it is useful to divide the insulation distance with several insulating barriers instead of leaving a space filled by a thick pressboard wall or a wide oil channel (Figure 4.29).

In addition, the above formulas assume an electric field gradient always perpendicular to the insulating barriers. It is not always possible to meet this requirement, particularly in structures with complicated geometries or small radii of curvature.

When the electric field is parallel to the surface of the insulating barrier, the phenomenon of *creepage current* occurs, which causes progressive destruction of the surface insulation by forming branching erosion structures called *tracking* (Figure 4.30), until a total breakdown happens, when the distance between the two parts at different potentials (so-called *creepage distance*) is not sustainable by the remaining insulation.

Withstand curves to creepage current are usually provided by the solid insulation manufacturer.

For this reason, it is essential to insert multiple barriers with the correct shape in the most critical parts of the transformer, possibly employing pressboard laminated with aluminum foil. In these cases, it is useful to know the electric field lines through finite element (FEM) simulations (Figure 4.31).

The impregnation of paper and paperboard is very important, as partial impregnation can leave vacuoles that have a lower dielectric strength, in which partial discharges can be created until they evolve into total discharges. The same

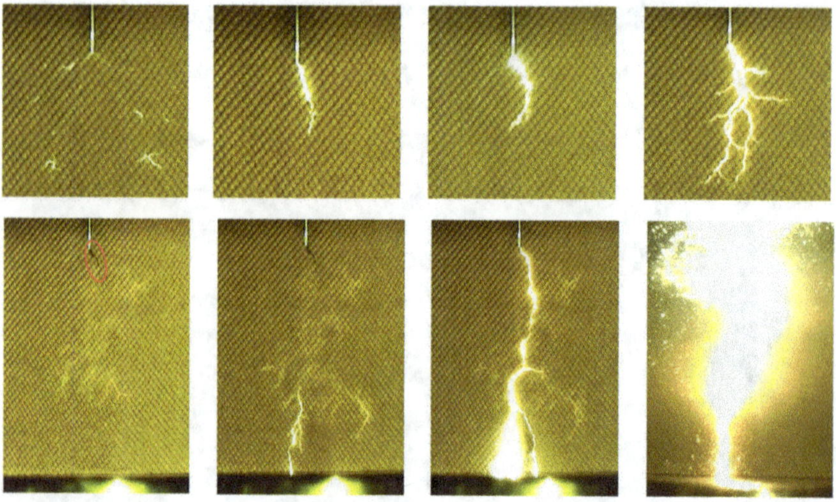

Figure 4.30 *Example of progressive erosion of insulation obtained with a live needle placed on the insulating surface of pressboard*

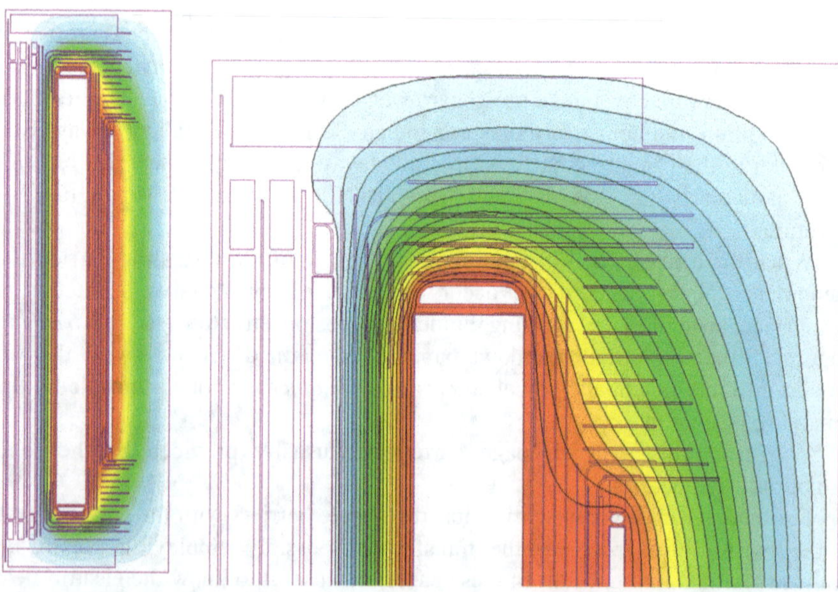

Figure 4.31 *FEM simulation of the electric field in the header of a winding (left) and detail of the highest gradient point (right)*

applies to residual moisture (Section 4.2.3) and extraneous bodies that can create areas with partial discharges and reduced dielectric withstand capability.

An additional factor to consider, particularly for transformers with forced oil circulation, is the phenomenon of *streaming electrification*, which consists of the accumulation in the oil of negative static charges due to friction between the oil flow and the walls of the circulation channels. These charges can reduce the dielectric withstand and cause partial or even total discharge. This is a poorly understood phenomenon because it is difficult to measure the static electricity contained in oil, but it is believed that very refined oils free of impurities have a lower tendency to develop streaming electrification. As a general rule, a circulation velocity of less than 0.5 m/s does not generate this phenomenon.

Consequently to what is stated above, the dielectric design must contain large margins of safety with respect to withstand values, so to account for all uncertainties in actual insulation conditions and constructive tolerances. Normally, electric field values for which the 1% probability of discharge results are taken as the reference. Having defined the required safety coefficient, the design is done iteratively, adding barriers or modifying the geometry where the calculation does not show that the required withstand margins are met.

4.3.3 Design for impulsive voltages

The dielectric withstand of a material is inversely proportional to the duration of test voltage application (Figure 4.32).

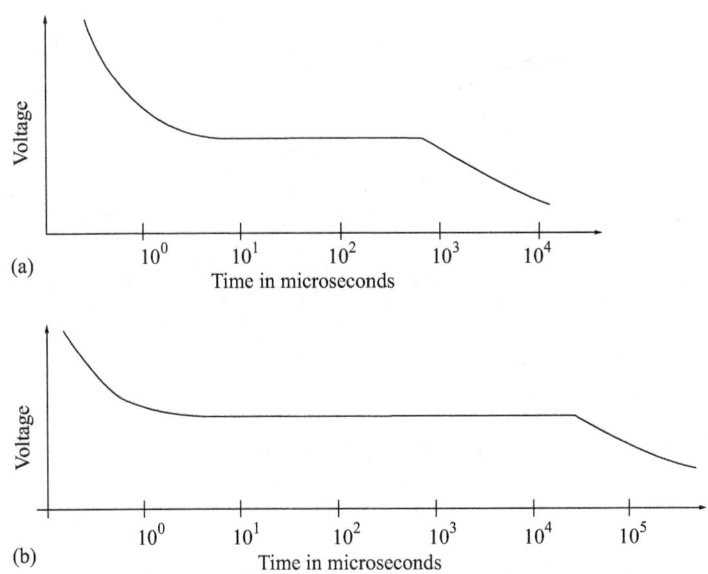

Figure 4.32 Qualitative trend of dielectric withstand as a function of voltage duration for oil (a) and oil-impregnated pressboard (b)

For impulsive overvoltages, with durations on the order of 1–10^3 μs, dielectric tightness is practically constant. Specific impulsive voltage withstand curves are normally provided by the solid insulation manufacturer depending on the analyzed geometry of the pressboard thickness and the size of the oil channel.

The application duration for sinusoidal periodic voltages can be assumed to be equal to a period.[8]

In contrast, for impulse voltages, the IEC 60060-1 standard [5] defines two types of curves (Figure 4.33):

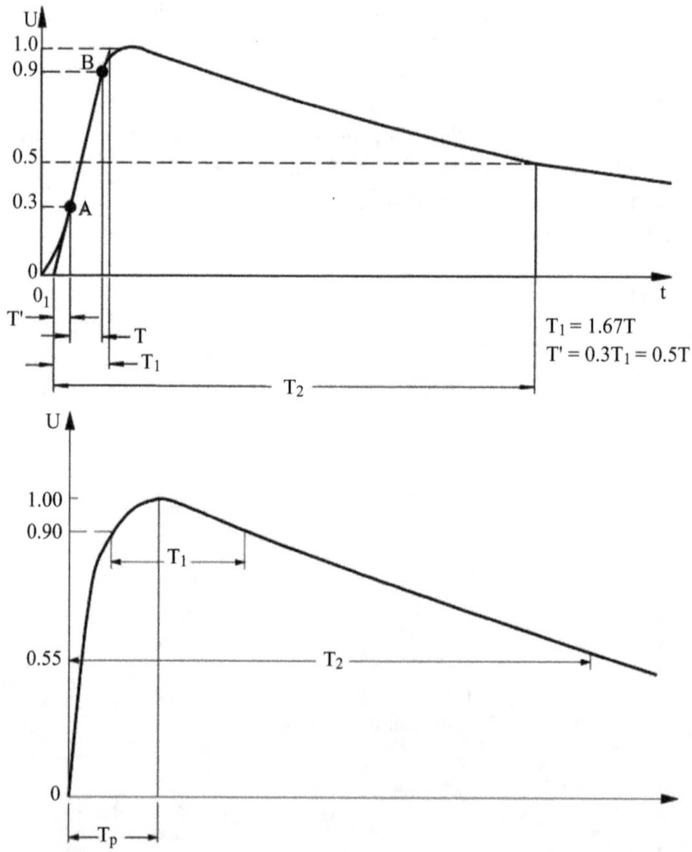

Figure 4.33 Conventional lightning impulse curve (top) and switching impulse curve (bottom)

[8]Non-sinusoidal periodic signals must be decomposed into the harmonic components to evaluate their respective frequencies; however, the combined effect of simultaneous application of signals at different frequencies cannot be calculated by simple superposition of effects, and requires case-by-case analysis, including experimental tests.

- the *lightning impulse*, representative of a lightning surge, characterized by the front time or the duration of the rising front (T_1), i.e., the conventional time it takes for the signal to reach the crest, and by a tail time or time to the half-value (T_2), that is the time it takes for the signal to reduce to half the peak value;
- the *switching impulse*, representative of an overvoltage due to the opening of a switch, characterized by the time to peak (T_p), the time above 90% of the crest value (T_d), and the time to zero (T_z), which is the time it takes for the signal to reach the first zero crossing. The signal typically then continues at opposite polarity gradually damping.

The IEC 60076-3 standard [3] provides the following values for impulse curves:

- lightning impulse: $T_1 = 1.2$ μs; $T_2 = 50$ μs;
- switching impulse: $T_p \geq 100$ μs; $T_d \geq 200$ μs; $T_z \geq 1000$ μs.

The tests are performed using: a pulse generator, consisting of a battery of resistors and capacitors that is charged in DC and then discharged via a spark gap on the test object, a truncator consisting of a series of hemispheres that act as a spark gap by shorting the pulse voltage to ground, and a capacitive voltage divider (Figure 4.34).

The standard requires that the lightning impulse be performed on all windings with $U_m > 72.5$ kV.

For windings with $U_m > 170$ kV, the impulse is truncated after 3–6 μs by abruptly nullifying the voltage to represent the triggering behavior of a spark gap.[9] The standard also requires that high voltage (HV) winding[10] is subjected to the switching impulse test if it has $U_m > 70$ kV. The duration of the shunting impulse, unlike the lightning impulse, allows a transduction of voltage between the windings, which are still subjected to an overvoltage as a result of the performance of the test on the HV winding. It follows that the switching impulse voltage can be applied to any winding as long as the voltage measured on the HV winding meets the normative requirements.

Instead, the voltage distribution for the lightning pulse in the winding must be calculated using a high-frequency model, in which the effects of stray capacitances are prevalent over inductive behavior (Figure 4.35).

Let us consider a winding whose turns have a capacitance per unit length between adjacent turns equal to c_s and a capacitance per unit length with respect to two walls at potential $U_a(x)$ and $U_b(x)$ equal, respectively, to c_a and c_b (Figure 4.36). With these assumptions, we intend to calculate the potential $U(x)$ of each node.

[9]Since high-voltage transformers are always associated with surge arresters rather than spark gaps, the truncated impulse test has minimal operational significance. The previous edition of the standard did not prescribe to routinely perform the truncated impulse test.

[10]Please note that the transformer standard defines HV winding as the one with the highest rated voltage, regardless of its actual rating value.

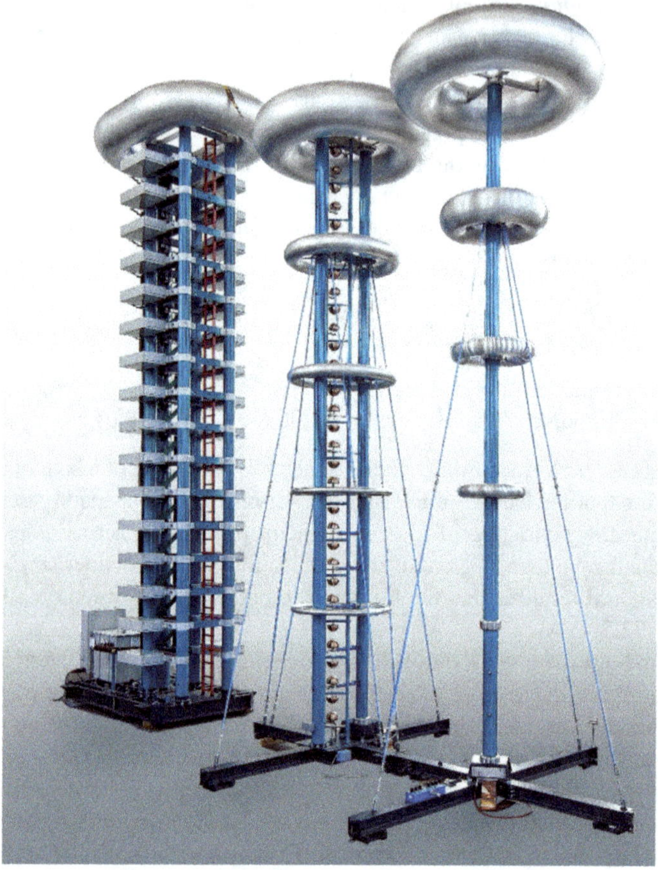

Figure 4.34 A pulse generator, a truncator, and a capacitive divider (left to right)

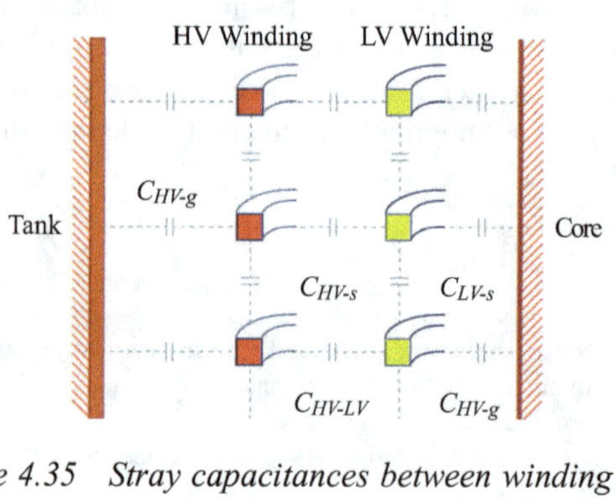

Figure 4.35 Stray capacitances between winding turns

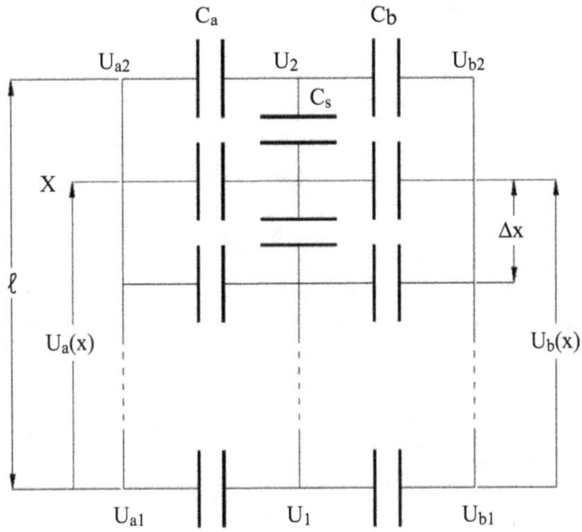

Figure 4.36 Schematic circuit of the parasitic capacitance of a winding

The following dimensionless stress distribution coefficients can be defined:

$$\alpha = \sqrt{\frac{c_a + c_b}{c_s}}, \tag{4.4}$$

$$\gamma_a = \frac{c_a}{c_a + c_b}, \ \gamma_b = \frac{c_b}{c_a + c_b}. \tag{4.5}$$

Said ℓ the overall height of the winding and $U(0) = U_1, U(\ell) = U_2$, and assumed constant potentials at the two walls, the intermediate voltage is defined:

$$U_\beta = \gamma_a U_a + \gamma_b U_b, \tag{4.6}$$

and the potential at the intermediate points by the following formula [6] (7.31):

$$U(x) = \frac{1}{\sinh \alpha}\left\{(U_2 - U_\beta)\sinh\left(\alpha\frac{x}{\ell}\right) + (U_1 - U_\beta)\sinh\left[\alpha\left(1 - \frac{x}{\ell}\right)\right]\right\} + U_\beta. \tag{4.7}$$

This formula, given the capacitances per unit length, makes it possible to estimate the trend of the potential, and thus the electrical gradient, for the cases of interest. In particular, in the case that equipotential walls are grounded ($U_a = U_b = 0$), as can be assumed for insulating winding cylinders, and that terminal 1 is pulsed with voltage $U_1 = U$ with the other terminal grounded, we obtain a potential distribution given by [6] (7.55)

$$U(x) = U\frac{\sinh\left[\alpha\left(1 - \frac{x}{\ell}\right)\right]}{\sinh \alpha}. \tag{4.8}$$

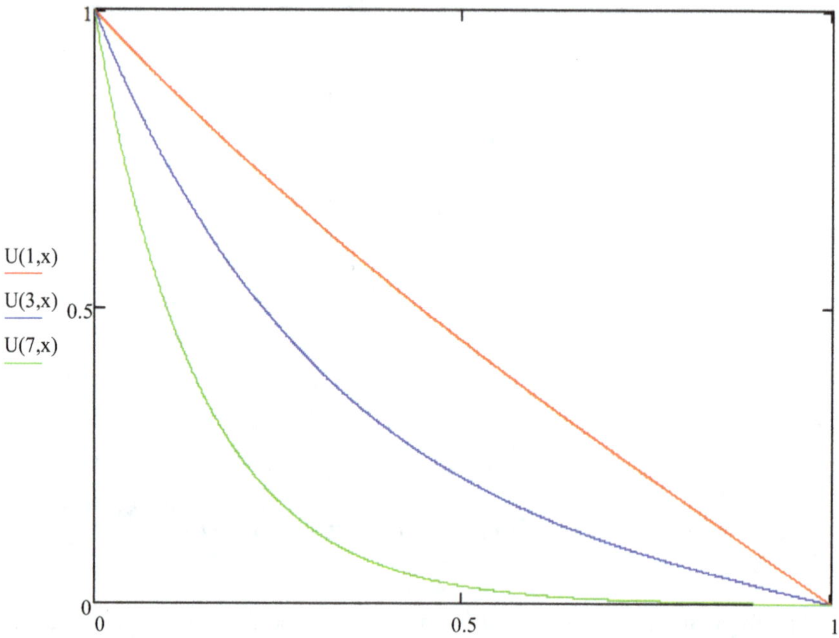

Figure 4.37 Trend of potential (α, x) for different values of α

Figure 4.37 shows that the trend of the potential is more uniform as α is closer to unity.

The maximum voltage gradient, i.e., the maximum value of electric field, is reached at the pulsed terminal ($x = 0$) and has value:

$$E(0) = \frac{\alpha U}{\ell \tanh \alpha}. \tag{4.9}$$

The most popular techniques for reducing the value of α are to increase the capacitance between turns c_s, through the transposition of conductors, creating an *interleaved disc* (see Section 3.5.1), or to reduce the capacitance to ground c_a, c_b through the interposition of conductors at ground potential (*shielded disc*). In addition, the use of equipotential rings (*shield/static rings*) at the ends of the windings, containing an aluminum foil that is placed at line or ground potential through an appropriate conductor (Figure 4.38), allows the winding potential to be adjusted at critical points. After the possible ring, the *clamping ring* made of laminated wood is installed, whose capacitance must be considered in the calculation of voltage distribution.

To allow the circulation of oil, *wedges* are inserted between the assembled phase, that is the cylinder consisting of all windings of the same phase, and the ring so to leave a sufficient gap (Figure 4.39).

Correspondingly, the bottom plane that allows the active part to rest on the lower yoke reinforcement also consists of a laminated wood structure (Figure 4.40).

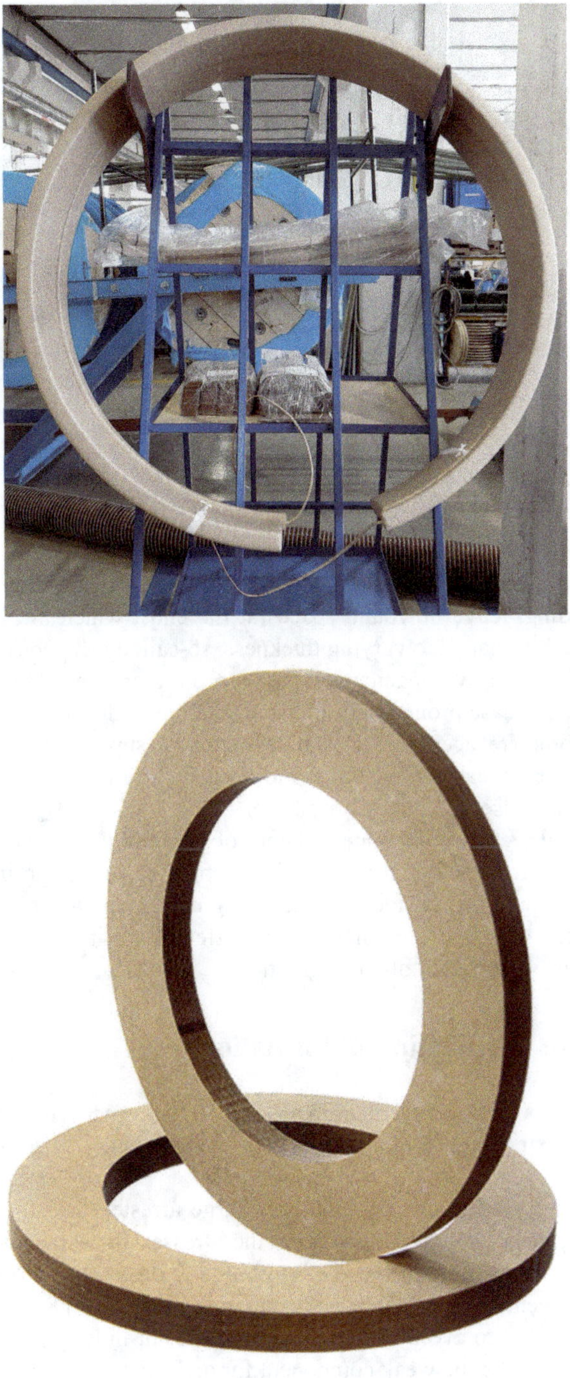

Figure 4.38 An equipotential ring (top) and a clamping ring (bottom)

Figure 4.39 Detail of the upper clamps with the ring and the wedges highlighted

Spiral windings, having an oblique arrangement of the turns, as is the case of regulating winding, leave unwound spaces at the ends, which are filled by insulating rings suitably shaped to varying thickness, so-called *false coils* (Figure 4.41).

Fully modeling stray capacitances between turns and insulation can be very onerous, and in any case prone to error since the value of dielectric constants is not known with complete accuracy. Design experience, supported by experimental verification and test measurements, leads to choosing the right degree of detail for calculating the impulse potential distribution.

Non-live coils, such as unconnected control taps and the tertiary winding, can also assume dangerous potential by capacitive coupling during an impulsive phenomenon. This should be considered both during design and testing by conducting the impulse test with the tap changer on different positions on the different impulsed phases, so to verify all configurations.

4.4 Tests for verification of insulation level

Dielectric design is a critical element in ensuring transformer reliability, especially for high and extra high voltage grids that experience various overvoltage phenomena.

Certainly, the systematic installation of surge arresters near the terminals of the autotransformers (ATRs) has reduced the stresses to which the devices are subjected in service, especially those of atmospheric origin.

However, given the importance and economic value of ATRs, coordination of insulation is essential to avoid failures that would result in long periods of outage. The ATR should not be the weak component for dielectric withstand of the system; instead, robustness should be achieved by devices installed for this purpose,

Figure 4.40 *Bottom plane (top) and its arrangement between winding and reinforcement (bottom)*

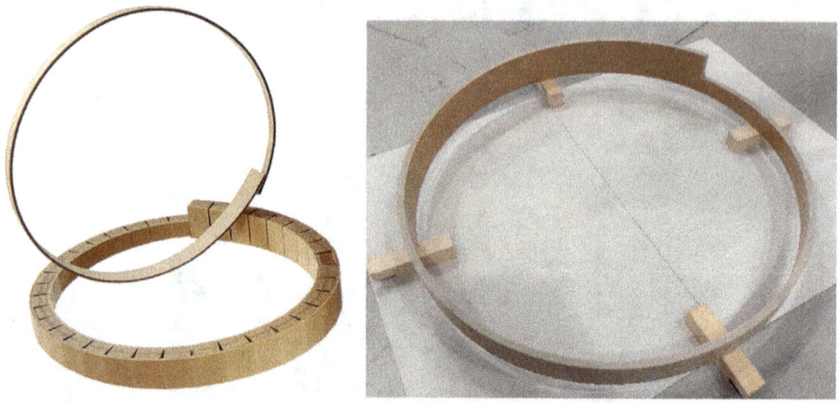

Figure 4.41 Two examples of false coil for spiral windings

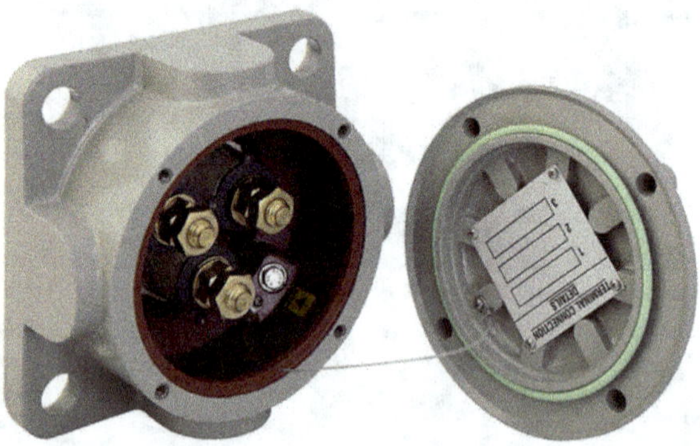

Figure 4.42 Terminals for connecting core, shields, and clamps to the tank

namely: arresters installed very close to the bushings, spark gaps at the entrance of the lines, and lightning masts in the substation.

The dielectric reliability of ATRs is not only ensured by impulsive insulation tests but also by the induced voltage test with partial discharge measurement, which is an important indicator of the quality of solid insulation.

In addition, the importance of industrial frequency withstand tests between the windings and the ground and between core and clamps with respect to the tank must not be overlooked. In fact, the core and armature are isolated from each other and normally connected to the tank at a single point via special terminals, usually accessible in a compartment located at the top of the case (Figure 4.42). If the insulation is compromised, a loop is established for the circulation of eddy currents that can create hot-spots and compromise the overall reliability. By removing the prearranged connection and checking the withstand of the insulation (routinely at 1 kV DC for 1 min), hidden defects can be detected.

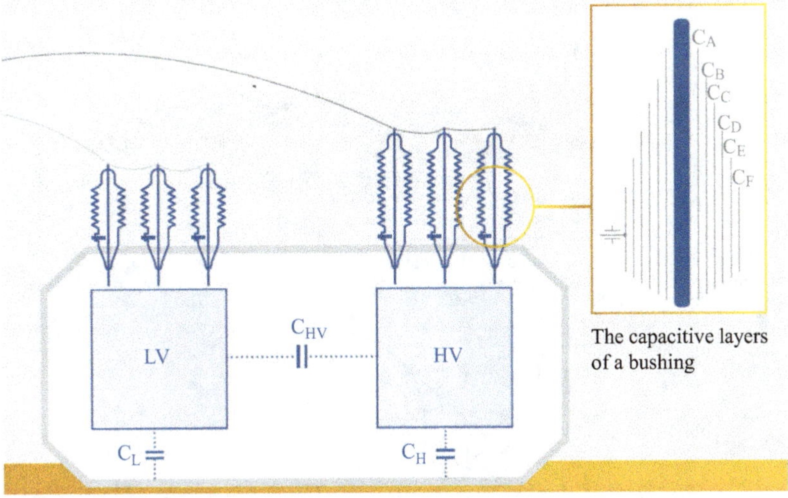

Figure 4.43 Measurement diagram of the stray capacitance of the windings toward the tank

Similar tests performed in the plant for insulation between windings and ground, through the grounded tank, performed routinely at 5 kV DC for 10 min (Figure 4.43), make it possible to find out possible insulation defects in the external structures of the active part (barriers, washers, etc.). Although there is no quantitative criterion described in the standards for passing the test, an insulation resistance measured at the end of the test of 2500 Ω for each volt of the rated voltage may be given as a reference value for checking the need for further investigations on the transformer.

Finally, measuring the capacitance and dissipation factor (tan δ) of the capacitive bushings, using the *test tap* (Figure 4.44), makes it possible to assess the state of the bushing's internal capacitance, with structure consisting of alternating layers of insulating foil and conductive aluminum foil.

Permanent installation of sensors on the bushing tap, for real-time monitoring of the status of the bushing, must be carefully evaluated, as a sensor defect could leave the test tap floating, which is connected to the first conductive layer of the internal capacitance, and thus it must always be grounded during the transformer's operation to maintain the capacitance at a stable voltage. Failure to do so may compromise the internal insulation of the bushing. Therefore, its widespread installation is not recommended, while it can be a useful investigative tool for critical or experimental bushings, as long as the system is periodically examined to ensure its functionality.

Instead, the periodic dielectric response check by means of the *frequency domain spectroscopy* is recommended, which allows measurement of the dissipation factor of windings at different frequencies (Figure 4.45) and provides a broader spectrum of information than simply measuring it at industrial frequencies.

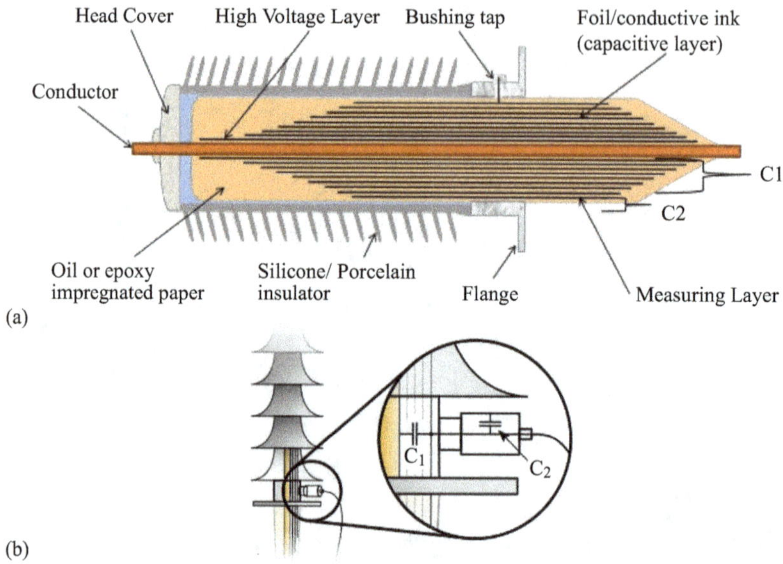

Figure 4.44 Schematic diagram of the internal capacitance of a bushing (a) and connection to the bushing's test tap (b)

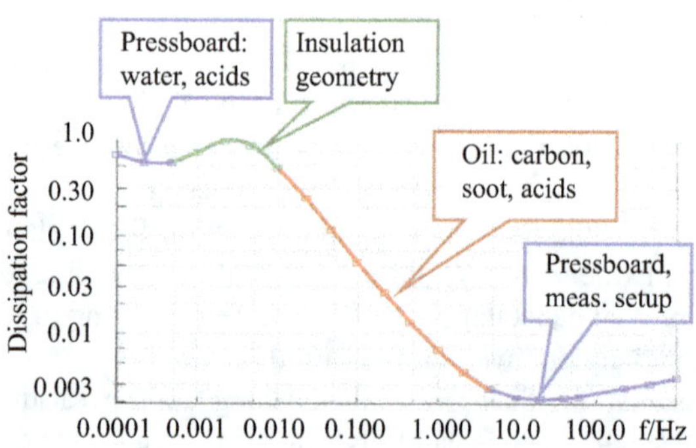

Figure 4.45 Trend of dissipation factor at different frequencies and possible inferable information

Although this measurement requires a longer test time of about 2 h, if done correctly, it allows additional information on solid insulation and internal transformer humidity to be obtained. For the interpretation of the measurements to be meaningful, it is essential to know the main geometric characteristics of the

Figure 4.46 A mobile laboratory for on-site HV testing

insulation (ratio of spacers to oil channels and barrier thickness), which should be requested when reviewing the design of the prototypal transformer of each type.

In cases where there is a real need for more thorough investigation of the insulation condition, the transformer must be brought to the factory for HV testing and inspection of the active part, or a mobile laboratory can be used (Figure 4.46) for HV testing directly at the substation.

Therefore, verification of transformer insulation is a continuous process, starting from design and passing through factory testing, and going on throughout the transformer's service life through periodic in-plant inspections.

Finally, *post-mortem* investigations of decommissioned transformers should not be overlooked, because they can provide, by visual examination and paper sampling for DP analysis, information on the state of insulation at the end of the transformer's life and possible critical points of degradation, which can be useful data for improving the design criteria of future transformers.

References

[1] Michael A. Franchek, and Aleksandr Levin: INSULutions® DPE, *Transformer Magazine*, vol. 3, issue 2, April 2016.

[2] Standard IEC 60076-7: *Power Transformers – Part 7: Loading Guide for Mineral-Oil-Immersed Power Transformers*, ed. 2.0, 2018.

[3] IEC standard EN 60076-3: *Power Transformers – Part 3: Insulation Levels, Dielectric Tests and Insulating Distances in Air*, 2014 (IEC 60076-3:2013).

[4] Bo Gao, Rui Yu, Guangcai Hu, Cheng Liu, Xin Zhuang, and Peng Zhou: *Development processes of surface trucking and partial discharge of pressboards immersed in mineral oil: Effect of tip curvatures*, Energies 2019, 12, 554 (https://www.mdpi.com/1996-1073/12/3/554).

[5] IEC standard EN 60060-1: *High-Voltage Test Techniques – Part 1: General Definitions and Test Requirements*, 2015 (IEC 60060-1:2010).

[6] Martin J. Heathcote: *The J&P Transformer Book*, Newnes, 12th ed. 1998.

[7] S.V. Kulkarni, and S.A. Khaparde: *Transformer Engineering*, Marcel Dekker, 2004.

[8] R.M. Del Vecchio, B. Poulin, P.T. Feghali, D.M. Shah, and R. Ahuja: *Transform Design Principles*, CRC Press, 2001.

[9] Lars E. Schmidt: *Transformer insulation materials – An introduction*, ABB Presentation at the "Life of a Transformer" Seminar, May 19–23, 2014, Padua (Italy).

[10] Christoph Krause: *Selection & processing of pressboard for prolonged transformer service life*, Weidmann Presentation at the MyTransfo Seminar, November 15–16, 2016, Turin (Italy).

[11] Tobias Wolfinger: *Lamination of Transformerboard using the structural elements of cellulosic fibers*, Transformer Magazine, vol. 2 issue 4, October 2015.

[12] Nynas: *Transformer Oil Handbook*, 2nd ed., 2009.

[13] Technical Brochure (TB) 349: *Moisture Equilibrium and Moisture Migration Within Transformer Insulation Systems*, Cigré, 2008.

[14] W. Ziomek, K. Vijayan, D. Boyd, K. Kuby, and M. Franchek: *High voltage power transformer insulation design*, Electrical Insulation Conference, June 5–8, 2011, Annapolis, MD, USA.

[15] Brian D. Sparling, and Jacques Aubin: *Assessing water content in insulating paper of power transformers*, Electric Energy T&D Magazine, no. 4-11, July/August 2007.

[16] Terna Document IO601MP: *Technical Inspections on Large Power Static Machines*, rev. 01 of 09/30/2018.

Chapter 5
Vegetable insulating fluids in transformers

5.1 Generalities

The active part of large transformers and autotransformers, composed of the windings and ferromagnetic core, is contained in a tank traditionally filled with a mineral oil composed of hydrocarbons, obtained by refining from petroleum. This is similar to a diesel oil which has been highly refined, so as to obtain a product for electrotechnical use with high dielectric and purity characteristics.

Oil performs the functions of insulating and cooling the active part. In fact, oil has a high dielectric withstand capability that increases the permittivity the solid insulation (the paper that wraps the winding conductors and the pressboard for the supporting structures) of the active part when it is impregnated by oil.

To ensure the insulating properties of the oil, it is necessary for it to maintain its dielectric capability, which can be reduced by the presence of moisture, high temperature, and oxidative process, which forms acidic compounds and solid precipitates called sludge.

In addition, the considerable heat capacity of the oil allows the heat produced in the active part because of the power losses to be extracted by convection through oil circulation, which can take place either naturally or through pumps. Dissipation of heat in the environment takes place by means of radiators or heat exchangers. Effective heat dissipation requires oil with low viscosity at all operating temperatures of the transformer. A lower viscosity allows, at a given thermal power to be dissipated, to reduce the flow rate and speed of the oil, thus the consumption of auxiliary services, and also prevents the formation of electrostatic charges on the contact surfaces, a phenomenon called *streaming electrification*, which occurs in the case of a high circulation speed (Section 4.3.2). The lowest temperature at which the oil can still flow is called the *pour point*. Mineral oil has little viscosity even at low temperatures.

The major disadvantages of mineral oil are as follows:

- flammability, whereby in the event of an internal arc within the transformer, the oil may ignite upon contact with oxygen;
- poor biodegradability, which becomes an environmental problem if the oil is accidentally spilled into the ground.

The lowest temperature at which oil ignites when approached by flame is called the *flash point*. Instead, the *fire point* is the temperature at which the oil spontaneously ignites and continues to burn for at least 5 seconds.

There are commercially available insulating liquids with a fire point above 300°C (class K according to IEC 60076-2), including silicone oils and vegetable oils; in particular, the latter, which also have greater biodegradability, are recently gaining more widespread use in new transformers. This chapter will describe the characteristics and conditions of use of vegetable oils.

5.2 Description of vegetable oils

5.2.1 Origin

Vegetable oils of natural origin are extracted from the seeds of several plants such as sunflower, soybean, castor oil plant, and rapeseed. In particular, the latter two have little food use and are mainly used for oil production.

Their introduction is not new. In fact, as early as about 1890, George Westinghouse had used linseed oil as an impregnating agent in insulation [1], and furthermore castor oil has been used as an insulator for capacitors as early as 1962 [2]. However, because of reliability problems with vegetable oil (mainly due to its tendency to oxidation), mineral oil was used for transformers over time.

Only recently has interest in vegetable oils as insulating fluids been rekindled, supported mainly by environmental issues. Since 1999, several products have been patented (Biotemp®, Envirotemp FR3®, and others), which perform better than the vegetable oils available in the past.

5.2.2 Chemical structure

A vegetable oil obtained from plants by a process of extraction and refining, but without changing the structure of the ester molecule, is called a *natural ester*. In contrast, if it undergoes additional treatments or is obtained through chemical reactions from organic compounds not directly derived from plants, it is called a *synthetic ester*.

Natural esters are triglycerides obtained from the esterification of glycerol (Figure 5.1) with three fatty acids [3]. Fatty acids are carboxylic acids composed of linear chains of alkyl groups (Figure 5.2) ending in a carboxylic group (Figure 5.3).

The esterification reaction replaces a hydroxyl group (–OH) of glycerol with the carboxyl group. If the linear chains all have simple bonds, the acid is said to be saturated; if double or triple bonds are present, it is said to be unsaturated.

Normally, synthetic esters consist of an alcohol in which the hydroxyl group is replaced by a carboxyl group (Figure 5.4). However, their composition and structure may vary according to the manufacturer and the type of performance required.

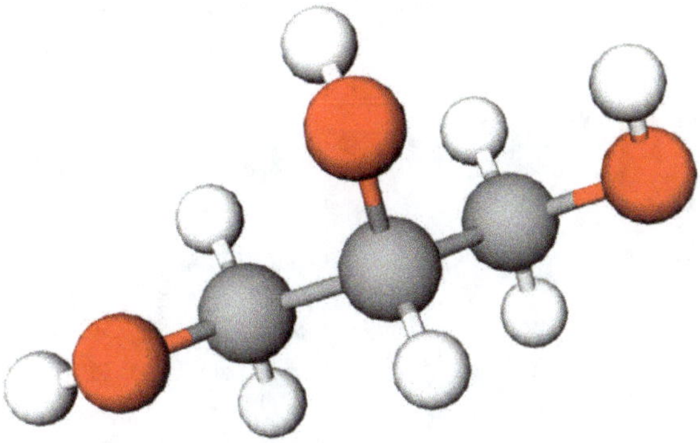

Figure 5.1 Glycerol molecule (O in red, C in gray, H in white)

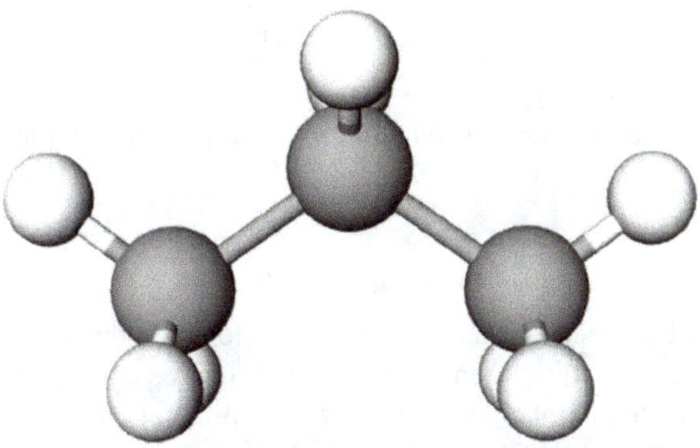

Figure 5.2 Generic alkyl group; chain length is variable and double or triple bonds may be present

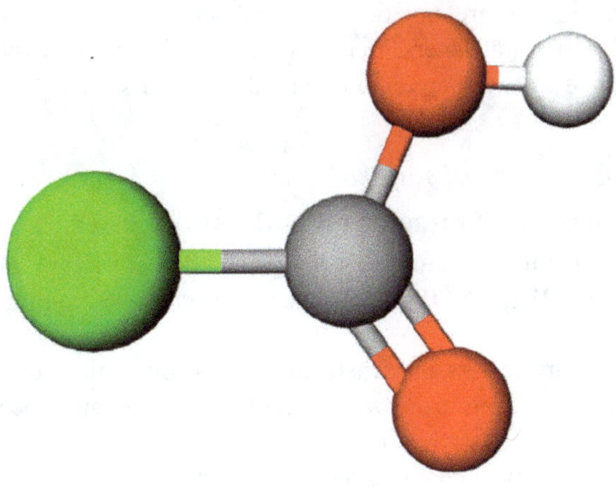

Figure 5.3 Carboxylic acid (the alkyl group is represented by the green sphere)

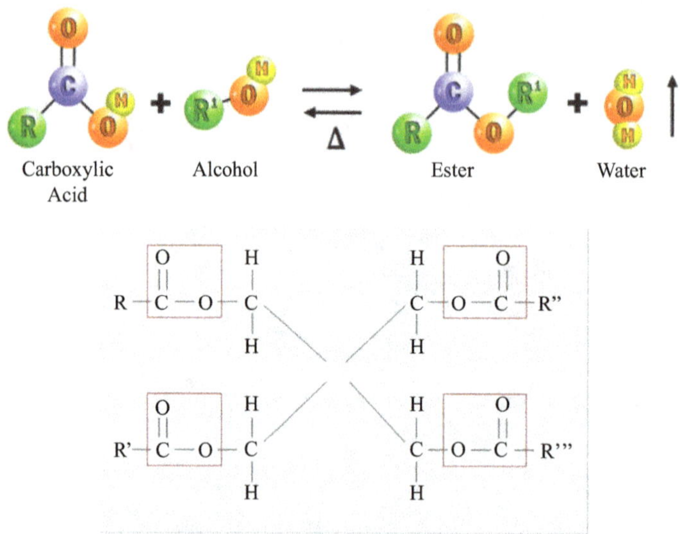

Figure 5.4 Esterification reaction and structure of a synthetic ester

5.2.3 *Characteristics of synthetic esters*

Synthetic esters have been commercially available since the 1970s [4] and have excellent performance at high temperatures, high hygroscopicity (higher than that of natural esters), pour point comparable to that of mineral oils, and good resistance to oxidation (better than that of natural esters).

They represent a product on which there is a fair amount of operating experience and established standards about them exist, both for the new products (IEC 61099) and for the maintenance of used products (IEC 61203).

Their characteristics are not described in detail in the following, which is instead devoted to natural esters, that nowadays are gaining wider interest and attract attention due to environmental compatibility and sustainability reasons.

5.2.4 *Characteristics of natural esters*

5.2.4.1 Comparison of mineral oil and natural ester

The chemical structure of natural esters gives them different properties from mineral oils. In fact, IEC 62770 defines specific normative requirements for these liquids.

The use of natural esters in transformers must be evaluated considering the differences with mineral oils from which result the design and operational advantages and disadvantages described later.

Table 5.1 shows typical characteristics and normative requirements for the two categories of liquids.

Table 5.1 Main characteristics of mineral oils and natural esters [3]

Characteristic	**Mineral oils**		**Natural esters**	
	Typical value	**Requirement by IEC 60296**	**Typical value**	**Requirement by IEC 62770**
Discharge voltage (kV)	30–85	>30	82–97	>35
Relative dielectric permittivity (ε_r)	2.2	–	3.1–3.3	–
Saturation humidity at room temperature (ppm)	40	–	750	–
Viscosity at 40°C (mm^2 s)	3–16	<12	16–37	<50
Pour point (°C)	−40 to −60	<−40	−19 to −33	<−10
Flash point (°C)	135–170	>135	315–328	>250
Density at 20°C (kg/dm^3)	0.83–0.89	<0.895	0.87–0.92	<1
Specific heat (kJ/kg K)	1.6–2.0	–	1.5–2.1	–
Thermal conductivity (W/m K)	0.11–0.12	–	0.16–0.17	–
Coefficient of thermal expansion (10^{-4} K^{-1})	7–9	–	5.5–5.9	–

5.2.4.2 Ecological compatibility

The most obvious advantage of natural esters is their *biodegradability*. According to the CEC-L-33 test [2], the biodegradability of natural esters is more than 90% (although the actual value depends on the amount of additives added to the commercial product), compared with about 30% of mineral oils. Therefore, in the case of accidental liquid spillage from the transformer's collection tub in the plant, as a result of a tank rupture, dispersion into the soil is not a major problem of environmental pollution. However, the spillage of natural ester into groundwater is still considered pollution for contamination of drinking water and water basins, so it should be avoided.

The lower life-cycle environmental impact depends on the fact that natural esters come from *renewable sources*, as opposed to the fossil sources from which mineral oils are derived. This characteristic is sometimes criticized by environmentalists who object to the cultivation of agricultural products for industrial use as taking resources away from food use. However, normally the use of renewable resources is viewed favorably with regard to *environmental sustainability*, and this can be valued in the sustainability report of utilities.

Moreover, natural ester extraction techniques are a process that requires less use of energy and chemical resources than oil refining [5], so they are more environmental friendly.

5.2.4.3 Relative dielectric permittivity

The relative dielectric permittivity of natural ester is higher than that of mineral oil. It follows that the use of ester makes the electric field distribution more uniform between liquid and solid insulation, since solid insulation has a higher permittivity than liquid insulation and the electric field distribution is inversely proportional to the permittivity.

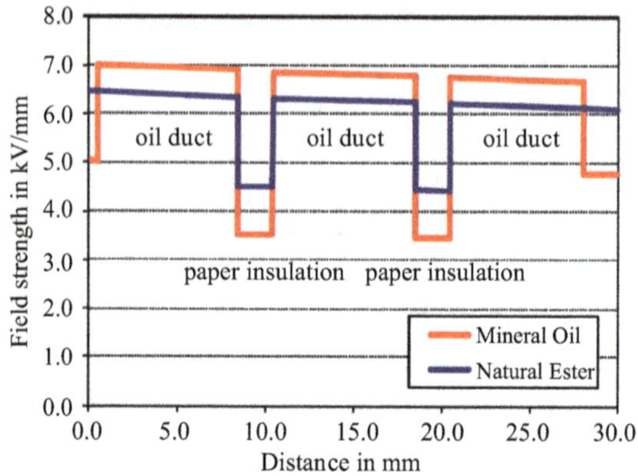

Figure 5.5 Electric field distribution between solid and liquid insulations

This means lower dielectric stress on the liquid, but higher stress on the solid insulation, i.e., paper and pressboard (Figure 5.5).

5.2.4.4 Breakdown voltage

Natural esters have a dielectric withstand capability, measured by the breakdown voltage, at industrial frequency (50–60 Hz) comparable to that of mineral oils [6].

However, natural ester has higher conductivity in AC when contaminated with particles than mineral oil [7]; therefore, the presence of impurities is more critical for natural esters. The dielectric strength drops dramatically in the presence of oxidation, when a viscous sol[1] is formed [8]; then upon complete oxidation, the ester becomes a semisolid gel. Therefore, ester oxidation presents a risk to transformer insulation.

5.2.4.5 Saturation moisture

In transformers, there is a certain amount of water dissolved in the solid and liquid insulations. Since the presence of water reduces the dielectric properties of the insulation, efforts are made to limit the entry of moisture into the tank. The method used in large transformers consists of an expansion bag separating oil from air in the conservator, and a desiccator containing hygroscopic salts connected to the air inlet pipe of the conservator (Section 4.2.2).

Natural esters have much higher saturation humidity[2] than mineral oils. Since the equilibrium of water concentrations between solid and liquid insulations occurs at the same relative humidity, this implies that—all other things being equal—natural ester contains a greater absolute amount of water [9] than mineral oil, which is subtracted from the moisture content of the paper, resulting in a drier paper.

[1]A sol is a type of colloid in which a solid material is dispersed in a liquid environment.
[2]Saturation humidity is the maximum amount of soluble water in a liquid mixture or of water vapor in a gas (typically air); this value varies with temperature.

Since the degradation of solid insulation is affected by moisture, the use of natural ester results in a longer nominal life of the insulation than mineral oil.

Moreover, since the dielectric withstand of the liquid depends on relative humidity, not absolute humidity, the higher amount of water in the natural ester does not affect its dielectric withstand.

5.2.4.6 Viscosity and pour point

The viscosity of natural ester at 40°C is much higher than the viscosity of mineral oil. This is due to the chemical structure of triglycerides, especially the saturated ones. However, it is preferable to reduce the percentage of unsaturated fatty acids in natural esters because, although they have lower viscosity, they also have lower oxidation stability, which is a very penalizing feature. This—given the same number and size of circulation channels—causes a lower flowing speed in the windings, thus a lower flow rate of coolant, than in mineral oil.

In addition, the viscosity of the natural ester increases rapidly with decreasing temperature, as indicated by the fairly high value of the pour point. This characteristic discourages its use in locations with ambient temperatures below −20°C.

5.2.4.7 Flash point

The flash point of natural ester is much higher than that of mineral oil, exceeding 300°C. Therefore, in the case of internal flashover, more energy of the electrical discharge in the liquid is required to raise its temperature enough to ignite it. As a result, natural esters provide greater protection from the risk of fire.

In addition, experiments [10] have shown that natural ester has self-extinguishing properties so that once the flame is removed, it tends not to sustain fire. Paper impregnated with natural ester also tends to ignite with greater difficulty.

Regulations recognize these characteristics and classify natural esters as less flammable liquids (category K). This characteristic is an advantage in substations within an urban area because it allows for shorter separation distances from residential buildings.

However, the natural ester does not eliminate the risk of explosion in the case of a total dielectric breakdown inside the tank. In fact, the rupture of the tank is caused by the overpressure resulting from the vaporization of the liquid by the intense heat of the electric arc, which is a phenomenon occurring with esters as well.

5.2.4.8 Density

The density of natural ester is 5% higher than that of mineral oil; therefore, for the same volume, the weight of a transformer filled with natural ester is higher.

5.2.4.9 Specific heat and thermal conductivity

The specific heat of natural ester is similar to that of mineral oil, while the thermal conductivity is higher. Therefore, under the same conditions of flow rate and thermal jump, natural ester has the advantage of removing heat more efficiently.

5.2.4.10 Oxidation stability

Poor oxidation stability is the major disadvantage of natural esters [5]. These tend to decompose and develop acids and hydroperoxides when exposed to oxygen even at moderate temperatures (40°C) [11], also increasing viscosity so as to make circulation in windings difficult. Unlike mineral oils, which upon oxidation form sludges (i.e., acids with a gelatinous appearance that tend to settle on the bottom of the tank), natural esters form polyester films on the contact surfaces, which are very adhesive and persistent.

Therefore, they must contain some antioxidant additives to improve oxidation stability. These inhibitors are rapidly consumed in the presence of oxygen, and special procedures must be taken to avoid exposing the natural ester to air during maintenance activities.

5.2.4.11 Aging

Laboratory studies show that paper dipped in natural ester has a longer service life based on mechanical strength [12] and degree of polymerization [13]. This advantage appears to be due to the higher moisture absorption capacity (hygroscopicity) of natural ester, which reduces paper moisture, an accelerating factor in aging.

In addition, the natural ester—like all esters—develops less acid at high temperature and is therefore particularly suitable for transformers intended for high load. According to a study [14], paper aging is significantly lower at temperatures above 130°C.

However, the segregation of the ester from air is critical to ensure its durability. At high temperatures in the presence of oxygen, the antioxidant additive is quickly consumed, and the ester loses its oxidation stability. In such a case, it is imperative to restore the additive immediately, otherwise the ester develops acids that can damage the paper and compromise the solid insulation.

It should be kept in mind that so far experience in the operation of devices with natural esters is limited, given their recent introduction, so there is no confirmed information on the state of the paper over time. In fact, its assessment can only be made at the end of the transformer's life, when the device is decommissioned and disassembled.

However, it can be said that under controlled conditions, the durability of natural ester is at least as long as that of mineral oil.

5.2.4.12 Insulation impedance

Natural esters have a greater tendency to the phenomenon of streaming electrification [5], which manifests itself at excessive circulation speed with the formation of electric charges on the paper. Therefore, it must be avoided in the transformer's design, but this requires larger circulation channels inside the windings.

The volumetric resistivity of natural ester is much lower than that of mineral oil, being $30 \cdot 10^{12}$ Ω cm instead of $250 \cdot 10^{12}$ Ω cm [15]. This causes the insulation resistance of a transformer in natural ester to be about half that of a transformer immersed in mineral oil. The loss factor[3], dependent on the ratio of conductivity to oil capacitance, is much higher for natural esters than for mineral oils (0.035 vs. 0.0017), increasing the dissipation factor of the transformer by up to 100% [16].

These characteristics do not affect the performance of the transformer because the insulation resistance is high in any case.

5.2.4.13 Gas production

In oil-immersed transformers, high temperatures and partial or total discharges cause oil molecules to split into combustible gases: hydrogen, methane, ethane, ethylene, and acetylene.

These gases, if the production rate is low, remain dissolved in the oil and can be analyzed by oil sampling from the tank and gas extraction (*dissolved gas analysis*). Conversely, if production is high, the gases form bubbles that flow into the Buchholz relay causing an alarm or the trip of the transformer.

Techniques for interpreting dissolved gases in mineral oil are well established by long experience and are described in standards, for example, by IEC 60599.

The studies carried out on transformers with natural ester [1] have shown a greater tendency for gas formation in the first few months of operation, particularly hydrogen and ethane. This phenomenon, which is also present to a lesser extent in mineral oils, is called *stray gassing* and tends to wear off over time. The causes are not fully known and could be due to the reaction of some additive catalyzed by the materials in the transformer.

Higher content of ethane as well as carbon monoxide and carbon dioxide are found to be produced during operation of transformers with natural ester, compared with mineral oil.

All these differences must be considered in the analysis of dissolved gases. In particular, the parameters of interpretations based on dissolved gas ratios developed for mineral oils may not be directly applicable to natural esters.

5.3 Design implications

Table 5.2 summarizes the advantages and disadvantages of using natural ester for transformers. All these have consequences for transformer design [17].

5.3.1 Thermal design

Regarding the thermal design of the machine, the higher viscosity of the natural ester reduces the flow velocity and thus its flow rate. This problem is partially offset by the higher thermal conductivity, but to maintain equivalent thermal

[3]The loss factor, used in ASTM D924-08, is the ratio of the resistive part of the isolation impedance to the impedance itself; it is normally referred to as $\cos\phi$ or PF (*power factor*); the dissipation factor, used in IEC 60247, is the ratio of the resistive to the capacitive part of the isolation impedance; it is normally referred to as $\tan\delta$ or *D* (*dissipation factor*). The relationship between the two quantities is $\tan\delta = \frac{\cos\phi}{\sqrt{1-\cos^2\phi}}$; for values up to 5% of $\cos\phi$, the difference between the two quantities is less than 0.1%, and therefore, they can be considered equivalent.

Table 5.2 Summary of the advantages and disadvantages of natural esters

Benefits	Disadvantages
Higher saturation moisture → higher water absorption capacity → reduced moisture in paper → less paper degradation	Lower oxidation stability → addition of antioxidants and segregation of liquid from air
Higher flash point → fire resistance	Increased ethane gas production, stray gassing → revising models of dissolved gas interpretation
Self-extinguishing capacity in the case of fire	Increased tendency for flow electrification → avoidance of high oil circulation velocities
Higher thermal conductivity → partially compensates for higher viscosity	Higher viscosity → difficult sliding in winding channels, lower heat transfer
Higher dielectric permittivity → better electric field sharing between oil and paper	Higher pour point → unsuitable for low-temperature applications
Biodegradability	Higher cost
Sourcing from renewable sources	Increased tendency to form acidity and solid deposits → possible impairment of the active part
Lower production energy and easier extraction	Lower resistivity → increase in dissipation factor
Better behavior at high temperatures	Lower dielectric strength if contaminated

disposal with the same temperature gradients, it is necessary to widen the circulation channels.

For large transformers, it is advisable to adopt a KDAF cooling system, i.e., with forced and guided liquid circulation, rather than KFAF, i.e., with forced but unguided liquid circulation, to ensure better performance, because otherwise an increase in liquid speed with a more powerful pumping system could cause the phenomenon of streaming electrification.

Consequently, the design of a transformer insulated with natural ester involves a larger volume of insulating liquid than the corresponding design with mineral oil.

5.3.2 Dielectric design

Regarding the dielectric design, the higher dielectric permittivity of natural ester results in a more uniform distribution of the electric field throughout the insulation, with a reduction in the electric field in the circulation channels and an increase in the electric field in the paper.

Given the higher dielectric stress on the solid insulation, with a 10% reduction in the overall withstand voltage at industrial frequency [18], and considering that the lightning impulse withstand of natural esters is slightly lower than that of mineral oils and very sensitive to liquid conditions [19], the design of a transformer with natural ester normally requires more solid insulation than the corresponding design with mineral oil.

It has been verified experimentally [20] that partial discharges in natural ester evolve into total discharges at a lower voltage (called acceleration voltage) than in mineral oil. As a consequence, the presence of electrical potential gradients that can cause the initiation of partial discharges should be avoided. Hence the geometry of the active part should avoid edges, spikes, and sharp points that can instead be normally tolerated in transformers designed for mineral oil.

The on-load tap changer, having electrodes with special configurations, may have a lower dielectric withstand in natural ester than that verified in mineral oil. Considering the lower arc interruption capability of esters [21], it is preferable to use switches with vacuum interrupting cells, and it may be necessary to oversize the selector and the preselector, both in voltage and current.

5.3.3 Mechanical design

Considering the problem of lower oxidation stability of the natural ester, the mechanical design necessarily requires a separation membrane in the conservator or an airtight tank with a nitrogen cushion.

Large power transformers already have membrane separation (the expansion bag), but it is important to implement a *membrane rupture* alarm that is remotely connected to the control center, which should be associated with the "urgent maintenance" alert.

Finally, due to the larger size of the circulation channels and the higher density of the natural ester, the weight of the transformer is higher and therefore the support structure must be assessed consequently.

5.3.4 Cost

Natural ester has a cost 2–3 times than mineral oil [22]; moreover, a larger quantity is needed to be used for the reasons explained before. Therefore, the choice to use natural ester is normally driven by environmental and safety considerations that must outweigh the higher economic burden.

Finally, because of the different characteristics of esters and mineral oil, the replacement in an existing transformer of mineral oil with a natural ester (*retro-fitting*) should be considered with caution, because the expected rating and reliability of the device may be significantly reduced.

5.3.5 Service considerations

The transformer designed for natural ester has no different operating requirements from transformers containing mineral oil at normal operating temperatures.

5.3.5.1 Low-temperature operation

At low temperatures, the higher viscosity of the natural ester can become a problem for liquid circulation and cooling of the active part. Therefore, it is advisable [23] not to start the pumps at the energization of a natural ester-insulated transformer if the ambient temperature is very low. This should already be included in the logics of the substation control system, so to start the pumps only when a top liquid temperature threshold (e.g., 15°C) is exceeded. The dielectric properties of the ester are not affected by low temperatures [24].

In addition, at low temperatures, fan starting does not allow the ester contained in the cooling system to heat up and reach a sufficiently low viscosity to allow circulation [25], as a result heat transfer worsens and temperatures deeper inside the tank can reach dangerous values. These phenomena are significant for very low ambient temperatures (<−20°C). Consequently, the power utility should gain more experience in low-temperature operation so to verify the reliability of the design under such conditions.

The on-load tap changer, containing moving parts immersed in the insulating liquid, may have difficulties in operating at low temperatures [21]. Therefore, a minimum temperature protection relay should be installed to inhibit tap-changer operations at temperatures below those considered safe by the tap-changer manufacturer.

5.3.5.2 Air exposure of natural ester

The greatest risk of the natural ester is the possible degradation over time. However, if the liquid preservation system works properly and the liquid–air separation membrane remains intact, the overall performance should be equivalent to mineral oil.

Conversely, if the liquid storage system had problems resulting from, e.g., membrane rupture, the condition of the ester can degrade quickly.

Therefore, much attention must be paid to the presence of oxygen among the dissolved gases and, above all, to the level of acidity, which is an indicator of liquid degradation. The goal is to intervene quickly to replace the degraded liquid.

Oxidation causes oligomerization of the ester [11], a process by which the fatty acids produced by oxidation react with each other to form polymer chains that adhere to surfaces, particularly metal surfaces, and form over time—especially when exposed to sunlight—a sticky layer that is very difficult to remove. For this reason, it is not recommended to [26] oven-dry the active part after it has undergone impregnation with natural ester. Instead, pretreatment with a solvent that removes the ester from the surfaces of the active part is necessary. Vapor-phase treatment, which employs kerosene as a solvent, should avoid the problem, but the formation of polymerized residues on the active part cannot be completely ruled out.

In conclusion, any air exposure or oven treatment of the active part impregnated with natural ester is critical to the reliability of the transformer.

Transportation of the transformer insulated with natural ester is also preferably done with the tank filled. This is unlike mineral oil, which is transported separately by tanker (also for reasons of fiscal control, in countries where a taxation of fuels exists), which is not relevant to natural esters, as they are not classified as fuels.

5.3.5.3 On-site treatment of natural ester

The treatment of insulating oil on site is accomplished with special machines to remove impurities, degas, and dehumidify the oil through a system of filters, heating towers, and vacuum pumps (Figure 5.6).

These machines, used to treat mineral oil, may be inadequately designed for the treatment of natural esters.

Figure 5.6 *Oil treatment machine*

In fact, considering the higher hygroscopicity of natural ester, the vacuum unit needs to dispose of more water. Therefore, it should be equipped with a refrigerated condenser, which protects the vacuum pumps and allows the ester to be processed faster. In addition, it is appropriate to use a Roots vacuum pump downstream of the rotary pump, which works at very low pressures and does not require lubricating oil.

Considering the higher viscosity of the natural ester, the back pressure of the filters is higher, and therefore, the flow rate must be reduced or filters with a larger surface must be employed. In contrast, it is not advisable to employ filters with a larger mesh, because they would allow particles to pass through them and compromise the dielectric behavior of the ester.

Higher viscosity also reduces the efficiency of degassing towers, because lamination may not be complete. As a result, it is preferable to increase the lamination area and to associate spray technology.

In-line residual dissolved gas measurement technology may also be unreliable for natural esters.

Finally, it should be kept in mind that the residual moisture of natural esters is not comparable with that of mineral oils (5 ppm) but is at least 10 times higher.

5.3.5.4 Interpretation of dissolved gases

The diagnostic models of dissolved gas interpretation need to be updated according to the different gas production propensities of the natural ester, which changes the proportions of the same as compared to the models valid for mineral oil. The model may not be unique, but dependent on the type of ester used, i.e., on the specific vegetable source of the ester.

5.3.5.5 Measurement of dissipation factor

The insulating characteristics of mineral ester are different from those of mineral oil (Section 5.2.4.12), so the acceptability criteria of transformer insulation impedance measurement must be reviewed.

Given the higher dissipations of the dielectric medium containing natural ester, the threshold of acceptability of the dissipation factor has to be increased. Instead of the typical value of 0.005 for transformers immersed in mineral oil, the limit for transformers immersed in natural ester should be around the value of 0.01, which is twice the previous value [16]. Since the measurement of the dissipation factor depends on the temperature of the dielectric, these limits are referred to the standard temperature of 20°C; the measured value should be brought back to the standard temperature by a correction coefficient that is different for natural ester than that valid for mineral oil (see IEEE C57.12.90-2006 Section 10.10.5).

In contrast, the limiting insulation resistance for transformers immersed in natural ester is half that considered acceptable for transformers immersed in mineral oil.

The measure of dielectric response (FDS/PDC) is basically a measure of the variable frequency dissipation factor and is used to estimate the moisture concentration in the solid insulator. The proportion of water content between liquid and solid insulators is different for natural esters than for mineral oil, so diagnostic models must be changed. Therefore, the power utility should ask the measuring instrument's manufacturer for valid algorithms for natural ester, which may be specific to the type of ester used.

5.4 Security implications

5.4.1 Fire regulations

Each country has a fire safety regulation for electrical devices with the presence of combustible insulating liquids in quantities exceeding a certain volume (usually a fairly small threshold, e.g., 1 m^3). The regulation prescribes precautions for installation and operation.

The regulation makes a distinction between indoor and outdoor installations (the latter being of greater interest to the power utilities), in which minimum distances between the transformer and other objects or buildings are given. If these distances cannot be met, walls with fire protection must be installed.

In addition, the regulation distinguishes the flash point of the liquid, which can be less than or equal to 300°C (type O) or greater than 300°C (type K), the latter being attributed to the natural ester. Table 5.3 shows the safety distances for different types and quantities of oil taken from the regulation valid in Italy. The distances to transformers filled with low flammability liquid are significantly lower. In addition, if increased protections are present (e.g., overpressure resistant

Table 5.3 Safety distances according to IEC EN 61936-1

Transformer type	Liquid volume (l)	Clearance G to	
		Other transformers or non-combustible building surface (m)	**Combustible building surface (m)**
Oil insulated transformers (O)	1,000 <... < 2,000	3	7.5
	2,000 ≤ ... < 20,000	5	10
	20,000 ≤ ... < 45,000	10	20
	≥ 45,000	15	30
Less flammable liquid insulated transformers (K) without enhanced protection	1,000 <... < 3,800	1.5	7.5
	≥ 3,800	4.5	15

Table 5.4 Safety distances according to Ministerial Decree dated 15/07/2014

Single machine liquid volume (L)	Distance to outside buildings (m)	Distance to interior buildings and other TRs (m)	Distance to fence (m)
$1000 < V \leq 2000$	7.5	3	3
$2000 < V \leq 20{,}000$	10	5	3
$20{,}000 < V \leq 45{,}000$	20	10	5
>45,000	30	15	5

tank, overpressure relief valves, and electrical protections), the distances to transformers can be further reduced.

Another regulation applicable to installations in Italy prescribes the safety distances shown in Table 5.4 for the external installation of electrical devices filled with combustible liquids.

Other requirements of the regulation for existing facilities are as follows:

- Installation of electrical protections, provision of maintenance procedures, securing and safety signage, and a containment system, i.e., the oil collection tub, as well as an exterior fence.
- For devices of more than 150 MVA installed near populated areas, an oil collection tub of 36 m^3 capacity or, if more than 250 MVA, 72 m^3 capacity is required. Finally, a reserve of foaming liquid or equivalent extinguishing agent, in sufficient quantity to produce medium-expansion foam depending on the characteristics of the product, must be made available to firefighting teams.

Some utilities retrofit the substation with a firewater system [27], consisting of a water supply of suitable liquid and an electric pump of adequate capacity.

This constitutes an additional economic and site aggravation, as well as a maintenance burden because the pump must be checked periodically as it normally stands idle. It is presumable that in the future the regulations will favor transformers filled with high flash point liquids, and it is therefore appropriate for the utilities to keep a watchful eye on the development of the regulations.

5.4.2 Spillage of natural esters

5.4.2.1 Spillage into the oil collection tub

A typical installation for large power transformers includes a tub under the foundation of the transformer bay, connected to a basin, called *oil collection tub*, so to prevent leakage of the insulating liquid into the ground in the case of leakage from the tank (Figure 5.7).

The basin is emptied of rainwater by a pump connected to a deoiler that separates it from suspended oil; a sensor stops the pump if the transformer tank breaks. The oil separator separates the oil from the water through a coalescing filter (Figure 5.8), which is based on the non-miscibility of the oil in the water

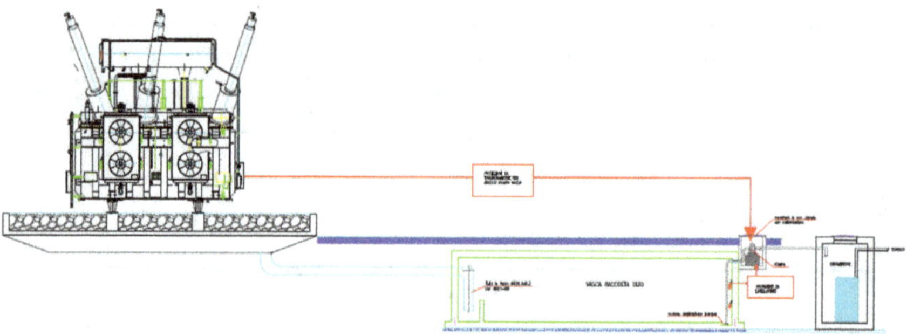

Figure 5.7 Oil collection tub

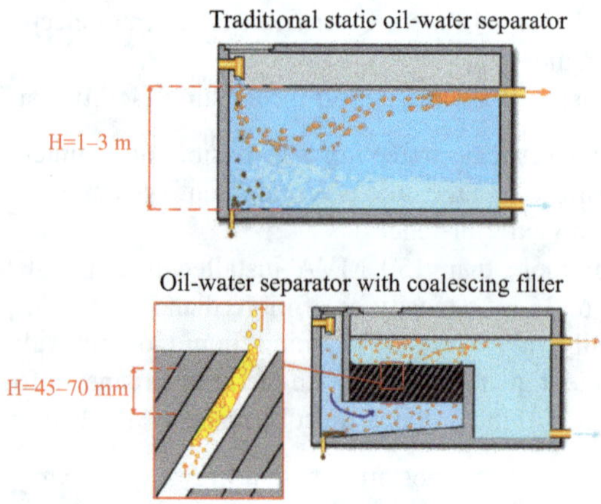

Figure 5.8 Schematic diagram of a coalescing filter

(as the oil forms an emulsion on the surface of the water). Since the natural ester biodegrades quickly, and the degradation products are miscible in water as early as a few days after exposure, current coalescing filters are ineffective for this kind of fluids.

5.4.2.2 Ground spillage

If the spillage is small, ground contamination is negligible, and there is usually no impact on groundwater. The plant operator implements emergency operations that already solve the problem, i.e., soil flaying, disposal of a few cubic meters of soil as waste, and sampling and analysis of the excavated terrain. The regulations may provide for different procedures depending on the size of the spillage, which for large volumes may include an environmental risk assessment and damage monitoring for a predetermined duration.

5.5 Performed experiments

5.5.1 Prototype "Demetra"

Tamini built for the Italian TSO TERNA a 250 MVA 400/135 kV autotransformer filled with natural ester (Cargill Envirotemp FR3®), named "Demetra". The transformer's nameplate data envisage the possibility of overloading the device up to 375 MVA, considering the favorable thermal performance of the insulating liquid (however, this capability has yet to be assessed by service experience).

This transformer required a KDAF-type cooling system with special arrangements of circulation channel inside the windings, which were suitably enlarged.

The transformer is equipped with special sensors for monitoring operating parameters. These sensors include:

- a dissolved gas analyzer, valid for eight combustible gases;
- an oil permittivity sensor;
- winding temperature sensors, with optical fiber technology;
- the monitoring of partial discharges through the bushing test taps.

In addition, to increase safety, this transformer was also equipped with overpressure valves on the bushings' domes (Figure 5.9).

Due to the peculiarity of this type of design and the lack of service experience in the use of natural ester, the online monitoring of the quantities recorded by the sensors has been established, which is:

- periodic, by means of recording by an operator of the measured values;
- continuous, by recording sensors connected to a logical memory, that are interfaced with a data acquisition and storage system (the additional recording of load current allows the measurements to be correlated with service duty).

It is also essential to carry out periodic physical-chemical analyses of the oil to check the oxidation stability.

Figure 5.9 The DEMETRA transformer installed in a TERNA substation

The autotransformer has been in regular service since 2017, and the information relating to the monitoring performed was the topic of a subsequent CIGRE paper [28].

Besides that, it would be useful to assess the performance during overloading operation if compatible with grid operation, with the aim of analyzing the performance at higher than rated power and the impact of high temperatures on the natural ester and solid insulation.

5.5.2 Flammability testing

The Italian TSO TERNA, together with the operational test center of the National Fire Brigade, carried out tests to verify the reduced flammability of natural esters.

Initially, a liquid atomizer and a pilot flame were used to ignite the ester (Figure 5.10). The self-extinguishing ability of the liquid was verified, because the fire faded out as expected when the pilot flame was removed.

Next, two MV/LV transformers containing 120 L of insulating liquid (mineral oil or natural ester) were used, from which two MV bushings were removed so that the liquid could be exposed to flame (Figure 5.11), generated via a propane cylinder connected to nozzles and placed on a metal pipe adjacent to the transformer.

Figure 5.10 Flammability test

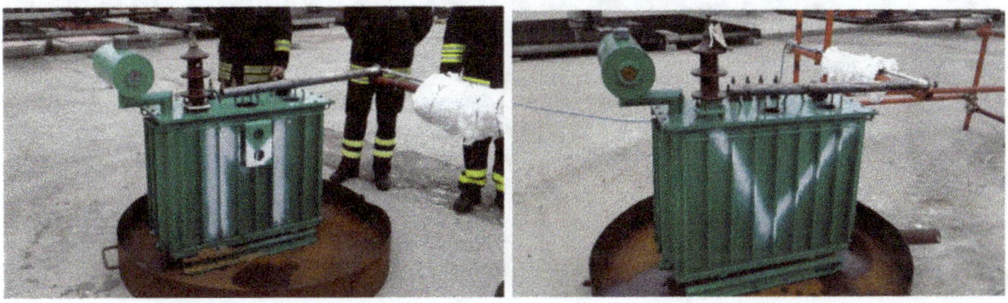

Figure 5.11 Test transformers filled with mineral oil (left) and vegetable ester (right)

Two tests were carried out, one with the flame directly above the liquid surface and the other with the flame near the transformer tank. In the first test, both transformers caught fire, but in the one filled with natural ester, the fire faded out by removing the flame, confirming its self-extinguishing ability. In the second test, only the transformer filled with mineral oil caught fire after about 20 min, while the other filled with natural ester did not catch fire (Figure 5.12); the test on it was stopped after about 40 minutes.

Next, two open containers of about 25 L containing pieces of winding wires were prepared (Figure 5.13), by filling them with natural ester and mineral oil, respectively.

The container was heated with a stove below until the fire ignition temperature was reached (Figure 5.14). It was verified that natural ester ignites at about 400°C, while mineral oil ignites at about 210°C. In both cases, the fire persists even after

Figure 5.12 Test of flammability

Figure 5.13 Container with wire chunks

Figure 5.14 Container in the ignition test before (left) and after (right) fire ignition

removing the heat source, as combustible liquids lose their self-extinguishing capacity at temperatures above the ignition temperature.

Extinguishing the fire was rather difficult after ignition (as superheated smoke tends to restore the fire); however, this situation is not representative of a fault in a transformer, because the internal arc is unable to heat the entire mass of insulating liquid beyond the ignition temperature.

5.6 Conclusions

The characteristics of natural esters (popularly known as "vegetable oils") allow, under certain constrained installation conditions, prospects for transformer use with advantages of smaller footprint (if rating is reduced by taking advantage of the possible overloading capability) and greater environmental compatibility.

However, special requirements for operation, additional provisions for their control, maintenance, and frequent and detailed monitoring of the state of the ester are necessary. Moreover, the long-term aging of natural esters, especially under high-temperature operations, has not been assessed yet.

Considering the warnings described before (which should be kept in mind during design, operation, and maintenance), and if the operational experience should confirm the advantageous features thus balancing the higher cost of this technology, installation of natural ester-filled transformers will increase, which can become advisable particularly in fire-sensitive sites or in protected ecological areas, where it is essential to avoid the risk of hydrocarbon pollutant dispersion into the soil.

The advantages of natural ester oil could be further enhanced by the regulators with an update of the regulations about fire safety, so to take into account the improved physical and chemical characteristics of natural esters, in order to reduce the fire and environmental safety measures currently prescribed for large electrical devices insulated with combustible fluids.

References

[1] Martin, *Preliminary results for dissolved gas levels in a vegetable oil-filled power transformer*, IEEE Electrical Insulation Magazine, 2010.

[2] T.V. Oommen, *Vegetable oils for liquid-filled transformer*, IEEE Electrical Insulation Magazine, 2003.

[3] Y. Bertrand and L.C. Hoang, *Vegetal oils as substitute for mineral oils*, Proceedings of the 7th International Conference on Properties and Applications of Dielectric Materials, 2003, Nagoya.

[4] Brochure CIGRE 436: *Experience in Service with New Liquids*, October 2010.

[5] C. Perrier and A. Beroual, *Experimental investigations on insulating liquids for power transformers: mineral, ester, and silicone oils*, IEEE Electrical Insulation Magazine, 2009.

[6] K.J. Rapp, C.P. McShane, J. Vandermaar, D. Vukovic, and S. Tenbohlen, *Long gap breakdown of natural ester fluid*, IEEE International Conference of HV Engineering and Application, 2010.

[7] J. Hao, M. Dan, Y. Li, R. Liao, and G. Chen, *Cellulosic particles accumulation and conductivity characteristics of natural ester used in power transformer*, Transformer Colloquium, 2017.
[8] X. Li, S. Qian, and X. Xu, *Breakdown characteristic of natural ester gelling process under oxidation and its impact on fresh oil*, 19th ICDL, 2017.
[9] C. Perrier, M.-L. Coulibaly, J. Lukic, and V. Mandic, *Ageing phenomena of cellulose/oil insulation in natural ester and mineral oil*, CIGRE 2014, paper A2-302.
[10] M. Lashbrook and M. Kuhn, *The use of ester transformer fluids for increased fire safety and reduced costs*, CIGRE 2012, paper A2-210.
[11] K.J. Rapp, A. Sbravati, J. Luksich, and D. Bingenheimer, *Oxidation aging and resulting dielectric performance of a natural ester insulation system for transformers*, CIRED 2015, paper 1533.
[12] A.R. Marulanda, M.A. Artigas, A. Gavidia, F. Labarca, and N. Paz, *Study of the vegetal oil as a substitute for mineral oils in distribution transformer*, IEEE Proceedings, 2008.
[13] Carcedo, *Aging assessment of dielectric vegetable oils*, IEEE Electrical Insulation Magazine, 2015.
[14] Martins, *Vegetable oils, an alternative to mineral oil for power transformers—experimental study of paper aging in vegetable oil versus mineral oil*, IEEE Electrical Insulation Magazine, 2010.
[15] J.G. Castellanos, *Experience with application of natural-ester filled HV transformers in Mexico*, CIGRE 2014, paper A2-306.
[16] Cargill reference document R2100: *Transformer Power Factor and Envirotemp FR3 Fluid*, 2013.
[17] R. Frotscher, D. Vukovic, M. Jovalekic, and S. Tenbohlen, *Behaviour of ester liquids under dielectric and thermal stress—from laboratory testing to practical use*, CIGRE 2012, paper D1-105.
[18] S. Haegele, F. Vahidi, S. Tenbohlen, K. Rapp, and A. Sbravati, *Investigation of interfacial surface creep breakdown at oil-pressboard interfaces in natural ester liquid and mineral oil*, 19th ICDL, 2017.
[19] K.J. Rapp, J. Corkran, C.P. McShane, and T.A. Prevost, *Lightning impulse testing of natural ester fluid gaps and insulation interfaces*, IEEE Transactions on Dielectrics and Electrical Insulation, 2009.
[20] M. Lashbrook, A. Gyore, and R. Martin, *A review of the fundamental dielectric characteristics of ester-based dielectric liquids*, Transformer Colloquium, 2017.
[21] R. Frotscher, *Alternative Liquids for Tap-Changers*, Maschinenfabrik Reinhausen, 2013.
[22] Suleiman, *Feasibility study on the use of vegetable oil (natural ester) in Malaysia power system transformers*, Transactions on Electrical and Electronic Materials, 2014.
[23] R. Wittmaack, *Simulation of cold start of an ester test transformer*, Transformer Colloquium, 2017.

[24] K.J. Rapp, G.A. Gauger, and J. Luksich, *Behavior of ester dielectric fluids near the pour point*, IEEE Conference on Electrical Insulation and Dielectric Phenomena, 1999.

[25] F. Bachinger and P. Hamberger, *Thermal measurement of an ester-filled power transformer at ultralow temperatures: steady state*, Transformer Colloquium, 2017.

[26] Cargill reference document R2080: *Dielectric Fluids—Thin Film Oxidation*, 2013.

[27] NT14-001-ING-IIM-IMP rev. 01 of 14/04/2014: Technical Guide "Design and Fire Permit Procedures for Electrical Machinery".

[28] Paper CIGRE A2-316: *Large Power Autotransformers Filled with Natural Ester—Working Parameters from the Field and Maintenance Notes*, TAMINI—TERNA, 2020.

[29] Martin, *Further studies of a vegetable-oil-filled power transformer*, IEEE Electrical Insulation Magazine, 2011.

[30] J. Li, *Preparation of a vegetable oil-based nanofluid and investigation of its breakdown and dielectric properties*, IEEE Electrical Insulation Magazine, 2012.

[31] Ministry of the Interior Decree of July 15, 2014: "Approval of the fire prevention technical rule for the design, installation and operation of substationary electrical machines with the presence of combustible insulating liquids in quantities exceeding 1 m^3."

[32] INAIL presentation at the Study Day on Natural Estates (Rome, 20/10/2016).

Chapter 6
Shell-type power transformers

6.1 Generalities

Shell-type transformer technology has significant differences from the more common core-type technology, and possesses interesting advantages for device operation, short circuit withstand, and the ability to change the turn ratio.

This chapter aims to describe the salient features of this technology, highlighting both its merits and shortcomings, in view of the possible increased use of this type of transformer in the electric transmission grid.

6.2 Introduction to shell-type transformers

Shell-type transformers[1] are the main technological alternative to *core-type* transformers. Figure 6.1 schematically shows the appearance of the active parts of transformers of these two technologies.

The main construction differences are as follows:

- the core of shell-type transformers surrounds most of the windings; instead in core-type transformers, it is the winding that surrounds the core's limb;
- the windings of core-type transformers are cylindrical, with the different coils (low and high voltage) coaxial but separate; in contrast, the windings of shell-type transformers are blunt rectangular, in which the turns of the different coils can be alternated with each other.

Figure 6.2 depicts how in principle it is possible to transition from core to shell technology by gradually changing the design of the active part.

Shell technology derives from the original idea of Lucien Gaulard and John Dixon Gibbs, who in 1882 proposed a transformer that they called *secondary generator*, consisting of two windings partially surrounded by two magnetic steel cylinders (Figure 6.3). They later sold the idea to the American firm Westinghouse, which patented the technology and licensed several firms, including Mitsubishi and Jeumont-Schneider, to build this type of transformer.

[1]Such transformers are sometimes improperly called *armored transformers*.

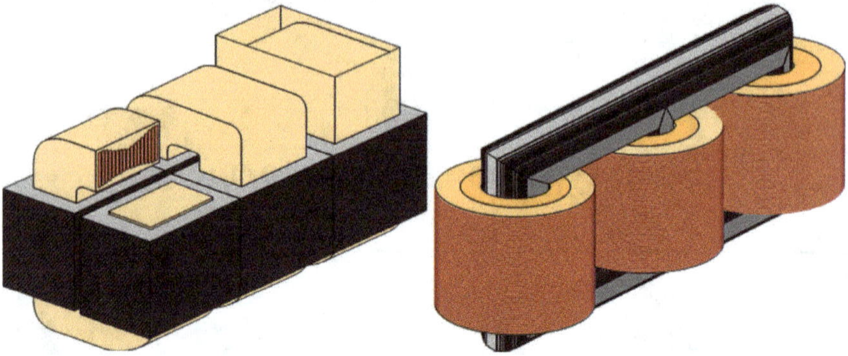

Figure 6.1 *Active part of a shell (left) and core (right) type transformer*

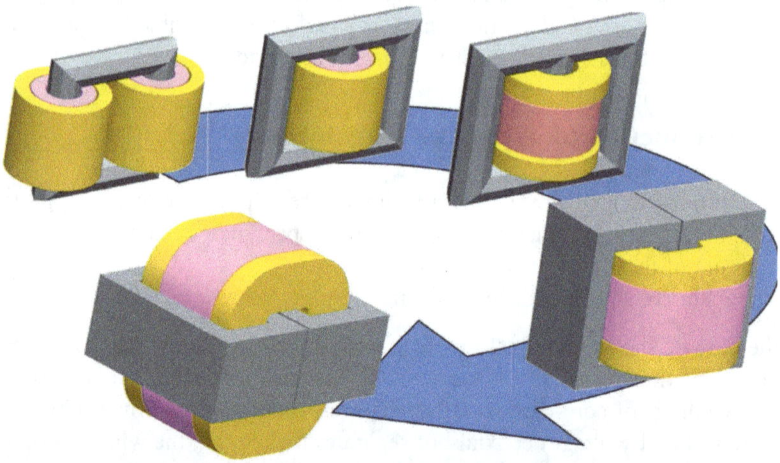

Figure 6.2 *Anamorphic transition from core to shell technology*

The industrial production chain for shell technology is different and incompatible with that of core technology, and only a few factories could operate under license of the Westinghouse patent. Thus, this technology had a lower diffusion than core technology, whose principle was not covered by patent exclusivity.

6.3 Construction characteristics of shell technology

6.3.1 Core

The core of shell-type transformers consists of rectangular windows, made from ferromagnetic steel sheets cut into trapezoid or lozenge shape (Figure 6.4), so its

Figure 6.3 Model of the original design by Gaulard and Gibbs (the winding is enclosed into the core)

construction is simpler than the core of core-type transformers, where the limbs are made of sheets of different widths, which must then be inserted into the yokes held by the clamps.

Apart from this, the construction of geometric figures of shell-type cores follows the same rules as core-type cores, such as staggered sheets and internal cooling channels.

The magnetic flux of each phase closes back into its own window and can circulate independently of the other phases (free flux). Thus, the zero-sequence equivalent circuit of three-phase shell-type transformers is identical to that of 5-limb core-type transformers.

Since in the three-window core model, as in Figure 6.4, the central window shares one side with each of the adjacent windows, the winding of the central phase has an opposite direction to that of the outer phases, so that the resulting flux on the shared columns is the same as on the other phases (Figure 6.5).

Alternatively, a 4-window core can be designed (Figure 6.6), where each phase is wrapped between two windows; in this way, the transport height can be reduced.

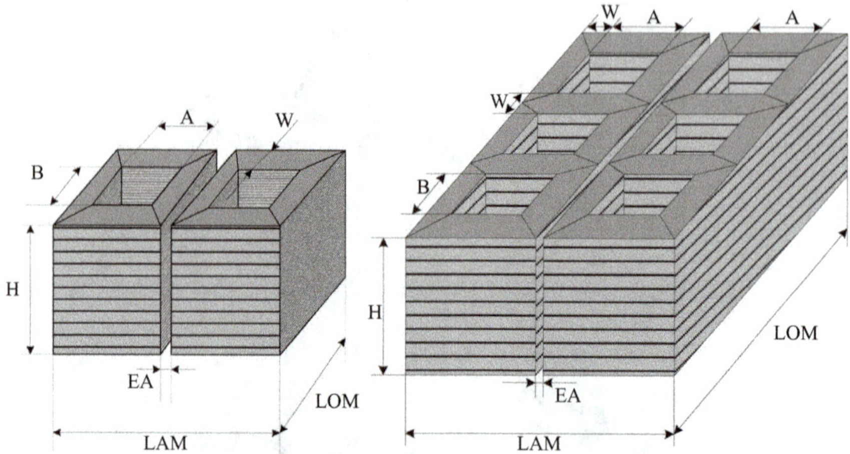

Figure 6.4 Representation of the shell-type core of a single-phase (left) and three-phase (right) transformer

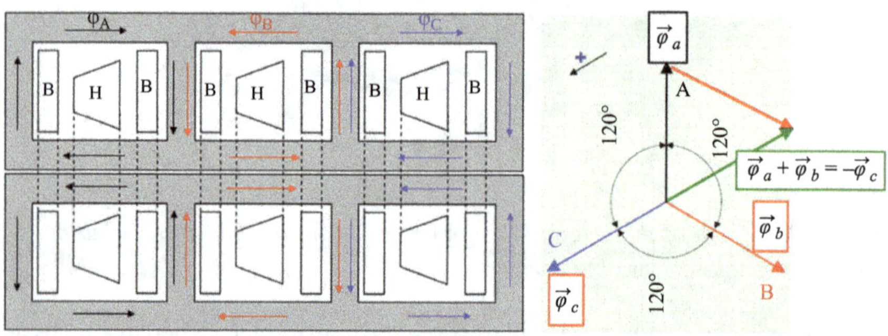

Figure 6.5 Directions of magnetic flux in a 3-window core and phasor diagram

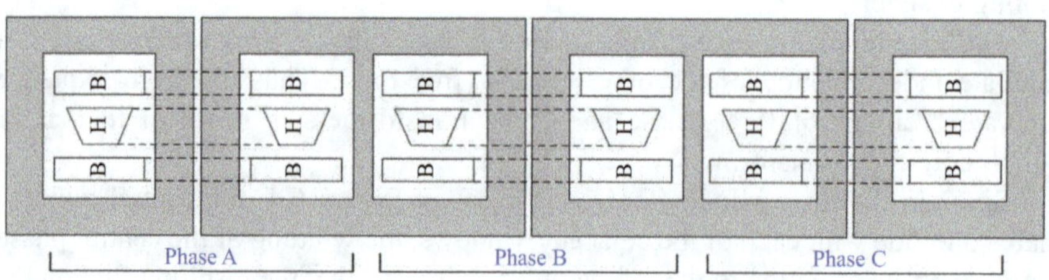

Figure 6.6 Design scheme with 4-window core

6.3.2 Windings

The windings of shell-type transformers are made of enameled and taped strands, either plain or transposed, as in core-type transformers. The difference lies in the different way the winding turns are formed and constructed.

In core-type transformers, the coils are built using special winding machines, but this does not apply to shell-type transformers, for which the coils are built on horizontal pallets (Figure 6.7), where the coils are wound in a geometric pattern formed by the underlying insulating surface, so as to achieve the characteristic blunt rectangular shape.

The individual coils are separated by layers of insulating material, usually pressboard, and contain appropriate spacers and channels for the passage of insulating liquid for cooling. The assembly of all coils constitutes the phase winding (Figure 6.8).

The complete winding takes on the characteristic *pancake* shape, as shown in Figure 6.9.

The winding, before it is fitted on the core, must be fixed by means of a special reinforcement with tie rods. Then, it is oven-dried and impregnated with oil.

Figure 6.10 shows the arrangement of coils in an autotransformer's winding. This arrangement makes possible to reduce the insulation to the core, because the lower voltage coils are the outmost ones. However, more complex arrangements, in which high- and low-voltage coils alternate, are also possible to reduce leakage reactance and thus short circuit impedance.

Figure 6.7 Different stages of winding construction

Figure 6.8 From left: single coil, insulating layer with spacers, complete phase winding with terminal conductors

Figure 6.9 Finished pancake winding, before assembly in the tank

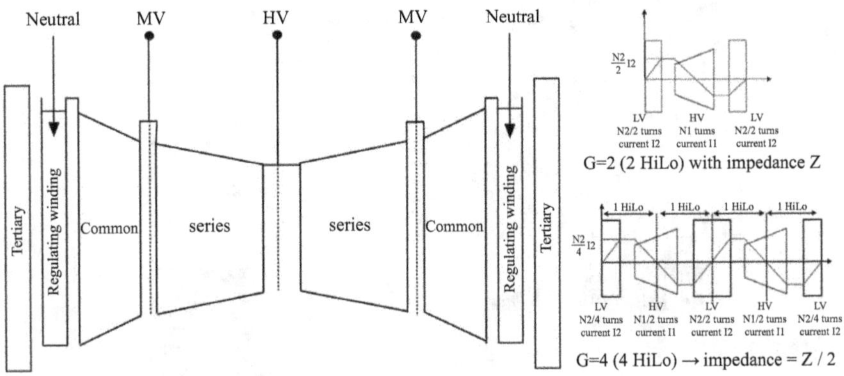

Figure 6.10 Left: typical arrangement of coils in an ATR winding; right: reduction of leakage reactance by alternating high- and low-voltage coils

6.3.3 Assembly

The assembly of the active part takes place directly in the tank, which consists of three pieces: the bottom, the middle part, and the cover.

Figure 6.11 depicts the different stages of assembly of the active part.

Additionally, the followings are included: wooden wedges to fix the winding to the case, metal supports to fix the core window to the winding, and springs between the core and tank to stiffen the active part against dynamic stresses. Figure 6.12 shows two factory steps in the assembly process.

6.3.4 Assembly and commissioning

Because the tank is designed to fit the size of the active part, the free space is reduced and thus so is the amount of insulating liquid needed to fill the tank. This has the advantage that the active part can also be tipped horizontally for transport, reducing its height. The obvious disadvantage is due to the larger size of the active part, and particularly the core, so the cost of a shell-type transformer is – for the same rating data – higher than the cost of a core-type transformer. Figure 6.13 shows the cross-sectional view of two shell-type transformers, one single-phase and one three-phase.

In case of transportation problems, it is also possible to place the phases in separate tanks, each with its own core, which are then brought together in the plant for assembly in three-phase arrangement (Figure 6.14).

6.3.5 Tap changer

Pancake coils can be connected to an on-load tap changer (OLTC), as in core-type transformers. In addition, because the pancake coils are connected to each other in the upper portion of the active part, they can be easily connected with a de-energized tap changer (DETC), which allows the turn ratio to be changed with a wider range of variation than can be achieved with the OLTC alone (Figure 6.15).

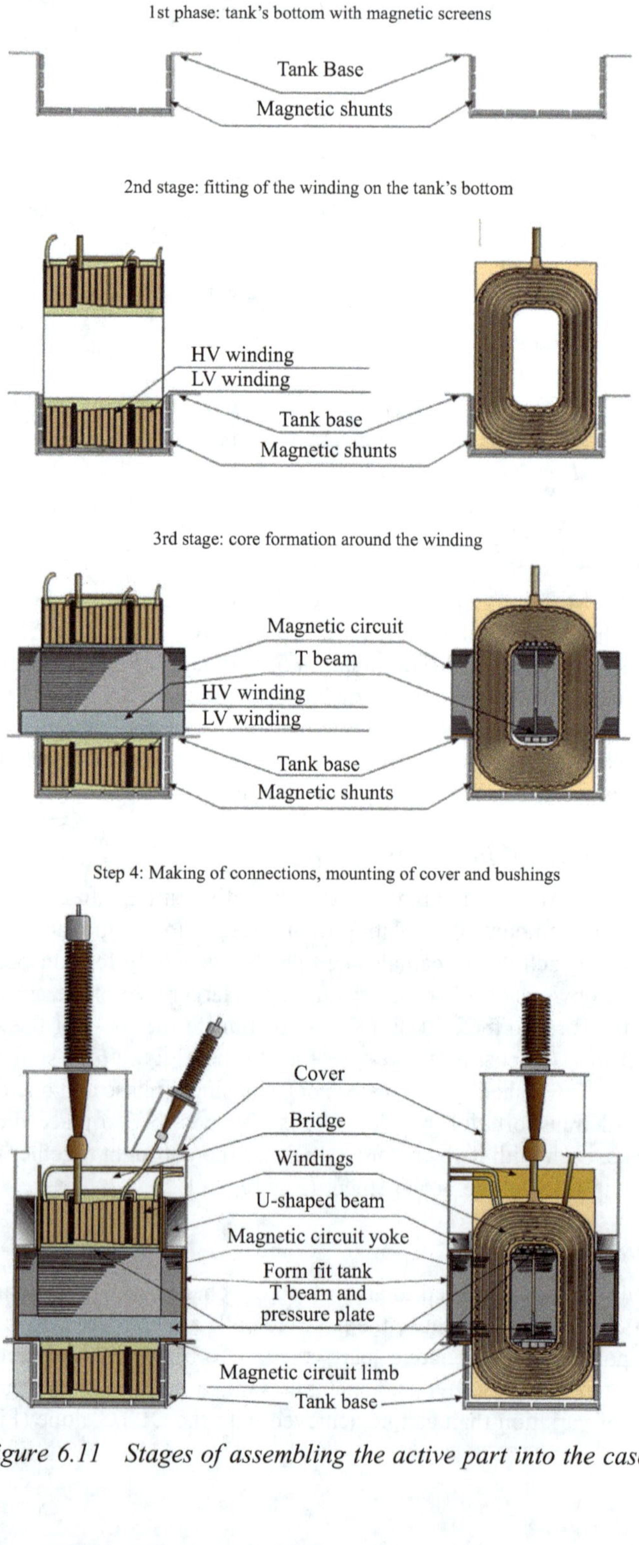

Figure 6.11 Stages of assembling the active part into the case

Figure 6.12 Assembly stages: winding fitting and core formation

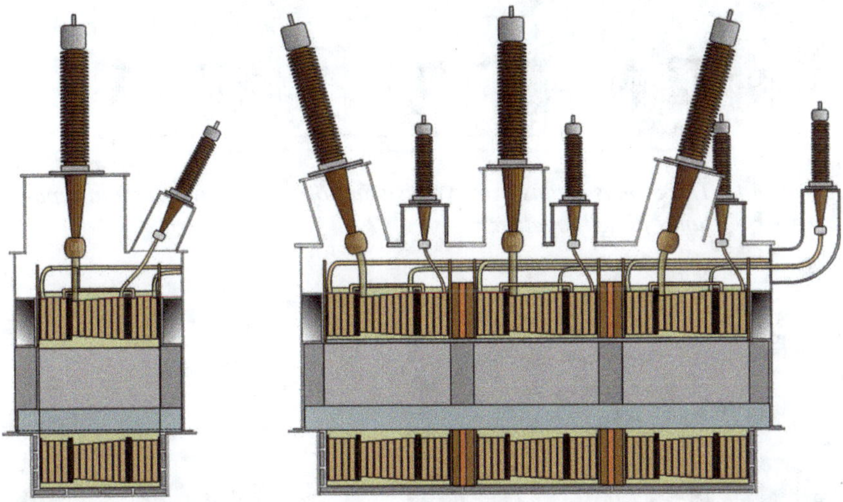

Figure 6.13 Cross-sectional view of a single-phase (left) and three-phase (right) shell-type transformer

This feature is useful for spare transformers, which can be sized to perform different transformation ratios to meet the most common ratios among transformers in service.

6.3.6 Cooling system

The insulating liquid is conveyed directly into the windings, requiring always guided circulation (OD), as depicted in Figure 6.16.

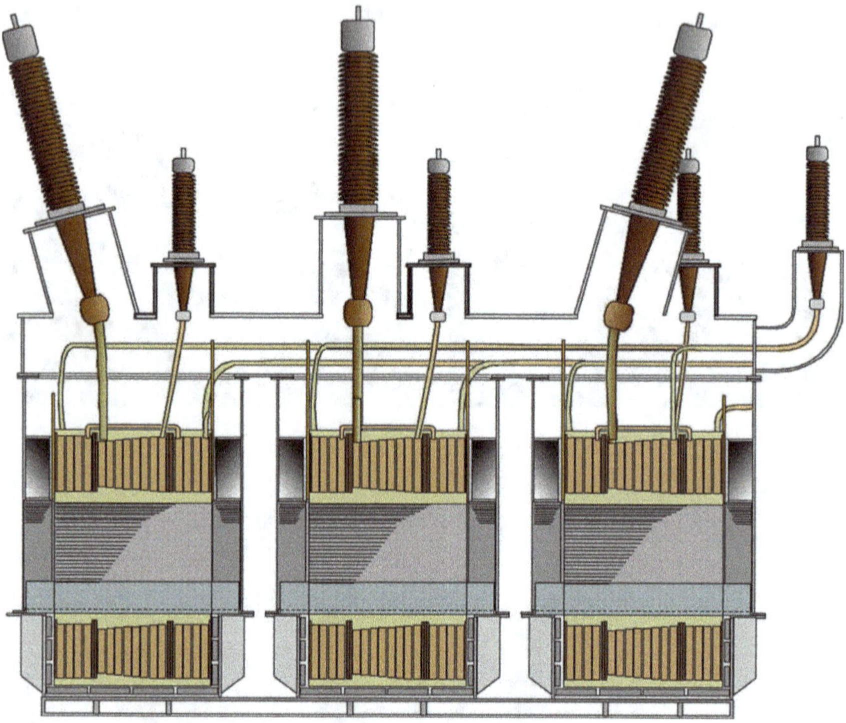

Figure 6.14 *Three-phase transformer consisting of three active parts in separate tanks joined together*

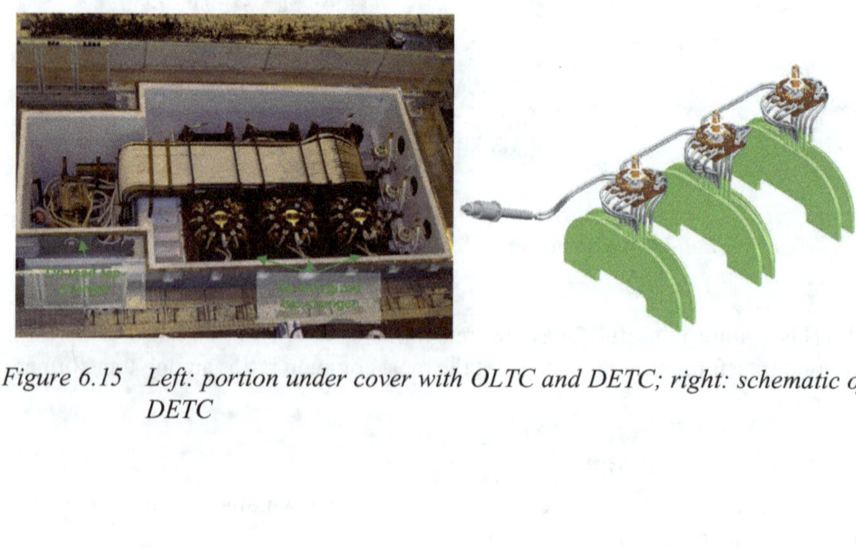

Figure 6.15 *Left: portion under cover with OLTC and DETC; right: schematic of DETC*

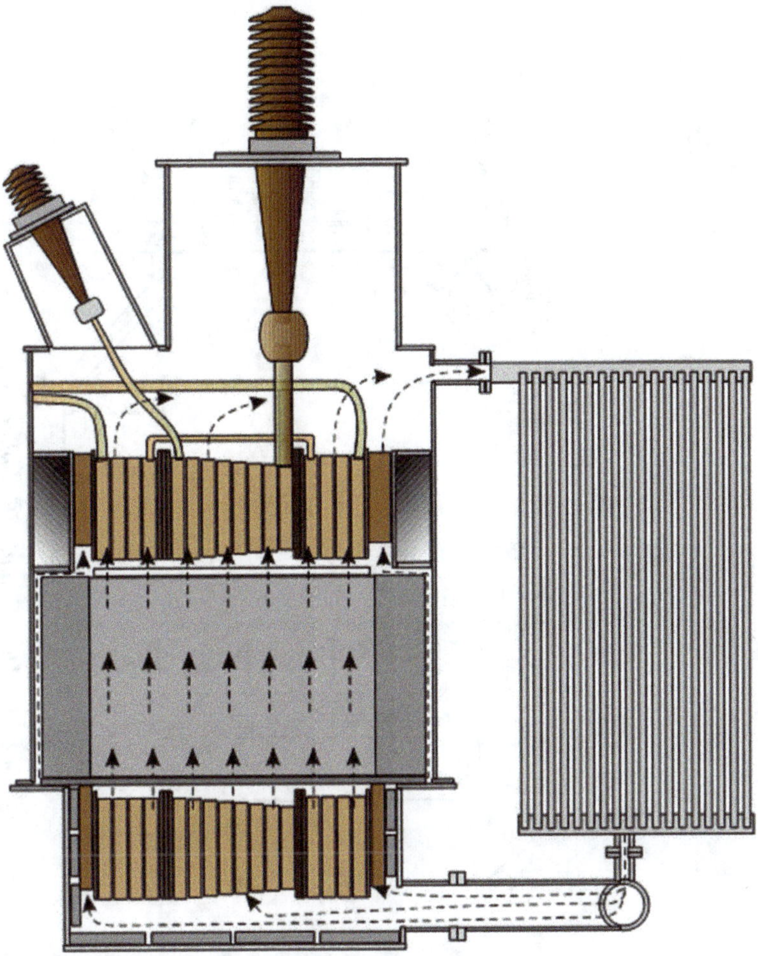

Figure 6.16 Guided circulation of insulating liquid in a shell-type transformer

Shell-type transformers can be equipped with all external cooling systems: AN, AF, WF, as depicted in Figure 6.17.

6.4 Design features of shell technology

6.4.1 Short circuit impedance

Shell transformers have free magnetic flux, and the leakage flux closes around each winding independently of the others. The flux lines are predominantly radial (Figure 6.18).

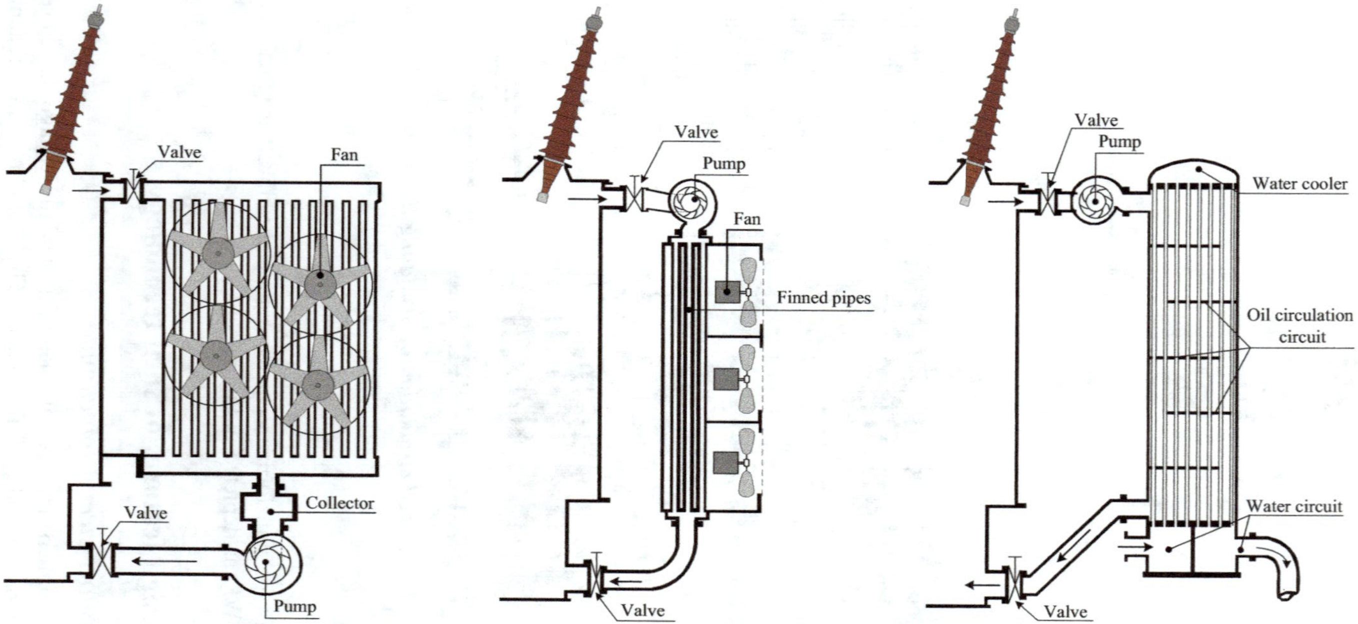

Figure 6.17 Schematic diagram of AN/AF, AF, and WF cooling systems

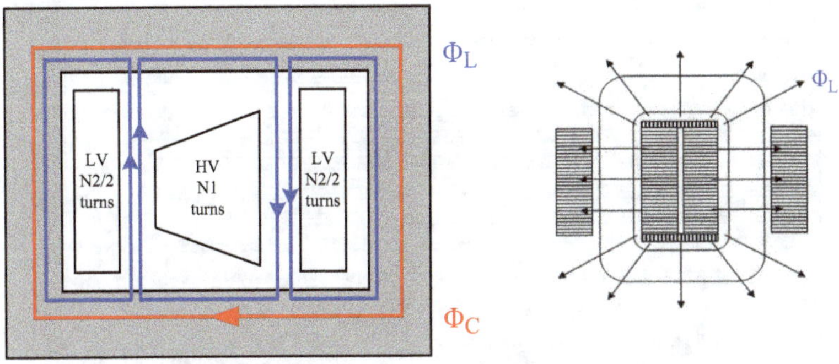

Figure 6.18 Magnetic flux pattern (left) and radial dispersion flux lines (right)

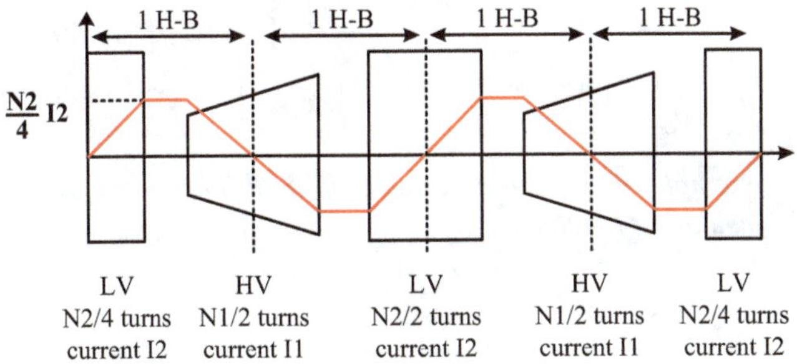

Figure 6.19 Arrangement of windings with alternating coils

The possibility of dividing the windings into several coils alternating with each other (Figure 6.19) allows for greater flexibility in determining the leakage reactance and thus the short circuit voltage.

6.4.2 Dielectric withstand

The dielectric withstand is calculated, as for core-type transformers, using finite element calculation computer programs (Figure 6.20).

To improve dielectric withstand, the pancake coils are arranged not exactly parallel to each other, but with the adjacent edges (to be connected together) closer than the non-adjacent edges, as depicted in Figure 6.21.

Because the coils have a large surface area and a small portion in contact with the case walls, the series stray capacitance is high and the parasitic capacitance to ground low. This results in a low α factor and thus a uniform distribution of impulse voltages (see Section 4.3.3).

Based on the arrangement of the coils and the results of the electric field distribution calculation, the arrangements of spacers and insulating barriers between coils are selected in order to adjust the electric field pattern appropriately (Figure 6.22).

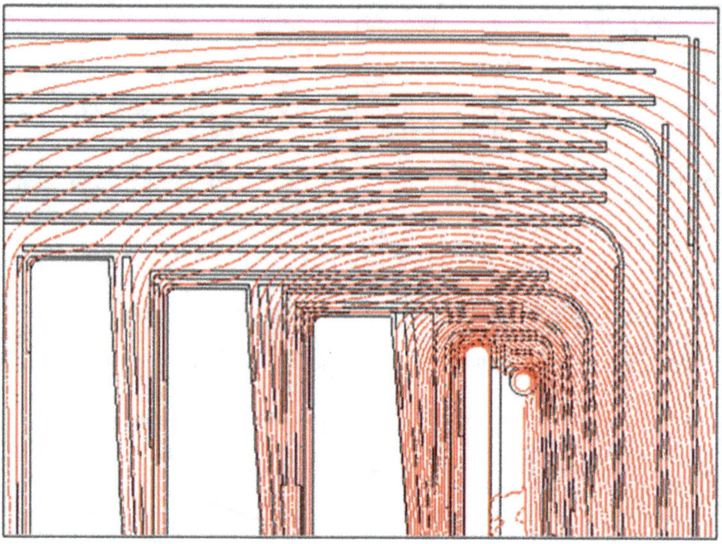

Figure 6.20 Example of calculation of electric field lines between windings and insulating barriers

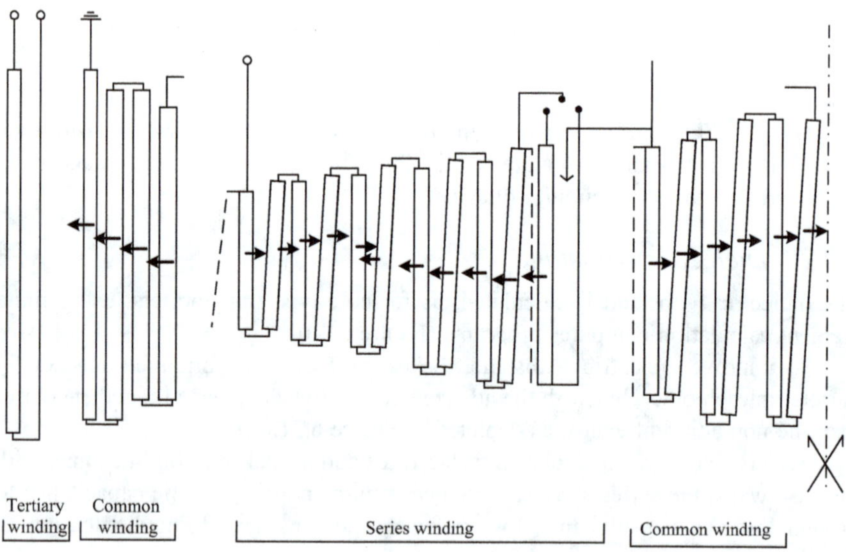

Figure 6.21 Arrangement and interconnection of pancake coils

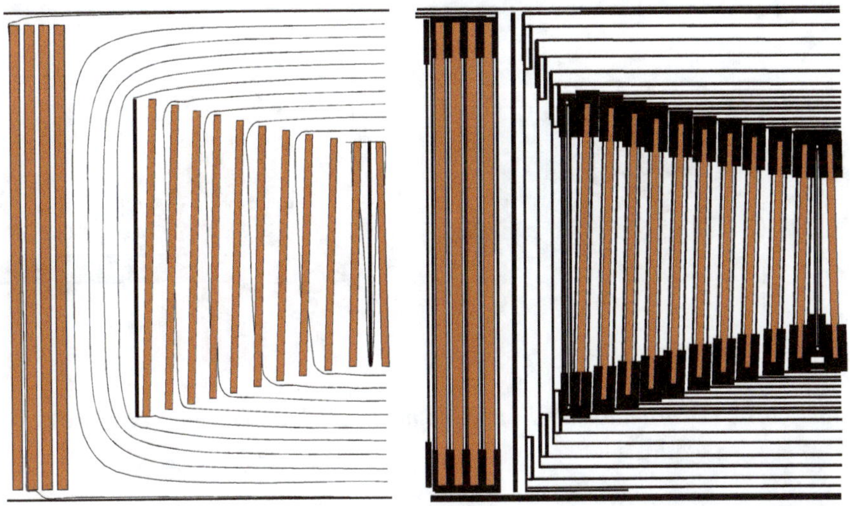

Figure 6.22 Placement of barriers (brown) and spacers (black) to equalize the electric field pattern

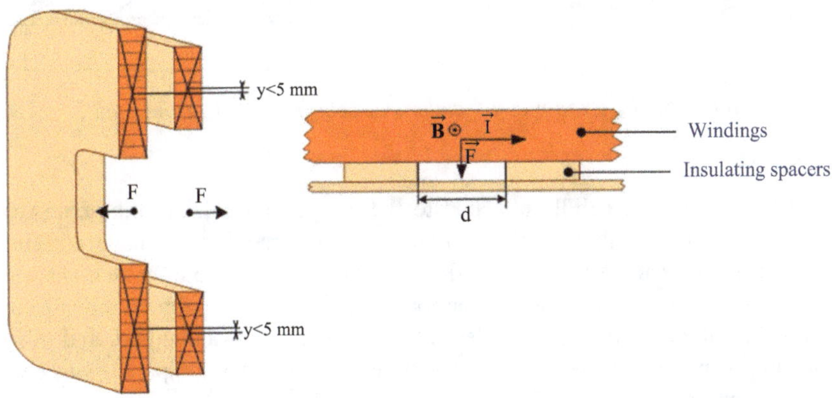

Figure 6.23 Diagram of electrodynamic forces between pancake coils

6.4.3 Short circuit withstand

Electrodynamic stresses in the windings of shell-type transformers are almost exclusively axial (Figure 6.23), because the coils are accurately centered with each other during assembly, thus eliminating any misalignment that would create the radial stresses.

These stresses, because the coils are fixed to the magnetic core and hence to the tank, are transmitted to the tank's walls and are balanced by the reinforcements in the tank (T-beam and springs), as shown in Figure 6.24.

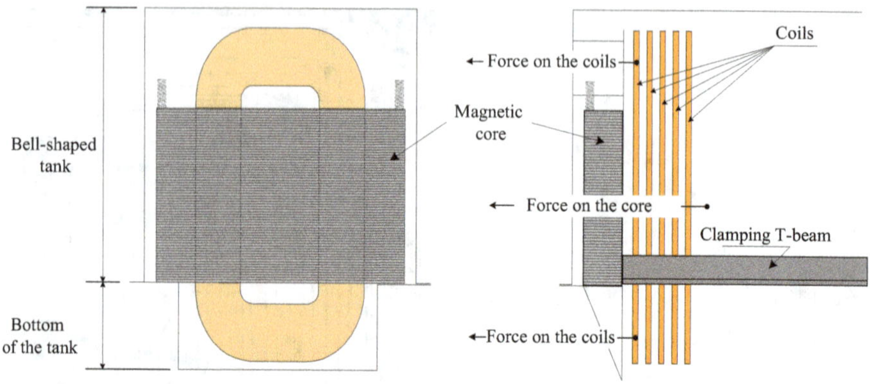

Figure 6.24 Distribution of electrodynamic forces between active part and tank

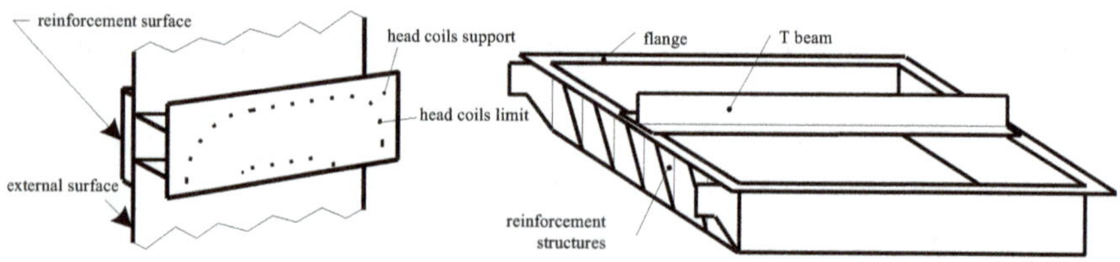

Figure 6.25 Examples of structural reinforcement of the tank

The active part is solidly attached to the tank on all sides (unlike core-type transformers), from which it follows that shell-type transformers have a natural resistance to electrodynamic stresses during short circuits. The only elements that require design attention are the connections, which are not rigidly clamped to the active part. All connections are made on the top side of the active part, and this is an advantage because fewer clamping systems are required compared with core-type transformers.

Otherwise, structural reinforcements to the tank, through metal support surfaces of the active part and beams on the tank's bottom (Figure 6.25), allow the desired mechanical strength to be achieved.

6.4.4 Thermal sizing

Because the circulation of the insulating liquid is guided (OD) within the windings, the distribution of overtemperatures is more uniform than unguided circulation (OF), with the advantage of reducing the hot-spot factor H.

However, guided circulation requires pumps with a high head, which causes higher cooling system's power consumption than core-type transformers.

As usual, the sizing of the cooling system and circulation channels is done by computer CFD simulations.

6.5 Conclusions

In summary, the main advantages of shell technology are as follows:

- more compact geometry with reduced insulating liquid's volume and possibility of horizontal transport;
- more efficient cooling mode with low hot-spot overtemperature;
- uniformity of distribution of impulse overvoltages, which reduces dielectric stress on insulation;
- possibility of obtaining very low short circuit voltages, thus reducing voltage drops under load;
- excellent resistance to dynamic short circuit stresses;
- ability to easily achieve off-circuit ratio change, with different transformation ratios available on the same transformers;
- possibility of manufacturing the three-phase units in three separate single-phase tanks, with smaller size and transport weight.

The main disadvantages compared with core-type technology are as follows:

- larger magnetic core size at the same rating, with higher production costs;
- higher cooling system consumption and criticality in case of pump's failure;
- very high no-load impedance at the zero sequence because the magnetic flux is free, as in 5-limb transformers (this can be a disadvantage in the presence of DC current components that saturate the core).

Because of what was previously described, shell transformers are cost-effective:

- for large devices, such as for power plant step-up transformers;
- when short circuit robustness is essential, such as for transformers in caverns;
- in cases where transportation requirements limit the geometry and weight of the transformer (e.g., for transformers in mountainous locations with impervious roads).

In addition to these cases, shell technology is particularly attractive for spare or emergency units; in fact, since such transformers can be equipped with a DETC for ratio change, a single three-phase spare unit (or a trio of single-phase units, for easier transport and emergency mounting) can meet wide needs for safety and resilience of the power system.

References

[1] Olivier Saint Paul: *Shell-form transformers*, Presentation at the Doble "Life of a Transformer" Seminar, May 19–23, 2014, Padua.

[2] JST: *JST Transformers Presentation*, November 2006.

[3] Olivier Saint Paul: *Shell-form technology*, Presentation for Terna, January 26, 2011.

[4] Gábor Gurszky: *Column and shell core transformers: Comparison of two technologies*, Presentation for Terna, January 26, 2011.

Chapter 7
Thermal aspects of power transformers

7.1 Generalities

During the operation of a transformer, the winding conductors and the ferromagnetic core are the main sources of electrical power losses and disperse heating. Therefore, in the design of a power transformer, the study of heat distribution and related cooling systems is crucial.

This chapter describes the main problems of thermal management of transformers and provides the technical justification of the relevant decisions for the effective specification of their cooling system.

7.2 Heating in power transformers

7.2.1 Power losses

The cause of internal temperature increase in transformers is the emission of heat generated within the metallic materials by the effect of electrical and magnetic phenomena during its operation. The emitted heat (*thermal load*) is subtracted from the transformed energy and is dissipated into the environment. This instantaneous *power losses* consists of two components:

- *no-load losses*, evidenced in the no-load operation of the transformer. They are due to the heat emitted by the core and other ferromagnetic components (clamps, shields, tank), due to the presence of the variable magnetic field generated by the no-load current. They are divided in hysteresis, eddy current, and anomalous no-load losses (see Section 2.2.3);
- *load losses*, evidenced in the short circuit operation of the transformer. They are the heat emitted by the Joule effect due to the circulation of current in the windings and the circulation of eddy currents generated by the magnetic field that does not close up in the magnetic circuit (*leakage flux*), instead investing other windings and ferromagnetic components. They are divided in ohmic losses and additional losses.

The ohmic losses can be calculated with Ohm's law:

$$P_{Ohm} = \rho \frac{\ell}{S} I^2, \tag{7.1}$$

where

- ρ is the resistivity of the conductor;
- S is the cross section of the conductor;
- ℓ is the length of the loop;
- I is the root mean square (RMS) value of the alternating current circulating in the loop.

The additional losses in the windings can be calculated by the following formula ([1] (10.22)):

$$P_{add} = \frac{\pi^2}{6}\frac{f^2 b^2 B_m^2}{\rho}, \tag{7.2}$$

where

- f is the frequency;
- b is the thickness of the conductor in the direction transverse to the magnetic field;
- B_m is the peak value of the magnetic induction, at the conductor.

Because of the dependence upon the square of thickness, it is convenient to divide the coils of the large-section winding into several conductors in parallel, so to limit the thickness and thus the additional losses.

The additional losses depend on the local values of magnetic induction. They are not uniform along the winding and are highest where the leakage flux is most intense. In core-type transformers, this occurs near the winding's top where the flux bends and frays to enter the yoke.

For the calculation of the average magnetic induction in the winding, an estimate based on the leakage flux and the winding cross section is sufficient. Instead, for the evaluation of additional losses, including those outside the windings and in the electrical connections, the peculiar magnetic field in the metal structures and connections must be calculated.

A reduction in losses can be achieved by using non-ferromagnetic materials, such as stainless steel, or by adding magnetic shields,[1] but in both cases the cost of manufacturing the transformer increases.

Ohmic losses, being dependent on current, are directly proportional to the resistivity of the metal, while additional losses, being dependent on voltage, are inversely proportional to the resistivity of the metal. Therefore, since resistivity depends on temperature, the temperature dependence of the two types of losses is opposite.

Losses in load operation are a very good approximation given by the sum of no-load and load losses. Normally, at or near to rated power operation, load losses are prevalent, and therefore, the cooling system's main purpose is to dissipate the heat generated by the windings.

[1]Magnetic shields are layers of magnetic sheets leaning against the inner wall of the tank, so that stray flux preferentially closes back on them rather than on the tank that would overheat the walls.

7.2.2 Cooling systems

The cooling method in transformers consists of a fluid circulating in the tank, lapping both the windings and the core, which heats up on contact with them and gives up thermal energy to another fluid outside the tank by means of a heat exchanger.

The transformer's cooling system is identified by four letters with the following meanings:

- 1st letter: cooling fluid in contact with the active part
  - insulating liquid with flash point below 300°C (typically, mineral oil);
  - K insulating liquid with flash point above 300°C;
  - L insulating liquid with unmeasurable flash point;
  - G insulating gas (typically SF_6).
- 2nd letter: circulation type for the internal cooling fluid
  - N: natural, i.e., by thermal convection and conduction;
  - F: forced, i.e., through pumps;
  - D: forced and guided, through pumps and conveyor pipes;
  - ND: natural and guided.
- 3rd letter: external cooling fluid
  - A: air;
  - W: water.
- 4th letter: circulation type for external cooling fluid
  - N: natural;
  - F: forced.

Heat exchangers with natural oil circulation are *radiators*, while in the case of forced circulation, *heat exchangers* are used, which differ from radiators by being more compact and having multiple channels in parallel for oil circulation to increase the heat exchange surface area (Figure 7.1).

For transformers and autotransformers in an electrical transmission grid, mineral oil is most frequently adopted as the internal fluid and air as the external fluid. However, the following considerations are also generally applicable to other insulating fluids such as natural or synthetic esters.

The above definitions do not specify the type of fluid circulation within the windings. In windings with crossover and spiral coils, there are no special arrangements for oil circulation. In contrast, for helical coils and especially for disc coils, there are channels in which oil circulation can be forced through special barriers (Figure 7.2), regardless of the presence of the conveyors in the OD system.

The heat exchange phenomena that allow cooling in transformers are *thermal conduction* between conductors and oil and between oil and air, through the wall of heat exchangers, and *convection* of the oil, which, in natural circulation systems, allows the liquid to move upward in the windings and downward in the heat exchangers.

Figure 7.1 Comparison (not to scale) between radiators with fans (ONAF system) and heat exchangers (OFAF)

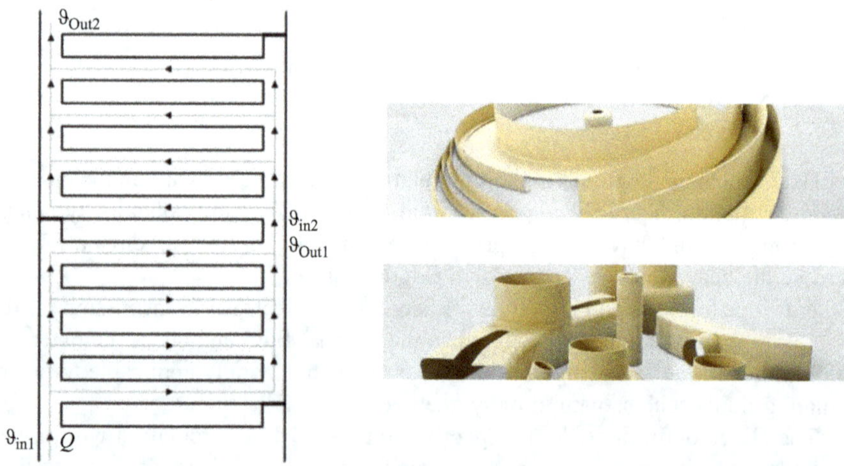

Figure 7.2 Pattern of forced circulation in oil channels (left) and barriers in pressboard (right)

The calculation of heat transfer by conduction can be done with a formula such as

$$P = \lambda \frac{A}{s} \Delta\theta, \tag{7.3}$$

where

- P is the heat power transferred by conduction;

- $\Delta\theta$ is the temperature difference between the two bodies (copper and oil, oil and air);
- A is the heat exchange surface area;
- s is the thickness of the separating wall between the two bodies (insulation paper, radiator pipe);
- λ is the heat conduction exchange constant.

In contrast, heat transfer by convection has greater complexity, because it depends on the type of fluid motion (laminar or turbulent) and the balance of upward force (buoyance) and viscous friction. These evaluations require computer simulation programs, generally described in a later section.

Heat exchange by radiation is normally negligible, except in small transformers where it is realized with fins welded to the tank, so to increase the area of the emitting surface. However, for installations in very hot climates, solar radiation on the tank contributes to the heating of the transformer, which will have to be oversized because its thermal load includes the external heating.

7.2.3 Thermal model

The most popular schematic model for defining temperatures inside oil-immersed transformers assumes a temperature dependence only as a function of height relative to the tank's bottom. This is a simplification because the temperature is not uniform among all sections at the same height. In particular, one must consider the parallel flows of oil between the different windings and that which does not circulate in the windings, but outside the active part.

Figure 7.3 highlights diagrams of temperatures of air (in light blue), coolant surface (in gray), oil in closed circuit between tank and coolant (in brown), conductor surface insulation (in yellow), and conductors (in red). Also shown in the figure are several temperature differences, also called *thermal jumps*, among which are notable the following (Figure 7.4):

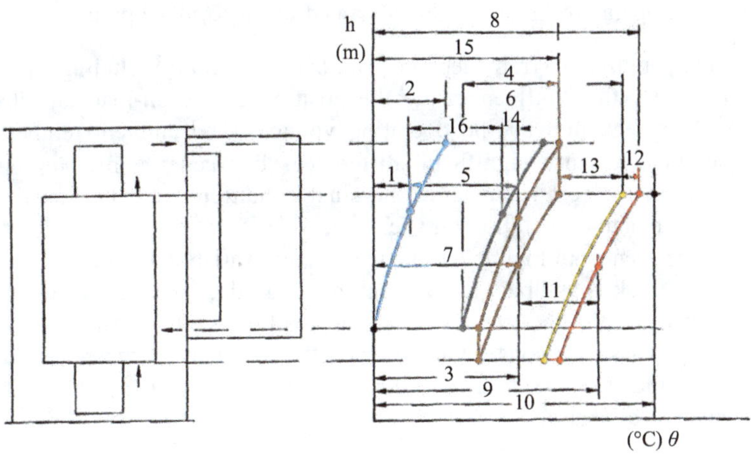

Figure 7.3 Temperature diagram in the transformer

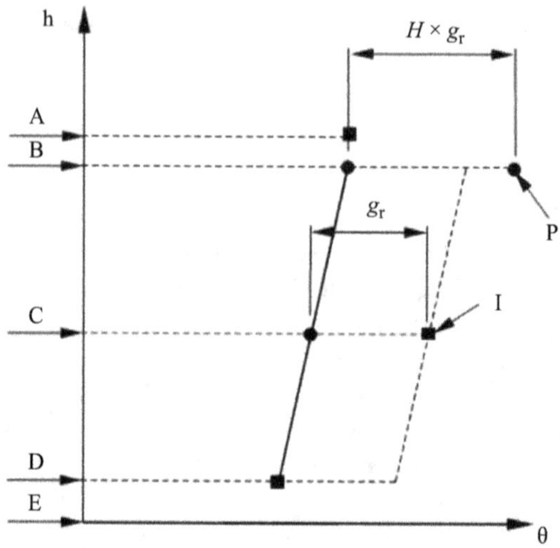

Figure 7.4 Thermal diagram with notable points of interest for temperatures highlighted

- oil temperature at the bottom of the tank (point E);
- oil temperature at the bottom outlet of the coolant system, which is considered equivalent to the temperature at the bottom of the windings (point D), named *bottom oil*;
- oil temperature at the top inlet of the cooling system, which is considered equivalent to the temperature at the top of the windings (point B);
- oil temperature under the tank's cover (point A), named *top oil*;
- average temperature of oil (point C) and conductors, named *copper temperature* (point I);
- highest temperature in the windings, named the *hot-spot* (P point).

The temperature diagram depends on several factors, including operating power, magnetic flux (influenced by the position of the on-load tap changer (OLTC), if present, and by the operating voltage), and ambient temperature. These factors can cause significant differences between the measured over-temperatures in the test room under nominal conditions and the actual over-temperatures in actual operation on site.

As can be seen from Figure 7.4, the temperature variation between the bottom and top of the tank is assumed to be linear, although this is an unrealistic simplification. Consequently, the average temperature difference between oil $\theta_{o(av)}$ and copper $\theta_{Cu(av)}$ is assumed constant and equal to the parameter g_r, where r stands for "rated," since the values refer to operation under nominal power conditions and at room temperature. That is

$$g_r = \theta_{Cu(av)} - \theta_{o(av)}. \tag{7.4}$$

To account for the fact that the temperature change in the windings is different from that of the oil and is not uniform, the *hot-spot temperature* (HST) θ_{hr} is defined, which is related to that of top oil θ_o by means of a parameter $H > 1$, called the *hot-spot factor*:

$$\theta_{hr} = H \cdot g_r + \theta_o. \tag{7.5}$$

In Figure 7.5, temperature diagrams for different cooling systems commonly used in oil-immersed transformers are schematized. The arrangement of the radiators raised above the case cover, called "gooseneck connection," allows a greater thermal jump in the radiators, and thus better circulation by convection. Also, in the ODAF system, the copper-oil heat jump is practically zero, given the higher efficiency of heat exchange. However, it should be kept in mind that it is not

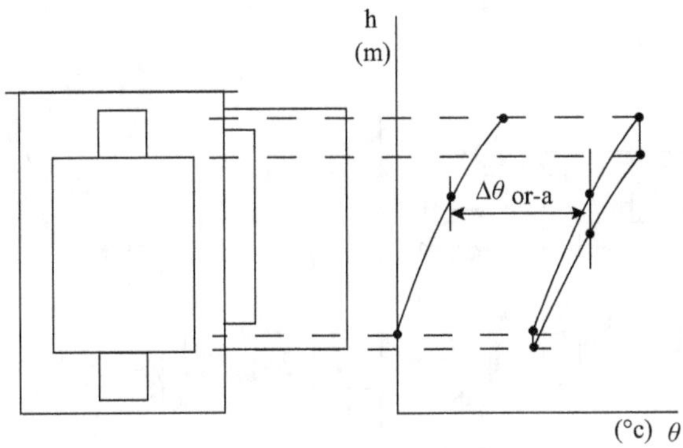

ONAN Cooling

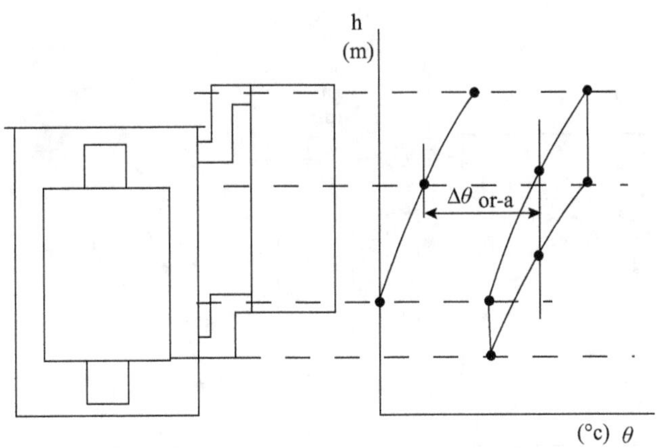

ONAN cooling with gooseneck connection

Figure 7.5 Temperature diagrams of different cooling systems

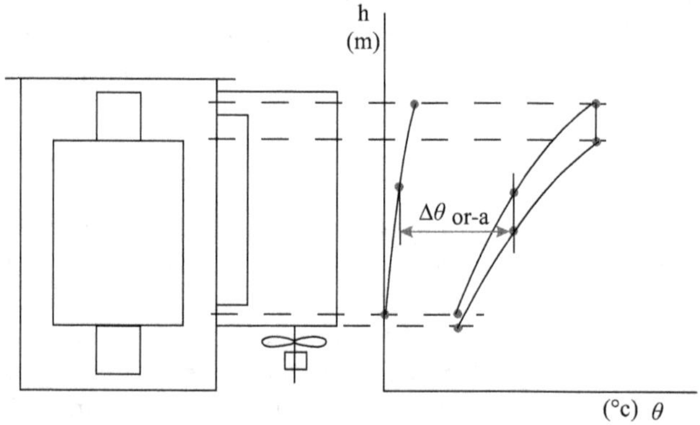

ONAF Cooling

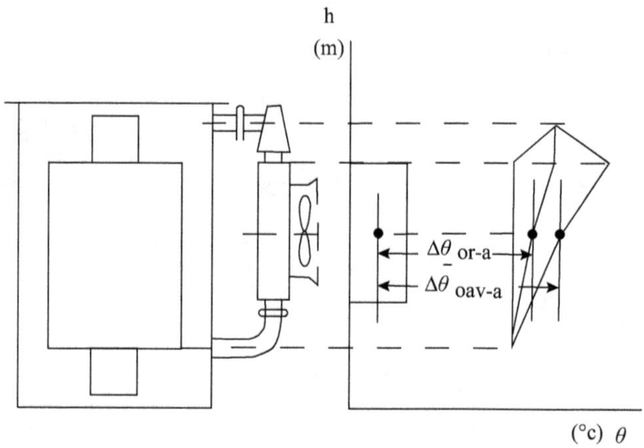

ONAF Cooling

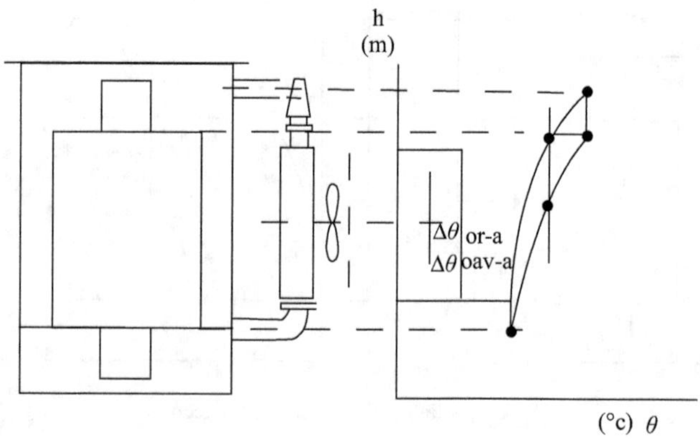

ODAF Cooling

Figure 7.5 (Continued)

possible to increase the speed of forced oil circulation beyond a certain limit, both because of its viscosity and because at high speeds the phenomenon of streaming electrification is generated (see Section 4.3.2), in which the oil takes on a static charge by rubbing against metal surfaces, which could generate partial discharges and compromise the insulation of the windings in the long term.

7.2.4 Aging of solid insulation

Solid insulation is the transformer's component most prone to aging, with the aggravating factor that it cannot be replaced. It consists of paper and pressboard in normal (not high-temperature) designs, both cellulose-based materials.

The phenomenon of degradation of cellulosic components in transformers has been extensively studied. The scheme of Figure 7.6 represents the known causes and includes the degradation of liquid insulation, which is a contributory cause of the solid insulation.

Oil degradation happens mainly by oxidation, and the decomposition products are semisolid acidic components and sludge that settle at the bottom of the tank. In

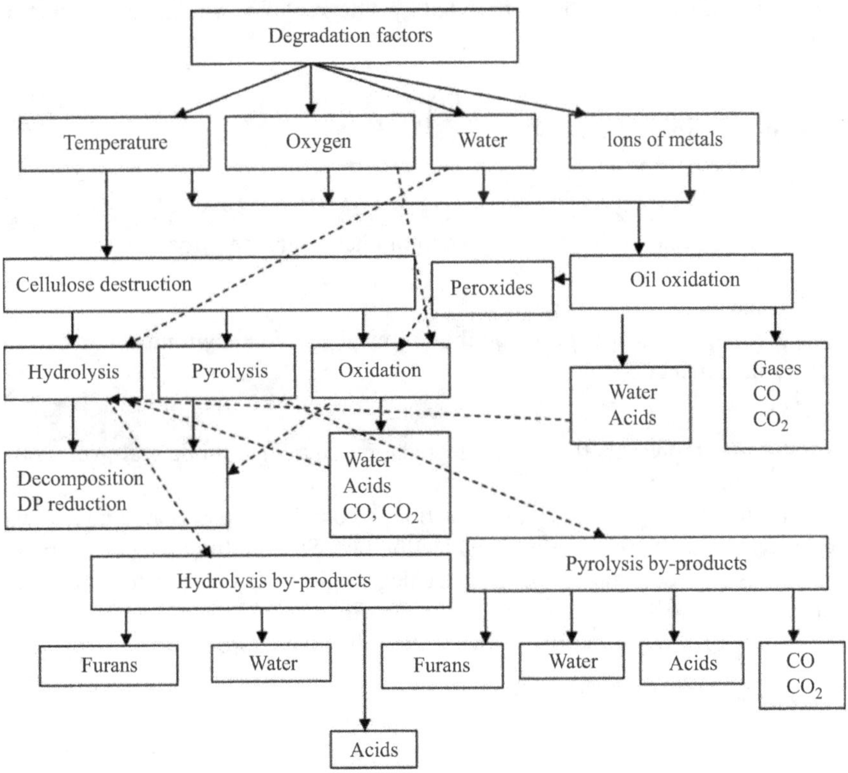

Figure 7.6 Paper and oil degradation phenomena

the presence of electrical discharges or at very high temperatures (>140°C), oil vaporization and combustible gas formation phenomena also happen.

Paper degradation happens in three ways:

- *oxidation*, through the reaction of dissolved oxygen in the oil that splits the cellulose molecules and forms water and CO_2 as end products. This phenomenon is prevalent at low temperatures and affects kraft and TUK paper equally. It happens most likely in transformers without the membrane in the conservator and without an airtight tank, where the oil is in contact with air.
- *hydrolysis*, by reaction with H^+ ions, then in an acidic environment with the presence of water. The decomposition products are low molecular weight acids and water, which induces a self-accelerating reaction. This reaction is prevalent at high temperatures and affects kraft paper more than TUK paper.
- *pyrolysis*, which is the thermal destruction of cellulose chains at elevated temperatures (>150°C), as can occur in the presence of hot-spots due to thermal defects in the active part due to defective welds and uncontrolled leakage flux.

All degradation phenomena depend exponentially on temperature, according to *Montsinger's law*, and can be represented by a degree of polymerization reduction law $DP(t)$:

$$\frac{1}{DP(t)} - \frac{1}{DP(0)} = \sum_{i=1}^{3} A_i \exp\left(\frac{-E_i}{R\theta}\right) t, \tag{7.6}$$

where

- i represents one of the three degradation phenomena described;
- R is the perfect gas constant;
- E_i is the activation energy;
- A_i a coefficient that depends on the chemical reaction environment;
- θ is the temperature;
- t is time.

Figure 7.7 represents the rate of degradation on a logarithmic scale versus the inverse of temperature.

Since the complete degradation reaction is complex to describe, conventional rules were adopted for calculating the reaction rate. Starting from the consideration that the HST θ_h determines the greatest degradation, all other conditions being equal, *aging laws* were derived for insulating paper.

The law of aging for kraft paper has the following expression:

$$V(\theta_h) = 2^{(\theta_h - 98^\circ C)/6^\circ C}, \tag{7.7}$$

where $V(\theta_h)$ is the aging rate. Set $V(98^\circ C) = 1$ as the nominal aging, it can be seen that the velocity doubles for every 6°C increase in HST.

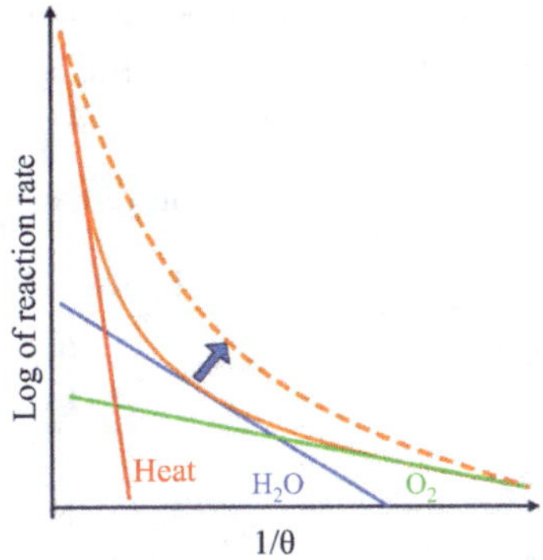

Figure 7.7 Rate of degradation as a function of the inverse of temperature

For the TUK paper, the law is more complicated:

$$V(\theta_h) = \exp\left(\frac{15{,}000°\text{C}}{110°\text{C} + 273°\text{C}} - \frac{15{,}000°\text{C}}{\theta_h + 273°\text{C}}\right) \approx 2^{\frac{\theta_h - 110°\text{C}}{6.8}°\text{C}}, \quad (7.8)$$

nominal aging occurs at 110°C, instead of 98°C, and the rate doubles every 6.8°C, instead of every 6°C.

These laws do not consider the possible presence of moisture, oxygen, and acids in the oil, which can greatly accelerate the degradation reaction. However, the dependence on these parameters is complex, and no consensus has yet been found among experts for a law that takes them into account.

7.2.5 Loading guide

The *loading guide* for oil-immersed transformers, as set out in the IEC 60076-7 standard [2], provides guidance and rules for calculating the HST, as a function of the top-oil temperature, detectable by thermoprobes inserted in special undercover wells, and the thermal characteristics of the transformer, known from thermal testing, or assumed (according to the rules of the loading guide) for each type of cooling.

Having defined the following parameters:

- θ_a: ambient temperature;
- $\Delta\theta_{oi}$: initial top-oil overtemperature;
- $\Delta\theta_{or}$: top-oil overtemperature at steady state at rated power;
- R: ratio of load to no-load losses under nominal conditions;
- K: load factor (ratio of load current to rated current);
- τ_o, τ_w: time constants of oil and copper, detectable from measurements of oil temperature and winding resistance at the end of the thermal test;
- k_{11}, k_{12}, k_{22}: thermal model constants that are not easy to determine; according to Ref. [3] § 4.6.2.3, additional thermal tests and measurements with fiber-optic sensors would be required. Normally, the conventional values given in the loading guide are adopted ([2] tab. 4);

- x, y: oil and copper exponents, which consider the nonlinear relationship between load and temperatures in both. Normally, the conventional values given in the loading guide are adopted ([2] tab. 4);

The laws of variation of the top-oil temperature are assumed to be[2] as follows:

$$(\text{increase})\ \theta_o(t) = \theta_a + \Delta\theta_{oi} + \left\{\Delta\theta_{or}\cdot\left[\frac{1+R\cdot K^2}{1+R}\right]^x - \Delta\theta_{oi}\right\}\cdot\left(1-e^{-t/k_{11}\tau_o}\right), \tag{7.9}$$

$$(\text{decrease})\ \theta_o(t) = \theta_a + \Delta\theta_{or}\cdot\left[\frac{1+R\cdot K^2}{1+R}\right]^x + \left\{\Delta\theta_{oi} - \Delta\theta_{or}\cdot\left[\frac{1+R\cdot K^2}{1+R}\right]^x\right\}\cdot e^{-t/k_{11}\tau_o}, \tag{7.10}$$

The laws of variation of the HST are assumed to be given by two contributions:

$$(\text{increase})\ \Delta\theta_{h1}(t) = \Delta\theta_{h1_i} + \{k_{21}Hg_rK^y - \Delta\theta_{h1_i}\}\cdot\left(1-e^{-t/k_{22}\tau_w}\right), \tag{7.11}$$

$$(\text{decrease})\ \Delta\theta_{h1}(t) = k_{21}Hg_rK^y + \{\Delta\theta_{h1_i} - k_{21}Hg_rK^y\}\cdot e^{-t/k_{22}\tau_w}, \tag{7.12}$$

and

$$(\text{increase})\Delta\theta_{h2}(t)=\Delta\theta_{h2_i}+\{(k_{21}-1)Hg_rK^y-\Delta\theta_{h2_i}\}\cdot\left(1-e^{-tk_{22}/\tau_o}\right), \tag{7.13}$$

$$(\text{decrease})\Delta\theta_{h2}(t)=(k_{21}-1)Hg_rK^y+\{\Delta\theta_{h2_i}-(k_{21}-1)Hg_rK^y\}\cdot e^{-tk_{22}/\tau_o}, \tag{7.14}$$

where the subscript i indicates the initial value of the overtemperature.

Overall, the HST results from

$$\theta_h(t) = \theta_o(t) + \Delta\theta_{h1}(t) - \Delta\theta_{h2}(t). \tag{7.15}$$

Figure 7.8 depicts the transient trend of the HST. Comparing the theoretical model with the actual measurement, it is observed that the standard parameters overestimate the temperature. The best results are obtained using parameters derived from measurements made with optical fiber sensors during thermal tests.

The loading guide provides the following estimate for service life under ideal conditions (moisture in oil $\leq$0.5% RH and oil segregated from air –, according to table A.1 in [2]):

- kraft paper with hot-spot at 98°C: 11.7 years;
- TUK paper with hot-spot at 110°C: 15.3 years.

[2]The two formulas – and subsequent pairs – are identical but written with a different order of addends to highlight the initial state and transient phenomena in situations of increasing and decreasing load.

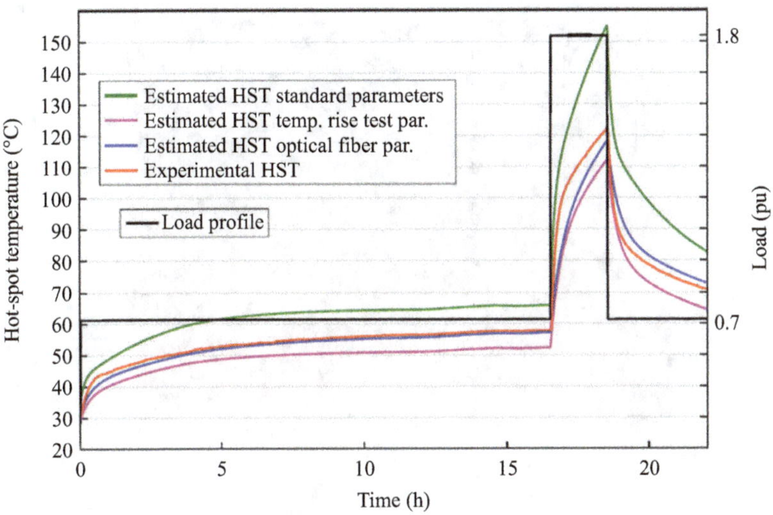

Figure 7.8 Estimates and measurements of hot-spot temperature trend

"End-of-life" is understood to be the condition where, starting with paper at DP = 1000, the paper reaches DP = 200; at this value the mechanical performance of the taping is considered inadequate, although the paper continues to perform its insulating function. However, destructive failure due to paper deterioration is not the predominant cause of transformer end-of-life. In fact, according to a survey by Cigré ([4] Tab. 9) only 5% of failures are due to insulation deterioration. Nevertheless, the loading guide model remains widely cited and applied for life estimates, because it is easily implemented and based on easily monitorable load and oil temperature data.

7.2.6 Thermal monitoring systems

There are two types of temperature measurement sensors used in transformers (Figure 7.9):

- thermostats that exploit the known change in the resistivity of a material as a function of temperature ("Pt100" sensors);
- optical sensors that exploit the known variation in the light reflection coefficient of a material (usually, GaAs) as a function of temperature ("optical fiber" sensors).

The former are the most common and are usually coupled with analog thermometers, for measuring oil temperatures via special thermoprobes under the cover or in the cooling system and core piping. The latter, which are more expensive and delicate, are instead dedicated to measuring temperatures inside the windings (Figure 7.10).

There are also sensors, called *thermal-imaging* sensors, consisting of a resistor inserted in an oil well connected to a circuit that injects a current proportional to the

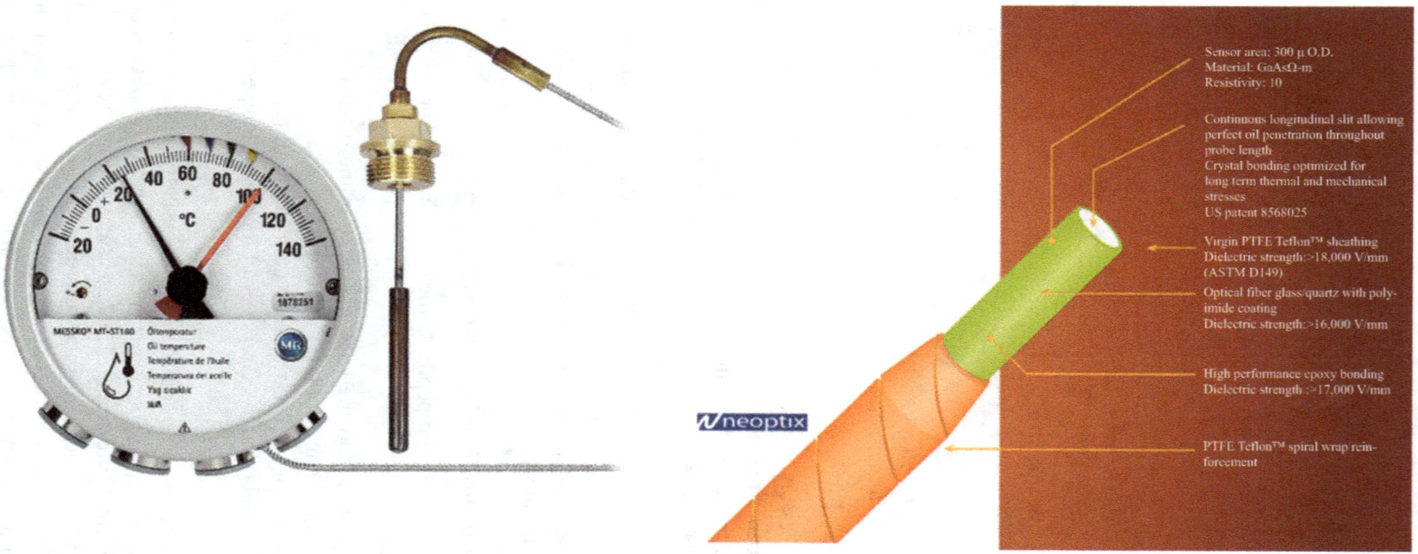

Figure 7.9 Oil temperature thermometer (left); optical fiber sensor (right)

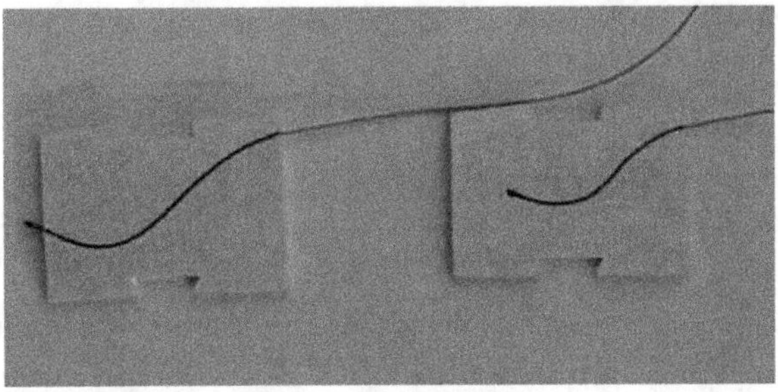

Figure 7.10 Examples of inserting optical fiber sensors into dovetail spacers

load current of the transformer. The oil temperature should then assume a value close to that at the windings. This system, popular in the past, has now been less widely used because of the growing popularity of the more accurate optical fiber sensors.

All these sensors provide an indication of the aging rate of the transformer, especially optical fiber sensors that can directly measure the HST. However, the following factors must be considered:

- aging monitoring is useful only if done from the beginning of transformer life. In fact, monitoring transformers that have already been in service for years does not allow estimating the remaining life, because the degradation prior to the installation of the monitoring system is unknown;
- aging depends on several causes (moisture, presence of oxygen and acids, etc.), all of which are not included in the loading guide model;
- sensor measurements in the plant are not always reliable, either for reasons of suboptimal installation [e.g., because the location of the optical fiber sensor may be different from the actual hot-spot, the determination of which requires a *computational fluid dynamics* (CFD) simulation], or because of the presence of possible environmental disturbances (the thermoprobes may be affected by solar radiation and wind).

For the previous reasons, monitoring temperatures with sensors is not completely reliable. Instead, it appears more promising to implement control of the cooling system pumps and fans based on predictive models of the load and expected temperatures in the transformer, using the time series of the operating parameter records. This would allow the cooling system to be started earlier than the sensor-controlled startup, thus preventing temporary high HSTs from being reached. Finally, by using the historical series of temperatures and loads, it would be possible to predict what the future temperatures will be, and thus to estimate the expectancy of life of the transformer.

Table 7.1 Comparison of OFAF and ONAF cooling systems

Feature	OFAF system	ONAF system
Efficiency of oil circulation	More uniform circulation due to pumps	Less uniform circulation resulting from convective motion alone
Dimensions	More compact, less oil in the exchangers	Less compact, more oil in the radiators
Noise	Noisier for pumps always running	Less noisy because fans run only if necessary
Reliability	Less reliable without auxiliary power: inoperative pumps and fans	More reliable for oil circulation even without auxiliary power
Energy consumption	Larger (18–20 kW)	Smaller (5 kW)

7.3 Possible improvements to the cooling system

7.3.1 Comparison of OFAF and ONAF systems

The increased efficiency of radiators and in the case of large transformers operated most of the time at low load factors ($K < 50\%$) suggests that the comparison between OFAF and ONAF cooling systems should be evaluated.

Both systems have advantages and disadvantages, summarized in Table 7.1.

Below are the most important characteristics for two types of autotransformers, 250 and 400 MVA, 150/400 kV (most common type in the Italian grid), calculated for the transition from OFAF to ONAF:

- dimensions compatible with the limits of existing bays;
- weight increments compatible with the static limits of the foundations:
  - 250 MVA: +25,000–35,000 kg total, including about 10,000 kg of oil,
  - 400 MVA: +18,000–23,000 kg total, including about 8000 kg of oil;
- active part virtually unchanged and thus no need to repeat all the type tests;
- no criticality for insulation distances of bushings and accessories due to the presence of radiators instead of unit heaters;
- higher reliability, energy saving, and lower noise.

Moreover, in the frequent cases of low-power operation at which ATRs are normally employed, the cooling system would operate practically with the fans off as if it were an ONAN system.

Alternatively, if there is a desire to provide a backup solution in case of very high outdoor temperatures, an ONAF/OFAF system can be used, with axial pumps instead of the centrifugal pumps normally used (Figure 7.11). In fact, although axial pumps have a lower head than centrifugal pumps, they allow the liquid to circulate even when they are stopped and thus would be in operation only when needed.

The solution of an ATR with an ONAF (or ONAF/OFAF) system should be tested with an installation in an area having high ambient temperatures, to monitor

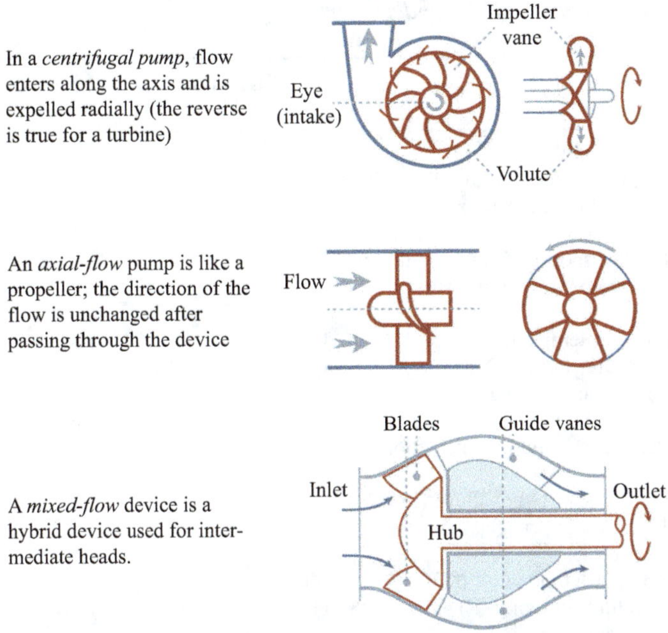

Figure 7.11 Types of pumps

the project during the most critical period. Observing positive results, the solution can be gradually extended to the existing fleet of transformers in the Italian grid.

Finally, it is possible to adopt the solution of pump and fan motors powered through an inverter, with electronic control of the rotation speed according to the temperatures measured under the cover. However, this system appears to be of considerable complexity and cost, not justified by the reduction in consumption, and above all it is inherently less reliable since it penalizes the useful life of the cooling system, moreover the electronic components have a service life of about 10 years and require frequent periodic checks.

7.3.2 Use of high-temperature materials

Transformers are traditionally designed for the maximum expected thermal load, i.e., losses at rated power. However, the loading guide introduced the concept of *cyclic overload*, according to which it is required that on *average* the rated power is met, even exceeding this value for some periods, as long as certain temperature limits are not exceeded during peak loads. With the same criterion, *emergency overloads of long and short duration* are defined, where accelerated aging is allowed, with higher limits than the cyclic condition (Table 7.2).

This possibility, which is not normally exploited in transformer operation, could allow temporary overloading during peak periods, which would be offset by periods of reduced load operation. This is the case when connecting renewable

Table 7.2 Temperature limits under overload [2]

Types of loading	Small transformers	Large and medium power transformers
Normal cyclic loading		
Winding hot-spot temperature and metallic parts in contact with cellulosic insulation material (°C)	120	120
Other metallic hot-spot temperature (in contact with oil, aramid paper, glass fiber materials) (°C)	140	140
Inner core hot-spot temperature (°C)	130	130
Top-oil temperature, in tank (°C)	105	105
Long-time emergency loading		
Winding hot-spot temperature and metallic parts in contact with cellulosic insulation material (°C)	140	140
Other metallic hot-spot temperature (in contact with oil, aramid paper, glass-fiber materials) (°C)	160	160
Inner core hot-spot temperature (°C)	140	140
Top-oil temperature, in tank (°C)	115	115
Short-time emergency loading		
Winding hot-spot temperature and metallic parts in contact with cellulosic insulation material (°C)	No limit set	160
Other metallic hot-spot temperature (in contact with oil, aramid paper, glass fiber materials) (°C)	No limit set	180
Inner core hot-spot temperature (°C)	No limit set	160
Top-oil temperature, in tank (°C)	No limit set	115

power plants. However, although this option is technically justifiable, a preliminary cost–benefit analysis is appropriate, considering the different aging rate and possible reduction in useful life, to be compared with the savings in purchasing a lower power machine.

The use of high-temperature materials offers a technological opportunity to reduce the size of transformers where constraints exist. Such materials allow transformer operation at very high temperatures while maintaining safe conditions and without the reduction of useful life. However, these materials have higher costs than traditional materials, so their use must be reasonably limited to those installations that are constrained in tight spaces or where transportation difficulties exist.

The standard IEC 60076-14 [1] provides operational guidance for the selection of high-temperature materials. For example, for the following three types of taping:

- kraft paper, which has a thermal class of 105°C (although for the hot-spot the maximum temperature is 98°C, that is, 20°C ambient + 78°C hot-spot overtemperature)[3];
- TUK paper, for which the thermal class can be increased from 105°C to 120°C, subject to agreement between supplier and customer;
- the synthetic tape (basically aramid), with thermal class 220°C.

As for solid insulation, the commercially available types are:

- the cellulosic pressboard, with thermal class 105°C;
- the PET-reinforced pressboard, with thermal class 130°C;
- the polyester (PET), with thermal class 155°C;
- the aramid in epoxy resin, with thermal class 180°C;
- the fiberglass in polyimide resin with thermal class 200°C;
- the phenolic resin or Bakelite with thermal class 300°C.

Regarding liquid insulation, the standard considers the following:

- mineral oil, with allowable overtemperature of 60°C;
- natural or synthetic ester, with allowable overtemperature of 90°C;
- silicone liquid, with permissible overtemperature of 115°C.

There are several possibilities for the use of high-temperature materials (Figure 7.12). Considering the use of mineral oil, the following options are available:

- *semi-hybrid* insulation, with only high-temperature taping;
- *mixed* insulation, with high-temperature taping and additionally high-temperature solid insulation only near the hot-spot;
- *fully hybrid* insulation, with high-temperature taping and solid insulation over the entire winding;
- finally, *fully high-temperature* insulation considers the use of ester or silicone liquid and high-temperature insulators for barriers and spacers outside the windings as well.

The permissible overtemperatures in the different application options for hot-spot range from 90°C for semi-hybrid insulation to 150°C for fully high-temperature insulation.

The actual allowable temperature value depends on the specific material used and the expected average ambient temperature.

7.3.3 Use of aluminum radiators

Although radiators are traditionally made of painted iron, a prototype aluminum radiator is being studied that would have substantial technical and environmental advantages [5].

[3]The value of 78°C was calculated by considering the copper-oil overtemperature limit $g_r = 65°C$ and a conventional hot-spot factor $H = 1.2$. Today this approximation appears oversimplified.

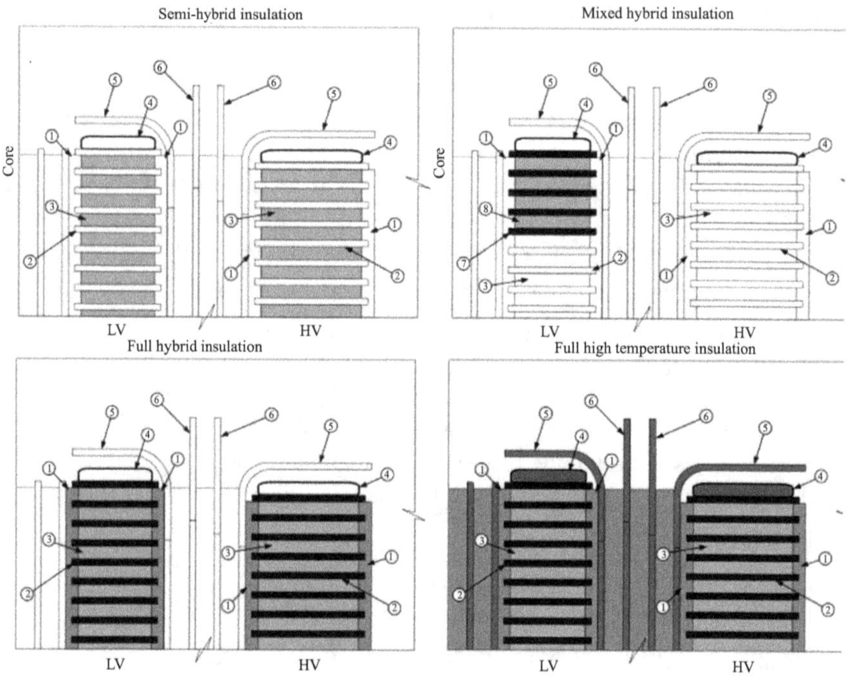

Figure 7.12 Types of insulation with high-temperature materials

In fact, aluminum has several advantageous characteristics compared with iron [6]:

- higher thermal conductivity, 236 W/m · K compared with 83.5 W/m · K, both at 0°C;
- larger surface of thermal exchange, equipping the pipes with fins;
- lower density, 2700 kg/m^3 compared with 7870 kg/m^3;
- increased resistance to corrosion.

As a result, for the same amount of heat to be disposed of, a more compact and lighter aluminum radiator battery can be obtained, with economic advantages as well, considering the lower amount of oil. In addition, the absence of the need for painting treatment of aluminum simplifies its maintenance and recycling at the end of its life.

One possible drawback is the lower mechanical strength of aluminum compared to iron, considering the fins on the pipes used to increase the heat transfer surface (Figure 7.13). This possible limitation can only be verified by testing a prototype transformer for a reasonable period (2–3 years). However, the use of aluminum radiators is not recommended for shunt reactors, because they are subject to considerable vibrations during operation that cause increased mechanical stress to the radiator tubes.

Figure 7.13 Iron (left) and aluminum (right) radiators; detail of fins on aluminum radiator's pipes (below)

7.4 Fluid dynamic model of transformers

7.4.1 Types of thermal models

Thermal models of transformers aim to calculate, more accurately than the loading guide allows, the transformer's internal temperatures [3].

The simplest model, and the only one employed before the advent of modern computer calculators, consists of correlating design values based on thermodynamics formulas with values measured in testing with special thermoprobes to derive correction coefficients for the design values of active parts.

A model of higher complexity is the *thermal hydraulic network* (THN) model, which consists of modeling the main thermal phenomena in a concentrated parameter network. The solution of such a network allows the main temperature surges to be derived.

The model of highest complexity is CFD simulation, which consists of geometric modeling (2D or 3D) of the active part and numerical solution of the fluid dynamics equations, i.e., the Navier–Stokes equations, considering the physical characteristics of the cooling fluid. This model can be coupled with a thermodynamic model that considers the distribution of losses in the windings as a function of eddy currents, which can be obtained from a separate magnetic simulation, and the effective temperature of the conductors, which influences the resistivity and thus the losses.

The characteristics, merits, and demerits of THN and CFD models will be briefly outlined below.

7.4.2 Thermodynamic network model

According to the THN model, an electrical network equivalent to the hydraulic phenomenon is constructed, in which current intensity corresponds to oil flow and potential difference corresponds to pressure difference. A thermal model that considers heat exchanges is associated with the hydraulic network. The different phenomena to be considered are:

- the pressure generated by the pumps (voltage generator);
- viscous friction (resistance), and other pressure losses in the various oil adduction channels;
- convection (voltage generator), due to the different densities of oil at different temperatures inside the machine, causes an upward force by difference with gravitational pressure. This is a pressure acting on the oil whose value can be represented in integral form as

$$p_T = \oint \rho(l)\vec{g} \cdot d\vec{l}, \tag{7.16}$$

where the integral is extended to the oil circuit path, $\rho(l)$ is the density as a function of position, and $\vec{g}$ is the acceleration of gravity. To a good approximation, the formula simplifies to

$$p_T = \rho_r \beta \Delta\theta_o g \Delta h, \tag{7.17}$$

where ρ_r is the density at the reference temperature, β is the coefficient of thermal expansion (K^{-1}), $\Delta\theta_o$ is the thermal jump with respect to the reference temperature, and Δh is the height difference between the centerline of the radiators and that of the winding.

The electrical network equivalent to the thermal model associates current with thermal power and potential difference with temperature difference. The phenomena to be considered are:

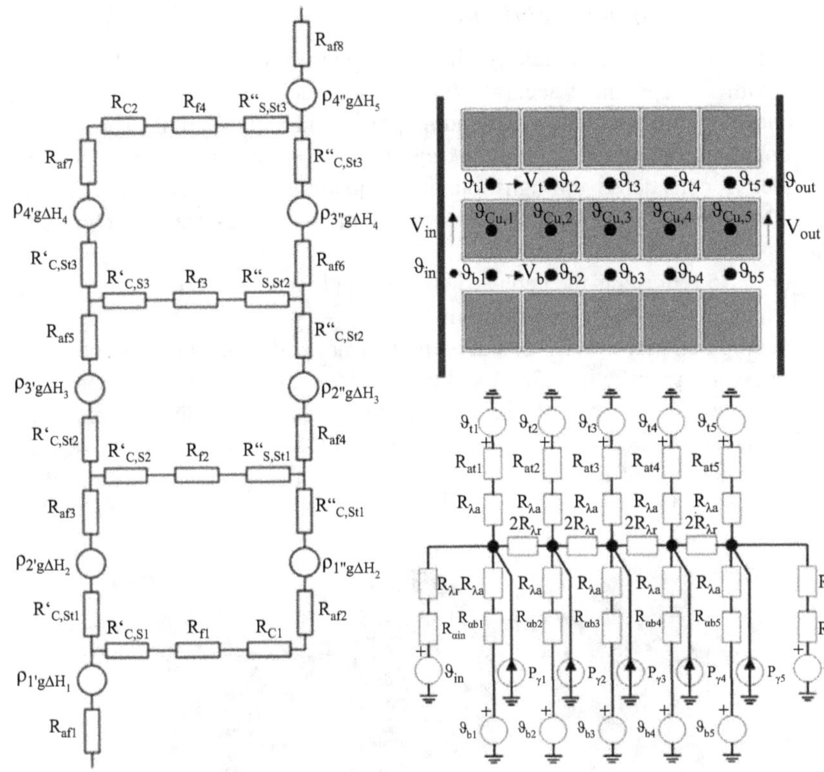

Figure 7.14 Equivalent hydraulic network (left) and equivalent thermal network (right)

- the production of heat from losses (current generator);
- the conduction and convection exchanges (resistances);
- the change in thermal energy by temperature change, as a function of the specific heat of the material (resistance);
- the thermal jump between the different parts of the transformer and the environment (voltage generator with respect to the reference ground point).

Figure 7.14 depicts a typical hydraulic network and a portion of a thermal network related to some discs of a winding; the two networks must be solved in a coupled fashion with successive iterations until convergence is achieved.

This model does not allow for the nonuniform distribution of losses within the discs, nor does it allow for consideration of the turbulent motion, consisting of vortices and nonlaminar flows of the oil. However, the complexity of constructing the equivalent networks and the calculation time make it suitable for rough estimates of design overtemperatures, the estimates of which can be refined by comparison with values measured during testing.

7.4.3 Fluid dynamics simulation

Fluid dynamics (CFD) simulations allow the geometry of the windings to be modeled with finite element accuracy, including consideration of conductor non-uniformities, for example, bulges at bending points and transpositions. Usually, due to the complexity of modeling, the tank and cooling system are not modeled, but their effect is considered through boundary equations, which consider the heat transfer characteristics of the coolant, whose operating curves are provided by the coolant manufacturer.

Figure 7.15 represents the geometric model made with a CAD software and the finite element discretization of the geometry.

The computational model considers the thermal and mechanical properties of the oil (viscosity, specific heat, etc.) and the precise state of the fluid (velocity and temperature), calculated at each node in the finite element mesh, to numerically solve the Navier–Stokes equations. The effectiveness of the simulation depends both on the accuracy of the geometric model and on the proper thickening of the

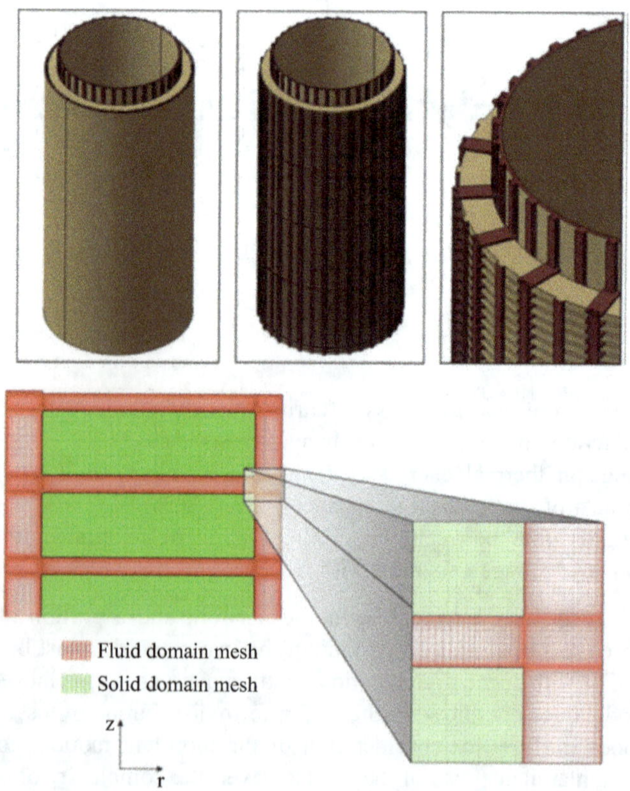

Figure 7.15 Model of a winding and finite element discretization of a channel between discs

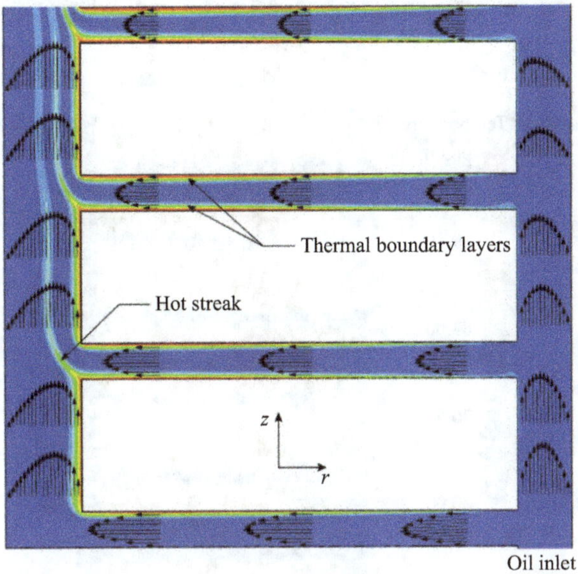

Figure 7.16 Distribution of oil velocities and temperatures

mesh at locations where higher computational accuracy is required. As a result, the number of points can reach enormous values, for example, 100 million nodes to solve, involving considerable computation time to reach convergence (on the order of days with current computers).

The solution of the model, which must consider the localized distribution of losses, preferably distinguishing the location of additional losses based on the magnetic simulation of the transformer, allows finding the velocity and temperature of the oil and solid parts, i.e., the conductors and insulator. For these, salient characteristics for thermal modeling, such as heat transfer coefficient, specific heat, variation of resistivity and viscosity with temperature, circulating current, etc., must be given.

Figure 7.16 represents the result of a CFD simulation, in which the laminar velocities of the oil and the temperature distribution of the oil are evident.

Figure 7.17 depicts the distribution of temperatures in the conductors, which allows the location of the winding hot-spot to be identified. The highest temperatures are reached in the disc immediately below the barrier because this prevents uniform circulation of oil around the winding. The absence of barriers would result in reduced oil circulation in the intermediate discs because the oil tends to follow the path at lower head between the winding inlet and outlet ducts.

Although it is possible to carry out simulations with 2D modeling and without considering the precise distribution of additional losses, i.e., assuming uniformly distributed losses, it has been demonstrated [3] that this causes an unacceptable computational error in estimating both the value and the location of the hot-spot (Figures 7.18 and 7.19).

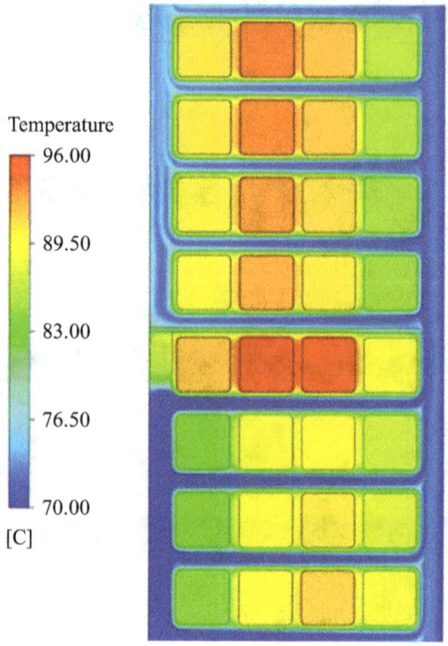

Figure 7.17 Temperature distribution in the conductors

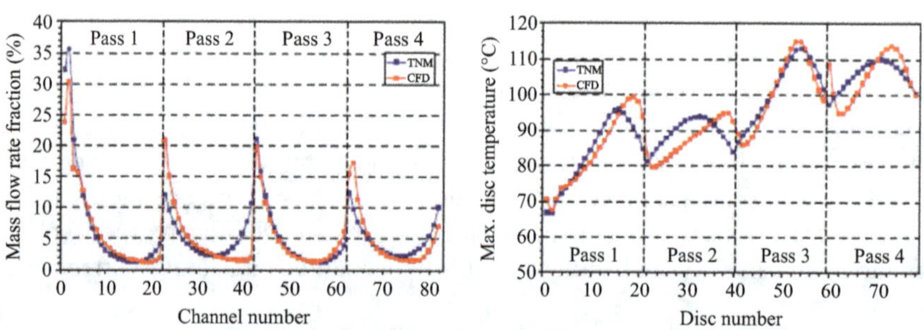

Figure 7.18 THN and CFD simulations with 2D geometry and uniform losses

CFD simulations can show unexpected distributions of the overtemperatures. In particular, the hot-spot is not always located in the top disc immediately below the yoke, as would be expected (Figure 7.20). This is not necessarily a symptom of a design error; on the contrary, it can mean efficient cooling in the upper discs and thus a hot-spot with lower temperature. This important result allows a more efficient installation of optical fiber sensors, which might otherwise be placed on discs incorrectly identified as the hottest.

CFD simulations are particularly useful for validating new design schemes, allowing for the detection of turbulent motions, vortices, and stagnant or reverse

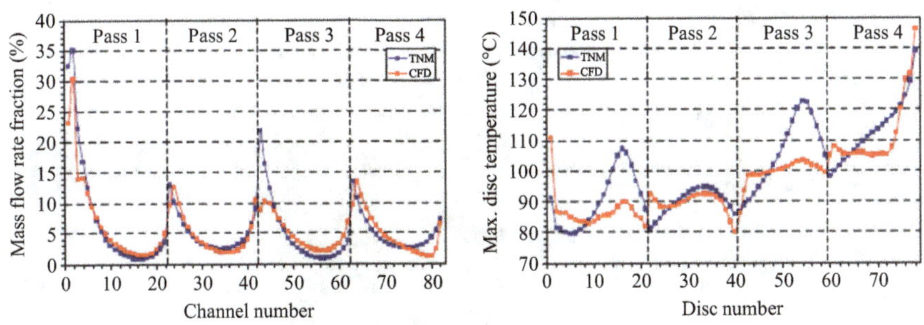

Figure 7.19 THN and CFD simulations with 3D geometry and nonuniform losses

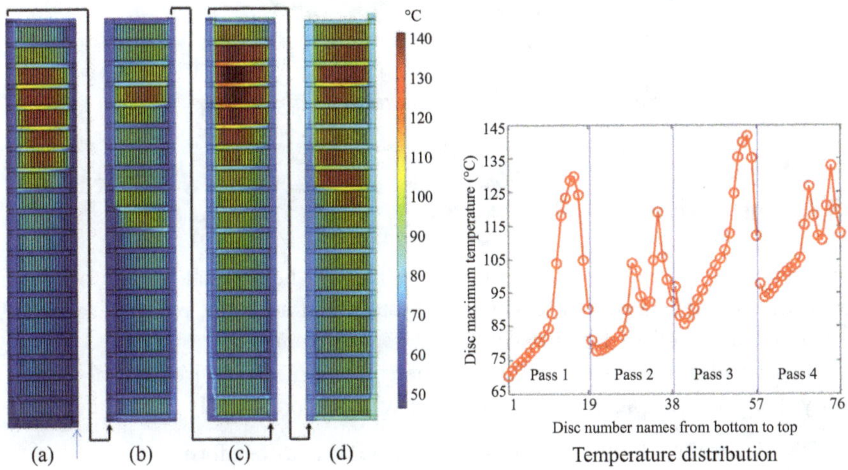

Figure 7.20 Thermal study of a transformer in which the hot-spot stands in an unusual position

flows, which are indications of incorrect preliminary design of channels between discs and barriers. However, given the long time required to prepare the model and run the simulation, the CFD model is not commonly used to design transformers with already adopted and verified rules.

CFD simulations are also used by cooling system manufacturers to validate the design of new radiators and heat exchangers (Figure 7.21). They are particularly effective because they allow both oil and air behavior to be modeled.

It has been observed that an accurate finite element model of the active part is no guarantee of a reliable result, either because of construction tolerances or because of approximations in the numerical calculation, including uncertainties in material properties. Therefore, it is always necessary to verify the results of a simulation by comparing them with measurements made in the test room, which can provide information to improve simulation techniques.

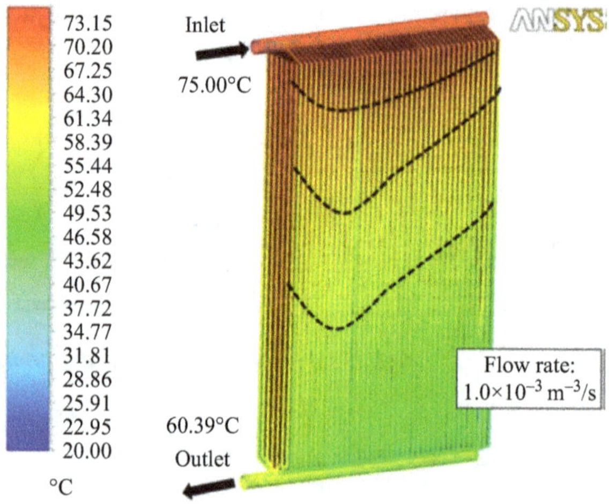

Figure 7.21 CFD simulation of a radiator

7.5 Conclusions

This chapter described the state of the art related to the thermal aspects of power transformers and related heat dissipation systems. In addition, possible technological innovations offered by market products were examined.

In particular, to pursue the goal of energy saving and environmental sustainability, the following solutions are technically and economically promising:

- ONAF (or ONAF/OFAF) cooling, for HV/HV autotransformers;
- aluminum radiator coils for smaller transformers such as HV/MV transformers.

The higher costs are to be compared with the savings that can be obtained from the higher efficiency of both the previous solutions. In contrast, the solution of powering pump and fan motors with inverters is not recommended due to lower overall reliability.

The option of using high-temperature materials, which is applicable to the case of ATRs with ester liquid, needs to be justified by a cost–benefit analysis, considering the projected usage of the devices, usually with variable load. With such materials, peak power can be met without increasing the power rating but using the higher temporary overloading.

The installation of fiber-optic HST sensors as a standard device should be avoided, given the considerable cost of the sensors themselves and their limited usefulness in service. Instead, it seems more useful to require additional thermal tests to be performed on the prototype transformers to determine all the parameters provided by the loading guide, which would allow an accurate thermal model of the design to be identified. In this way, estimates of internal temperatures at each point of interest would be possible even without the use of optical fiber sensors.

Thermal study with fluid dynamic models (Section 7.4) should be limited to special cases of new transformer designs with critical functions and design peculiarities, such as converter transformers and phase shifters, to accurately estimate the temperature and hot-spot location. The same estimates should be subsequently verified and validated by thermal testing.

Finally, a comprehensive revision of the IEC 60076 series of standards relating to thermal aspects of transformers (parts 2, 7, and 14) is underway. As a result, there is a need to continue to monitor the evolution of the standards in order to seize any new opportunities, including the definition of the most significant thermal tests and intercept possible critical issues, such as in the case of the use of high-temperature materials, the choice of which must comply with the criteria of consistency of the thermal classes with the limits provided for all materials involved.

References

[1] IEC 60076-14: *Power Transformers – Part 14: Liquid-Immersed Trans formers with High-Temperature Insulating Materials*, 2013.

[2] IEC 60076-7: *Power Transformers – Part 7: Loading Guide for Mineral-Oil-Immersed Power Transformers*, 2018, ed. 2.0.

[3] Brochure 659: *Transformer Thermal Modelling*, Cigré, 2016.

[4] Brochure 642: *Transformer Reliability Survey*, Cigré, 2015.

[5] G. Veller: *Aluminum Finned Elliptical Tube Radiator*, Convett paper, 2018.

[6] The Engineering Toolbox: https://www.engineeringtoolbox.com/thermal-conductivity-metals-d_858.html, https://www.engineeringtoolbox.com/density-solids-d_1265.html

[7] F. Coppadoro: *Thermal Behavior of Power Transformers – Volume 1: Transformer Heating: From Losses to Cooling*, CESI Handouts, 1993.

[8] Robert M. Del Vecchio, Bertrand Poulin, Pierre T. Feghali, Dilipkumar M. Shah, and Rajendra Ahuja: *Transformer Design Principles – With Applications to Core-Form Power Transformers*, CRC Press, 2002.

[9] IEC EN 60076-2: *Power Transformers – Part 2: Overtemperatures in Liquid-Immersed Transformers*, 2012 (IEC 60076-2:2011).

[10] Brochure 393: *Thermal Performance of Transformers*, Cigré, 2009.

[11] Terna technical note: *ATR Feasibility Study with ONAF Cooling*, 29 September 2008.

[12] Qiang Liu (University of Manchester): Transformer Thermal Management for future power networks, Presentation at CWIEME 2018, Berlin 20 June 2018.

Chapter 8

Parameters at the zero sequence of transformers

8.1 Generalities

Design choices leading to the construction of a transformer must consider the effects on its electrical characteristics that will affect the performance of the device in the electrical system. This chapter aims to analyze the equivalent circuit at the zero sequence of transformers, which is quite different from the circuits at the positive and negative sequence, that are essentially the same. The analysis is appropriate because the measurement of impedance at the zero sequence encounters difficulties and sometimes approximate values are used. Nonetheless, the correct knowledge of the circuit parameters at the zero sequence is indispensable for the precise analysis of ground faults and the consequent calibration of line protection relays.

8.2 Sequence impedances of transformers

8.2.1 Summary of single-phase circuit model

Transformers consist of two or more wire coils, which are galvanically separated in conventional transformers and instead are connected in series with each other in autotransformers. These coils are called *windings* because they are wound and fitted onto a structure of ferromagnetic material called *core*, which enhances the concatenation of magnetic flux between them (Figure 8.1). They are conventionally named *primary* and *secondary* windings.[1]

In alternating current, applying Lenz's law in the ideal condition where all flux is confined between the windings and the core has infinite magnetic permeability, the voltages and currents at the primary and secondary are in proportion to each other according to the ratio of the number of turns of the windings:

$$U_2 = \frac{N_2}{N_1} U_1, \tag{8.1}$$

$$I_2 = \frac{N_1}{N_2} I_1. \tag{8.2}$$

[1]The IEC standards employ the names HV and LV referring to the winding with the highest and lowest voltage, respectively. Primary and secondary refer to the normal flow of power; for transformers in the transmission grid, they are equivalent to HV and LV, and this correspondence is assumed throughout the chapter.

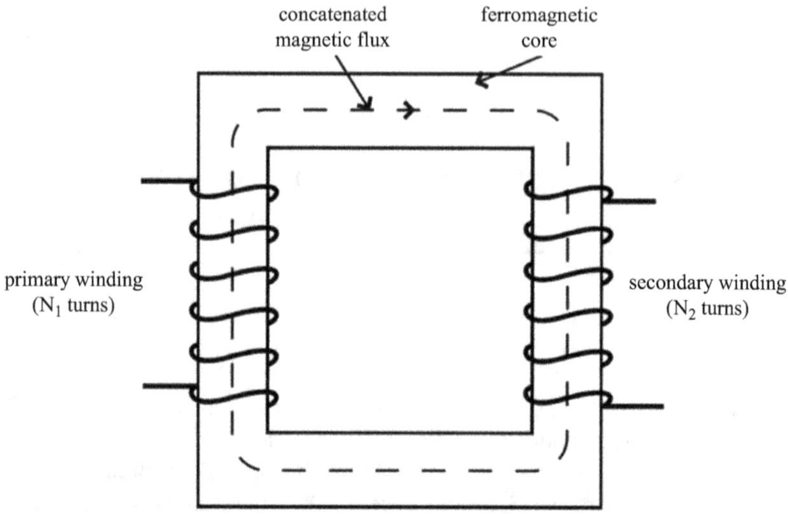

Figure 8.1 Schematic representation of a single-phase transformer with two windings

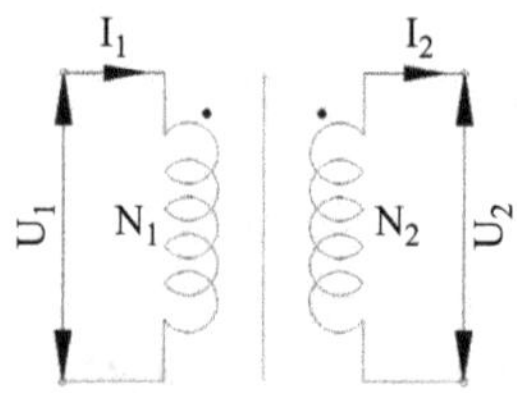

Figure 8.2 Circuital representation of the ideal transformer

The circuital component representing these constitutive equations is called the *ideal transformer* (Figure 8.2).

However, actually:

- part of the magnetic field inevitably flows out of the core (Figure 8.3) and concatenates with only one of the two windings, resulting in a self-inductance in series with the considered winding, called *leakage inductance*;
- the magnetic core does not have infinite magnetic permeability and a current circulation is required to maintain the concatenated flux, called *magnetization current*. This current is not transferred between the windings according to Equation (8.2). The phenomenon is represented by a branch in parallel with the ideal transformer having a nonlinear inductance, dependent on the reluctance of the ferromagnetic circuit.

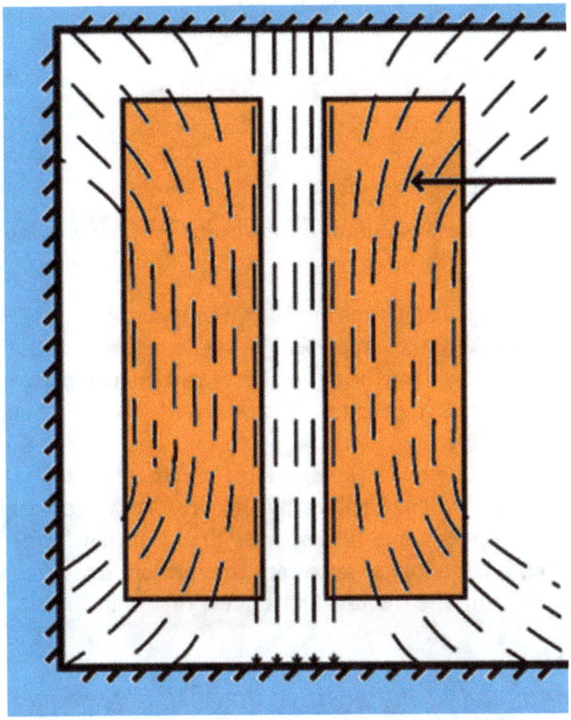

Figure 8.3 *Cross-section of the two concentric windings (orange) and the core (blue) of a transformer with the representation of stray magnetic flux lines*

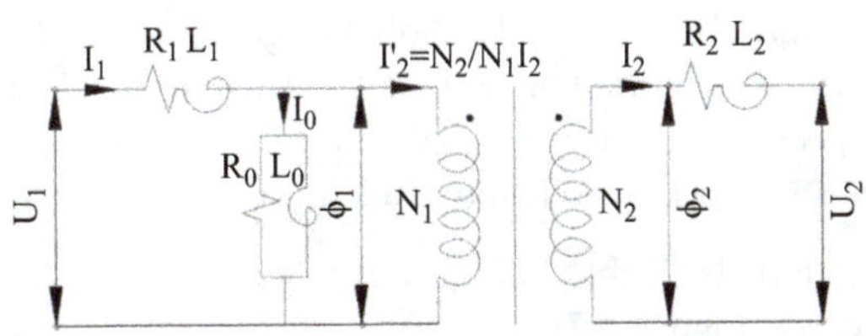

Figure 8.4 *Equivalent circuit of a real transformer*

Accordingly, a more representative electrical circuit of the *real* single-phase transformer is drawn in Figure 8.4 and also includes series resistors to represent losses in the windings and a parallel resistor to represent losses in the core.

Impedances can be represented indifferently at the primary or the secondary to facilitate calculations; the value of impedances referenced to the secondary is obtained from that referenced to the primary using the relationship:

$$Z_2' = \left(\frac{N_2}{N_1}\right)^2 Z_2. \tag{8.3}$$

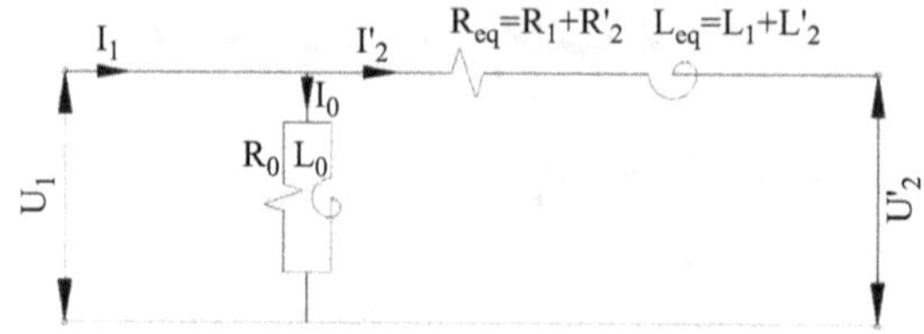

Figure 8.5 Equivalent circuit of a real transformer with relative values

Although the leakage reactances are shown in Figure 8.4 as X_{L1} at the primary and X_{L2} at the secondary, they represent a single phenomenon that cannot be separated. For this reason, and because the circuit model does not change by representing a single series impedance, it is preferred to refer all quantities to one side, normally the primary.

If the *relative* (or *per unit*) *values* of the electrical quantities, obtained by dividing the current and voltage values by their respective ratings, are used, the equivalent circuit is further simplified because the ideal transformer assumes unit ratio, and therefore, it is no longer necessary to represent it.[2]

Ultimately, the simplified equivalent circuit of a transformer is depicted in Figure 8.5.

8.2.2 Circuit model at the positive and negative sequence

Three-phase transformers[3] can be made with different magnetic core configurations, of which the most common ones are (Figure 8.6) as follows:

- the *shell-type*, in which the core surrounds the windings, which are made from several layers (*pancake coils*) that follow the geometry of the core windows.
- the *core-type*, in which the windings are fitted into the limbs. The magnetic circuit is closed by the upper and lower beams, called *yokes*. There may be 3 limbs, one per phase, or 5 limbs, one per phase and two additional side limbs that are not wound.

The winding terminals in three-phase transformers can be connected to each other in different ways (Figure 8.7):

- in *star* (or *wye*) *connections*, the initial terminal of each winding is made accessible for connection to the power grid, while the final terminal is connected with homologous ones of the other windings in the *star point*, which may or may not be accessible through a *neutral* terminal, used for ground connection;
- in *delta* (or *triangle*) *connections*, the initial terminal of each winding is connected to the final terminal of the next, and this common point is made accessible for connection to the power grid; there is no star point.

[2]In transformers equipped with a tap changer, the transformation ratio is not constant, and this must also be considered in the model with relative values.
[3]The case of three-phase transformers consisting of three single-phase units with separate ferromagnetic cores is excluded.

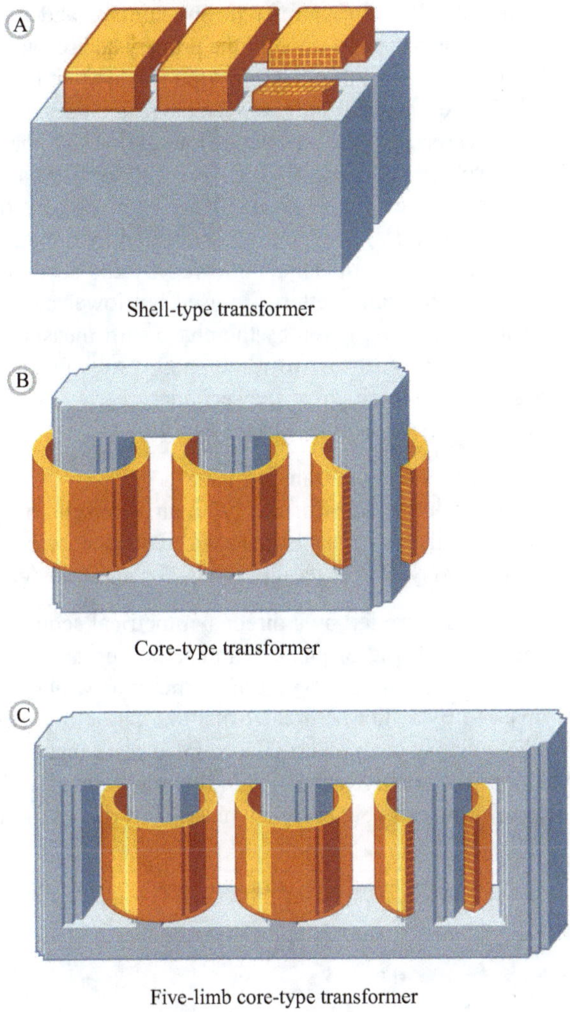

Figure 8.6 Types of active parts for three-phase transformers

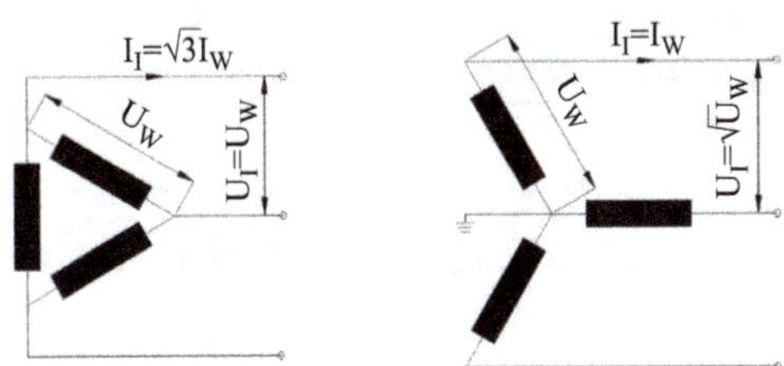

Figure 8.7 Winding connections: delta (left); star with neutral (right)

Transformer nameplate data always refer to line currents and concatenated line voltages, regardless of the internal connection. If the primary and secondary connections are not the same, this implies that the transformation ratio is different from the turns ratio by a factor $\sqrt{3}$ and that there is a phase shift between the magnitudes of the homologous phases at the primary and secondary. Conventions for naming the phases are various, usually A-B-C for high voltage (HV) and a-b-c for low voltage (LV) are employed.

The type of connection and its phase shift are indicated by the *connection symbol*. The type of primary (HV) connection is indicated by the letters: Y for star, YN for star with neutral, and D for delta, followed by the secondary (LV) connection symbol, for which the same letters are used but lower case. Finally, there follows the *vector* (or *hour*) *group* given by the phase shift measured clockwise in sexagesimal degrees divided by 30º between the phase “A” voltage of the primary and the homologous phase “a” voltage at the secondary.

The most common connection symbols are as follows:

- YNd11 for power plants step-up transformers;
- Yy0 for HV/LV distribution transformers without Petersen coil;
- Yyn0 for HV/LV distribution transformers with Petersen coil;
- YNa0, equivalent to YNyn0 at external effects, for autotransformers.

When the A-B-C phases are fed by a direct symmetrical sequence of voltages, the phase fluxes are also 120° out of phase with each other, and thus, their vector sum is null. Therefore, yokes in core-type transformers or windows in shell-type transformers are invested by a flux no greater than the phase flux (Figure 8.8).

The magnetic behavior of each phase is decoupled from that of the others, and the circuit equivalent to the positive sequence of the transformer is identical to that of Figure 8.5, except for the possible presence phase shift due to the vector group.[4]

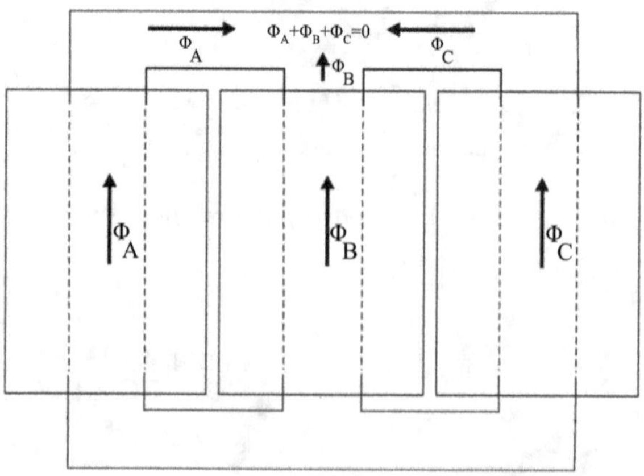

Figure 8.8 Flow diagram in a 3-limb transformer

[4]The transformer sequence circuit does not consider the phase shift of the vector group, which must be taken into account separately in the circuit analysis.

When phases A-B-C are fed by a negative symmetrical sequence of voltages, the same magnetic effects are obtained, but the vector group changes and becomes complementary to that at the positive sequence.

8.2.3 Circuit model at zero sequence

Unlike the positive or negative sequence, the phase voltages at the zero sequence are identical and therefore have no null sum. Consequently, the phase fluxes relative to the zero sequence also sum to each other to generate a net flux equal to $3\Phi_0$, which in 3-limb transformers saturates the yokes and then closes outside the ferromagnetic circuit (Figure 8.9) aiming at the path of least reluctance. This includes the steel tank of the transformer, which however is not sized to withstand high magnetic fields and therefore heats up quickly due to hysteresis and eddy currents, even with voltages having a small zero component.

In 5-limb transformers (Figure 8.10), the flux closes back into the side limbs, which however have a smaller cross section than the phase limbs. At high zero-sequence voltages (more than one-third of the nominal phase voltage), the side limbs saturate like yokes and the zero-sequence flux closes back into the tank.

In transformers with shell-type cores (Figure 8.11), the privileged reclosure of the zero-sequence flux is the outer sides of the windows, if these do not saturate as in the case of 5-limb transformers.

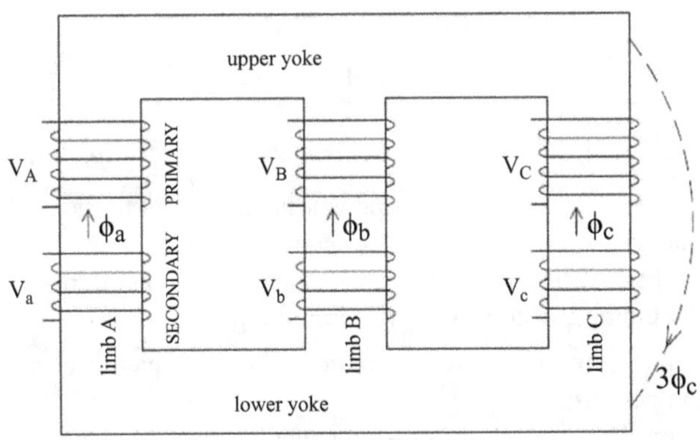

Figure 8.9 Zero-sequence flux trends in 3-limb transformer

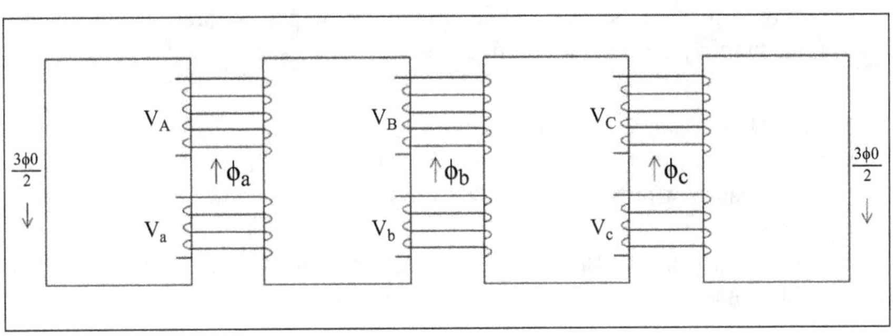

Figure 8.10 Zero-sequence flux trends in 5-limb transformer

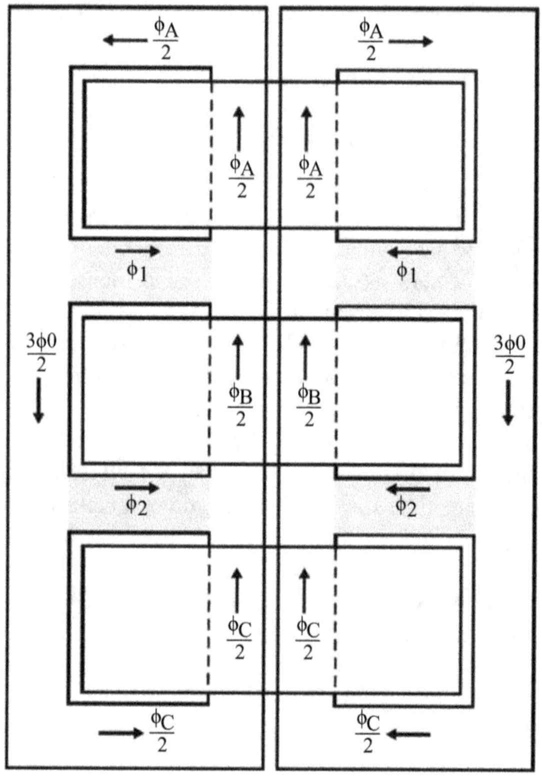

Figure 8.11 Zero-sequence flux trends in shell transformer

Ultimately, the values of circuit impedances equivalent to the zero sequence of a transformer depend on the type of ferromagnetic core.

The possibility of circulation of zero-sequence currents depends on the connection of the windings, according to the following cases:

- the YN connection[5] allows the circulation of zero-sequence current that closes through the neutral;
- Y connection does not allow zero-sequence current circulation;
- connection D allows zero-sequence current to flow in the windings, but not in the connection terminals.

Figure 8.12 represents the possible circuits to the zero sequence as a function of different winding connection configurations.

It should be noted that:

- in the YNyn0 connection (case 1), both zero-sequence voltage and current are transferred from the primary to the secondary according to the turns ratio and without phase shifting;

[5]It is understood, also in the following, that the YN connection is always with grounded neutral; if the neutral is floating, it is circuitally equivalent to the Y connection.

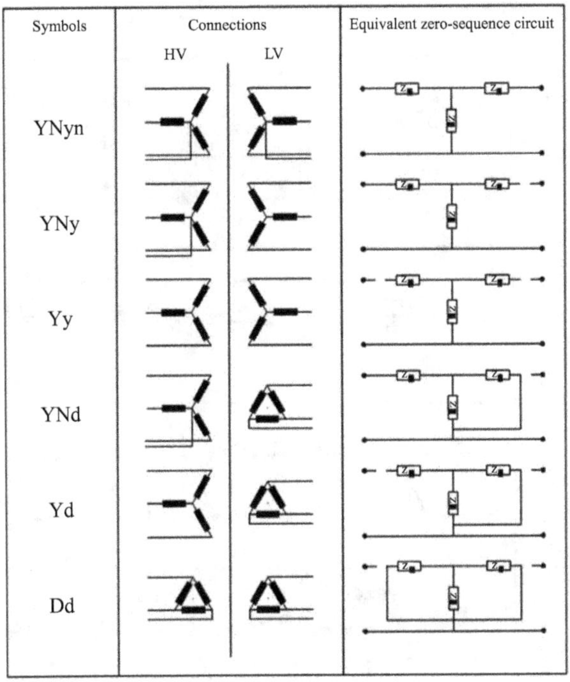

Figure 8.12 Circuits at the zero sequence

- in the YNy0 connection (case 2), the zero-sequence voltage at the primary creates a potential difference at the ends of the ground-connected windings, which allows the circulation of a magnetization current; the voltage is transferred to the secondary as if the transformer were in no-load operation, since there can be no circulation of zero-sequence current in this winding;
- in the rare Dd0 connection (case 3), zero-sequence current circulation in the windings cannot be induced by a zero-sequence line voltage, because all terminals are brought to the same voltage, and thus, there is no potential difference at the terminals of the windings that can cause current circulation;
- in the YNd11 connection (case 4), the zero-sequence voltage at the primary creates a circulating current at the secondary, which remains closed within the phases, and thus, at zero sequence, the transformer appears practically short circuited. A zero-sequence voltage is transferred to the secondary, but no zero-sequence current can circulate to the outside;
- in the Yd11 connection (case 5), there is no zero-sequence current circulation at either the primary or the secondary, and the zero-sequence no-load voltage is not transferred between the windings.

In three-winding transformers, the same criteria are followed, considering one branch for each winding (Figure 8.13). Other cases for different three-winding configurations are given in chapters ([1] Tab. 5.3 and [2] Tab. 1.4b).

TABELLA 1.

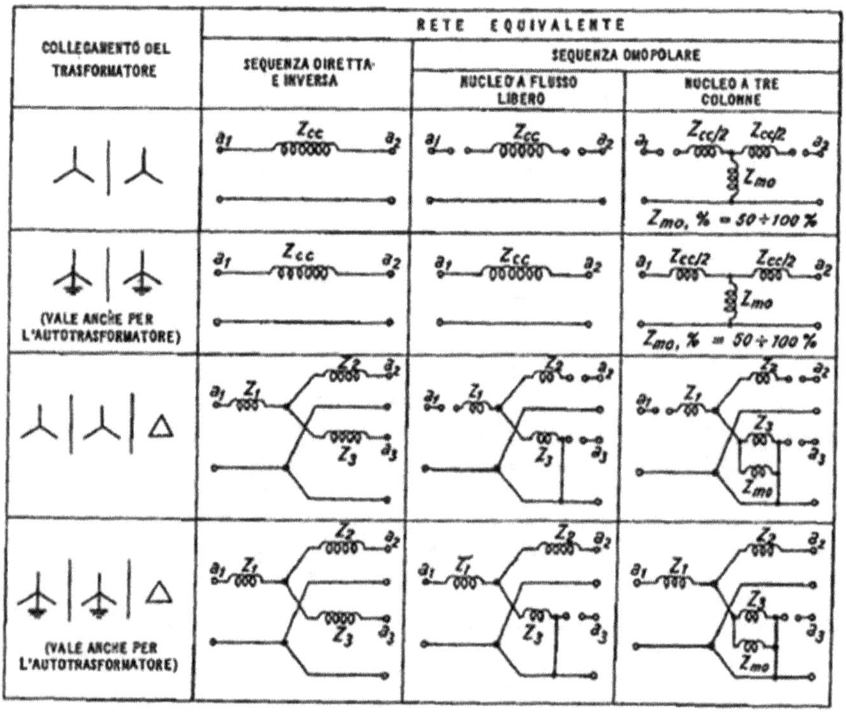

Figure 8.13 Equivalent circuits for two- and three-winding transformers

8.3 Measurement of sequence impedances

8.3.1 Positive sequence impedances

The series impedance at the positive sequence is measured during the short circuit test at rated current. The three-phase impedances are very similar, and normally the average of the three measurements is taken as the reference value.

The parallel impedance at the positive sequence is measured during the no-load test at nominal voltage. The three-phase impedances may differ significantly for 3-limb transformers, with a higher value for the center column phase, but the average of the three measurements is always taken as the reference value.

From the values of short circuit losses P_{sc} and no-load losses P_o, the series resistance $\dot{R}_{sc}$ and the parallel conductance $\dot{G}_o$ can be derived (both in relative value, represented by dotted symbols), with the well-known formulas:

$$\dot{R}_{sc} = \frac{P_{sc}}{S_r}, \tag{8.4}$$

$$\dot{G}_o = \frac{P_o}{S_r}, \tag{8.5}$$

where S_r is the rated apparent power. By vector difference, the series reactance $\dot{X}_{sc}$ and the parallel susceptance $\dot{B}_o$ are obtained in relative value:

$$\dot{X}_{sc} = \sqrt{\dot{V}_{sc}^2 - \dot{R}_{sc}^2}, \tag{8.6}$$

$$\dot{B}_o = -\sqrt{\dot{I}_o^2 - \dot{G}_o^2}, \tag{8.7}$$

where $\dot{V}_{sc}$ is the relative short circuit voltage (numerically equal to the relative short circuit impedance $\dot{Z}_{sc}$), and $\dot{I}_o$ the no-load relative current (numerically equal to the no-load relative admittance $\dot{Y}_o$).

8.3.2 Zero-sequence impedances

The following measurements with zero-sequence power supply are significant:

- YNyn0 or YNa0: short circuited and at no load, feeding either the primary or the secondary (usually the measurement from the primary is taken as a reference);
- YNy0: no-load supply from the primary;
- Dd0: no measurement possible;
- YNd11: no-load supply from the primary; short circuit supply from the primary is not necessary as it provides the same value as no-load supply (zero-sequence current does not circulate out of the delta connection to the secondary)[6];
- Yd11 or Yy0: no measurement possible.

The standard test setup consists of connecting the primary terminals together and applying voltage between them and the neutral, measuring voltage U and supply current I (Figure 8.14). The secondary terminals are left open in the no-load measurement, connected together with their neutral, if any, in the short circuit measurement.

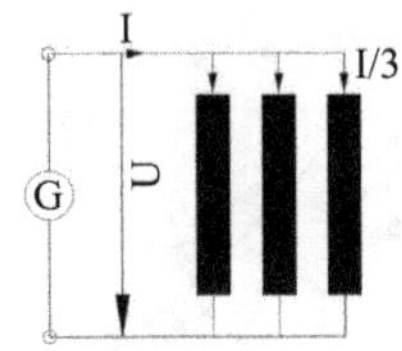

Figure 8.14 Power supply for zero-sequence impedance measurement

[6]According to one study [3], the impedance measured by short circuiting the triangle is slightly lower (−3%) than that measured by leaving it open. This is probably due to the mutual inductances that are cancelled by shorting the terminals. However, the standard measurement is that with the terminals open.

Since the supply current is divided among the three phases, it is equal to three times the zero-sequence current; hence, the zero-sequence impedance is obtained as

$$Z_0 = \frac{3U}{I}. \tag{8.8}$$

8.4 Typical values and calculation of impedances of zero sequence

8.4.1 No-load impedance

In no-load measurement, the value of zero-sequence impedance is affected by the type of magnetic circuit.

In three-limb cores, the zero-sequence flux mainly circulates off the yokes, invests the tank, and is affected by its saturation. Consequently, the measured zero-sequence impedance depends on the supply voltage: it increases up to 10%–15% U_r (U_r rated phase voltage), because the tank is not fully magnetized and then decreases until the saturation value is reached for voltages > 30%–35% U_r (Figure 8.15).

Consequently, for measurements made at a voltage lower than the rated voltage, to bring the measured value back to that at the nominal voltage, one must divide the result by an empirical correction factor (Figure 8.16). However, this factor has considerable uncertainties, so it is preferable to make the measurement at rated voltage, except in cases where terminal insulation limits would be exceeded.

The typical value of zero-sequence impedance at no-load rated voltage for 3-limb transformers is quoted in the literature ([4] § 2.7) to be in the range of 75%–200% Z_r ($Z_r = \frac{S_r}{V_r^2}$ reference impedance, where $V_r = \sqrt{3}U_r$ is the nominal phase-to-phase voltage). However, normative references ([2] Tab. 1) indicate values in the range of 50%–60% Z_r.

For an estimate of the zero-sequence impedance at no load, one can consider the tank as a short-circuited winding and employ the leakage reactance formula ([5] (3.66)):

$$\dot{X}_{0o} = \omega\mu_0 \frac{\pi N^2}{H_{eq} Z_r}\left(\frac{1}{3} T_w D_w + T_g D_g\right), \tag{8.9}$$

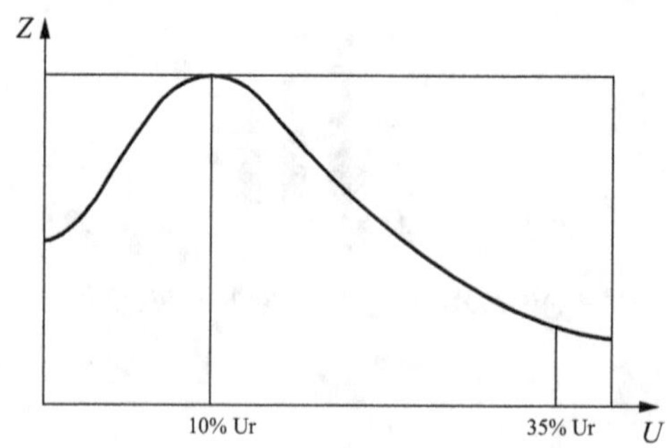

Figure 8.15 Typical trend of zero-sequence impedance at no load of a 3-limb transformer

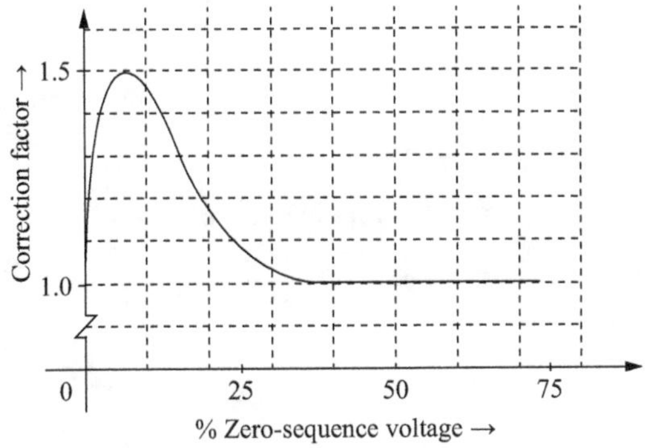

Figure 8.16 Zero-sequence no-load impedance correction factor for a 3-limb transformer

where

- H_{eq} is the equivalent winding height, that is the actual height divided by the Rogowski correction factor;
- T_w is the thickness of the fed winding;
- D_w is the average diameter (with respect to the column axis) of the fed winding;
- T_g is the distance between the fed winding and the tank;
- D_g is the average diameter of the space between the fed winding and the tank.

Therefore, the zero-sequence no-load impedance in 3-limb transformers depends on the fed winding.

In 5-limb or shell-type transformers, the zero-sequence impedance at no-load is similar to the positive sequence impedance, i.e., nonlinear on the order of magnitude ~100% Z_r; it may be slightly lower if the yokes or windows are partially saturated at the rated voltage.

8.4.2 Short circuit impedance

The measured short circuit impedance, including that measured on YNd0 transformers even with open delta winding, is very similar to that of the positive sequence, albeit slightly lower in the 3-limb core transformers.

Typical values of this impedance for 3-limb core transformers are ([4] § 2.7) as follows:

- $Z_{0sc} \simeq 85\%–95\%\ Z_{sc}$, if the fed winding is inside the short-circuited one;
- $Z_{0sc} \simeq 80\%–90\%\ Z_{sc}$, if the fed winding is outside the short-circuited one.

The difference is due to the fact that the leakage flux has a different pattern according to which winding is fed (Figure 8.17). In fact, the stray flux cannot invest the short-circuited winding, because otherwise it would create a potential

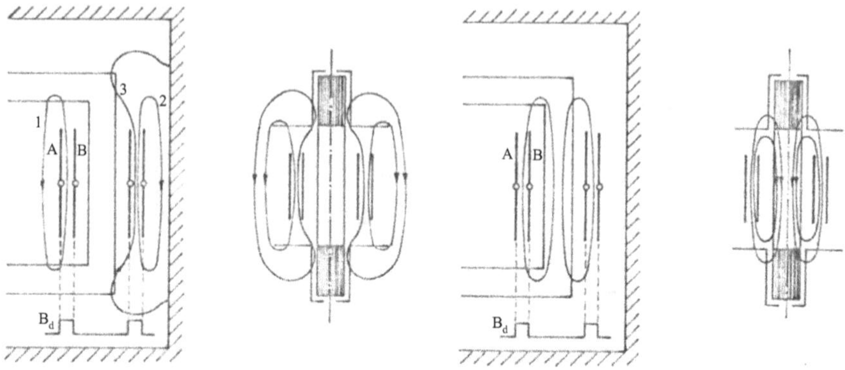

Figure 8.17 Leakage flux by short circuiting the inner (left) or outer (right) winding

difference at its ends (basically, the circulation of the short-circuited current exactly compensates for the magnetic flux that induces it).

A fairly reliable estimate of the value of the zero-sequence short circuit reactance is obtained by considering the tank as a third short circuit winding ([5] § 3.7.2). The reactance is equal, in relative value, to the absorbed reactive power $\dot{Q}$. By denoting with the letters *h*, *l*, *t* primary, secondary, and tank, respectively[7]:

$$\dot{X}_{0sc} = \dot{Q} = -\left[\dot{X}_{hl}\dot{I}_h\dot{I}_l + \dot{X}_{ht}\dot{I}_h\dot{I}_t + \dot{X}_{lt}\dot{I}_l\dot{I}_t\right], \tag{8.10}$$

where neglecting the magnetization current, the following relationship is respected:

$$\dot{I}_h + \dot{I}_l + \dot{I}_t = 0. \tag{8.11}$$

Assuming $\dot{I}_h = 1$, we can substitute $\dot{I}_t = -1 - \dot{I}_l$ and write

$$\dot{X}_{0sc} = -\dot{X}_{hl}\dot{I}_l + \dot{X}_{ht}(1 + \dot{I}_l) + \dot{X}_{lt}\dot{I}_l(1 + \dot{I}_l). \tag{8.12}$$

The distribution of currents between secondary and case is found by minimizing the expression of reactive power with respect to the variable $\dot{I}_l$, and this is obtained:

$$\dot{I}_l = \frac{\dot{X}_{hl} - \dot{X}_{ht} - \dot{X}_{lt}}{2\dot{X}_{lt}}. \tag{8.13}$$

[7]Since the currents are all taken with the same direction (incoming or outgoing), the reactive power is calculated with the convention of generators; therefore, the minus sign in front of the expression of the zero-sequence reactance is necessary. Kulkarni and Khaparde [5] use the opposite convention for currents at the secondary and in the tank, but this complication seems unnecessary.

Taking l as the shorted winding, if this is external, we can write

$$\dot{X}_{ht} \simeq \dot{X}_{hl} + \dot{X}_{lt}, \tag{8.14}$$

and therefore, we obtain $\dot{I}_l \simeq -1, \dot{I}_t \simeq 0$; consequently, the following relation holds

$$\dot{X}_{0sc} \simeq \dot{X}_{hl}, \tag{8.15}$$

so the zero-sequence reactance is equal to the leakage reactance between the primary and secondary.

If, on the other hand, the shorted winding is the inner winding, we can pose

$$\dot{X}_{lt} \simeq \dot{X}_{hl} + \dot{X}_{ht}, \tag{8.16}$$

and therefore, we get $\dot{I}_l \simeq -\dot{X}_{ht}/\dot{X}_{lt}, \dot{I}_t \simeq -\dot{X}_{hl}/\dot{X}_{lt}$.

Accordingly, the zero-sequence reactance is

$$\dot{X}_{0sc} \simeq \frac{\dot{X}_{ht}}{\dot{X}_{lt}} \dot{X}_{hl}, \tag{8.17}$$

and is therefore less than the leakage reactance between primary and secondary.

In 5-limb or shell-type transformers, since most of the flux affects the core, the short circuit impedance at the zero sequence is not significantly different from that at the positive sequence.

8.4.3 Equivalent model for autotransformers

Autotransformers, and in general transformers with YNyn0 connection, can be represented with a three-parameter model,[8] as shown in Figure 8.18.

From the possible four homopolar impedance measurements (short circuit and no-load on the primary and secondary sides), the three[9] equivalent parameters can

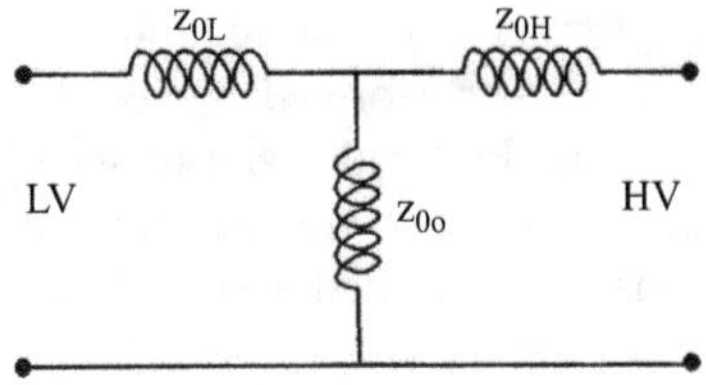

Figure 8.18 Circuit equivalent to the homopolar sequence of an ATR

[8]In Figure 8.13, the two series impedances of the 3-limb transformer model are depicted identically, but as far as shown this approximation is not valid.

[9]Three parameters would not be sufficient to represent four independent quantities, but the four measures are not independent of each other.

be derived, according to the circuit relationships:

$$\dot{Z}_{0sc,L} = \dot{Z}_{0L} + \dot{Z}_{0H} \| \dot{Z}_{0o}, \; \dot{Z}_{0o,L} = \dot{Z}_{0L} + \dot{Z}_{0o}, \tag{8.18}$$

and similarly for the primary. In this model, the series impedance of the outer winding may be negative, while the overall impedance is always positive.

The measurements also allow the values of short circuit and no-load losses at the zero sequence to be derived, which can be modeled using series and parallel resistors similarly to the direct sequence. Normally these parameters are not considered significant. If necessary, and in the absence of measured values, the series resistances will be taken equal to those of the positive sequence and the parallel conductance proportional to that of the positive sequence according to the ratio of no-load susceptances between the two sequences.

8.4.4 *Finite element model calculation*

Precise knowledge of transformer geometry allows finite element simulation to be employed for accurate estimation of homopolar impedance.

Studies carried out [6] have shown that calculation by evaluating the magnetic energy between pairs of windings, proportional to the leakage reactance between them, provides an accurate estimate of the zero-sequence reactance value, using formula (8.10).

8.5 Recommendations

8.5.1 *Measurement*

Impedance measurements at the zero sequence can be made during transformer testing in the test room and are inexpensive. Figure 8.19 shows the possible measurements for different types of transformers.

In the case of autotransformers, two no-load and one short circuit measurements, taken as a type test, may be performed. It would be appropriate to perform the short circuit measurement on both sides and to specify that the no-load measurements should be made at the rated voltage. On the other hand, it does not appear to be very useful to make the no-load measurements for the entire regulation range of the OLTC.

As for HV/MV transformers having a Yyn0 connection, only measurement on the no-load MV side is possible, to be carried out as a type test. Again, the measurement must be made at rated voltage and with the OLTC in the central position.

Measurement from the primary for autotransformers can also be considered an acceptance test, given its low onerousness and the usefulness of knowing the precise parameters at the zero sequence.

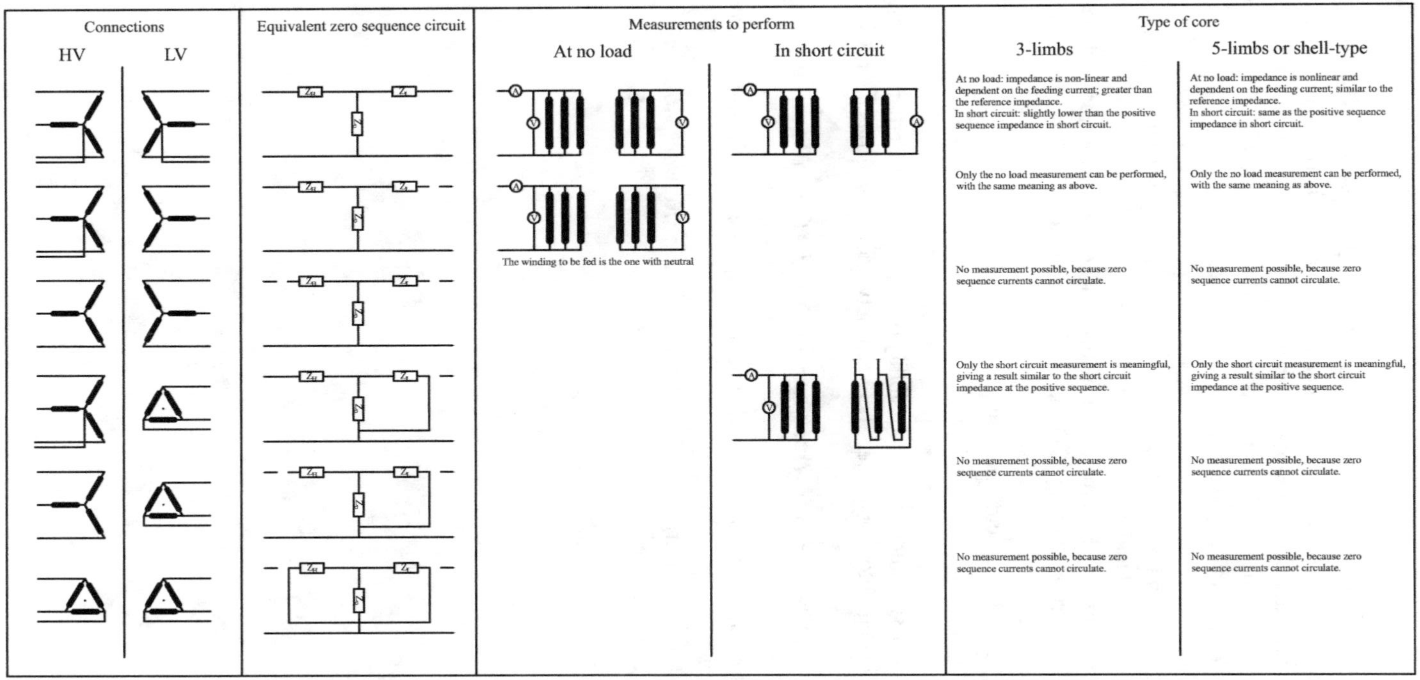

Figure 8.19 Summary diagram of zero-sequence impedances of transformers

References

[1] M. Alderson *et al.*: *Power Transformers – Vol. 2: Expertise*, Alstom, 2011.

[2] IEC 14-4/8: *Power Transformers-Application Guide*, 2001 (IEC 60076-8 ed. 1).

[3] E. Sorrentino, and J.C. Burgos: Comparison of methods for measuring zero sequence impedances in 3-phase core-type transformers, *47th International Universities Power Engineering Conference*, 2012.

[4] M. Heathcote: *The J & P Transformer Book*, 12th ed., Newnes, 1998.

[5] S.V. Kulkarni, and S.A. Khaparde: *Transformer Engineering – Design and Practice*, Marcel Dekker, 2004.

[6] D. Šegović, and F. Kelemen: Estimation of zero sequence impedance in power transformers using 2D and 3D FEM, *Cigré 3rd Transformer Colloquium*, 2014.

[7] J. Wang, and R. Lascu: Zero sequence circuit of three-legged core type transformers, *62nd Annual Conference for Protective Relay Engineers*, 2009.

[8] F. Iliceto: Considerations on the possibility of dispensing with tertiary winding in interconnecting autotransformers between networks, *L'elettrotecnica*, October 1965.

[9] IEC EN 60076-1: *Power Transformers-Part 1: Generalities*, 2012 (IEC 60076-1 ed. 3).

[10] Å. Carlson, J. Fuhr, G. Schemel, and F. Wegscheider: *Testing of Power Transformers*, ABB, 2003.

[11] M. Bamber, M. Bergstrom, A. Darby, *et al.*: *Network Protection & Automation Guide*, Alstom, 2011.

Chapter 9

Short circuit withstand test for autotransformers

9.1 Generalities

HV/HV autotransformers installed in electric transmission grids are subjected to various stresses during their operation. Among them, one of the most onerous is the feeding of a short circuit into the connected grid.

The importance of short circuit withstand capability in ensuring the reliability of autotransformers (ATRs) was not widely felt as long as grids were poorly meshed, as the currents resulting from a fault had small values. However, beginning in the 1970s, with the release of the first standard document for short circuit testing (IEC 76-5:1976), the demand for the demonstration of short circuit withstand of machines began. Initially limited to distribution transformers, short circuit testing has been extended to ATRs in the transmission grid since the late 1980s, when test laboratories were set up at electrical substations (such as the substation in Rondissone, Italy), built for the purpose of testing machines connected to a high-power grid node. Subsequently, other laboratories with high-voltage test capabilities were built, most notably the one in Arnhem (the Netherlands), which uses alternators capable of powering the transformer even after separation from the grid, allowing greater flexibility in test conditions.

The current IEC 60076-5 standard [1] also allows the possibility of proving withstand by calculation and comparison with transformers that have already been successfully tested.

Short circuit withstand is important to verify the correctness of the design and construction quality of ATRs. Knowledge gained from it can support experts in choosing the most appropriate design criteria and evaluating the results. Therefore, this chapter describes the state-of-the-art, requirements for testing and design verification, and the prospects for the evolution of technique and standard.

9.2 Short circuit stresses

9.2.1 Short circuit currents

Autotransformers are critical components for the interconnection between transmission grids and between them and electrical sub-transmission networks. Frequently power lines, and more rarely substation components, experience faults with the loss of insulation between phases or between a phase and ground, which cause the circulation of large *short circuit* currents, especially in networks operated with the neutral grounded.

The evaluation of the expected value of short circuit current, at each electrical node and for different fault configurations, is performed periodically by TSOs with dedicated calculations assuming conservative assumptions for generators, grid meshing, and operating voltage. The result of the calculation provides the maximum three-, two-, and single-phase short circuit current values for each node in the transmission grid.

The standards [2] give guidance for the calculation of the short circuit currents; the three-phase short circuit, under symmetrical voltage regime, is considered the reference case for the verification of the short circuit withstand capability of power transformers.

For the study of the parameters affecting the value of the short circuit current in ATRs, let us denote by Z_s the impedance of the positive sequence equivalent grid under short circuit that is feeding the transformer and with Z_t the series impedance of the transformer. For a three-phase fault, the RMS short circuit current is obtained as[1]

$$I = \frac{V_r}{\sqrt{3}(Z_t + Z_s)}, \tag{9.1}$$

where V_r is the phase-to-phase voltage of the theoretical infinitely powered (i.e., with fixed voltage under all conditions) grid feeding the fault. This value is called steady-state *symmetrical current* and represents the RMS value of the current that would occur if the short circuit's conditions were constant over time. Actually, in the first instants of the transient, the current takes on a maximum peak value that is called *asymmetrical current*. It is obtained by multiplying by $\sqrt{2}$ the symmetrical current and by a peak factor given by the following expression:

$$k = 1 + \exp\left[-\frac{(\phi + \frac{\pi}{2})}{\frac{X}{R}}\right] \cdot \sin\phi, \tag{9.2}$$

where X is the total reactance of the grid and transformer, R the total resistance of the network and transformer, and $\phi = \arctan\frac{X}{R}$ is the phase of the total impedance.

This factor varies from 1 (in the unrealistic case that $X = 0\Omega$) to 2 (in the case that $R = 0\Omega$); however, the standard limits its value to 1.9 for transformers with rated power over 100 MVA.

[1]The calculation is made under the assumption that the grid impedance and the series impedance of the transformer are in phase and thus can be summed arithmetically rather than through phasors. It should be noted that this current does not necessarily correspond to the current I_w circulating in the winding because it depends on the type of phase connection: for star connection $I_w = I$ and for the delta connection $I_w = I/\sqrt{3}$.

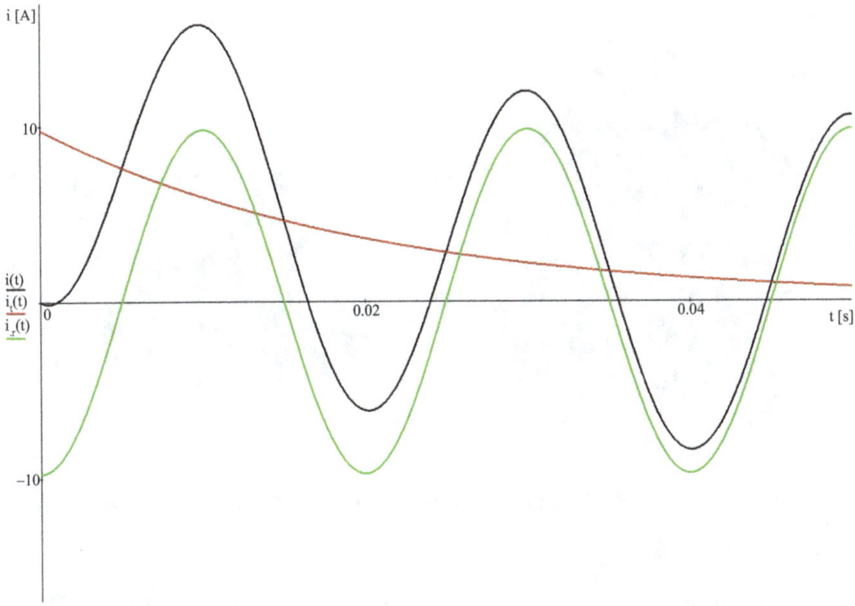

Figure 9.1 Closing transient of a resistive-inductive circuit, representative of a shorted ATR

This result is obtained by considering a sinusoidal voltage $v(t) = V\sin\omega t$, to which corresponds a closing transient at the instant $t = 0$ with asymmetric current:

$$i_t(t) = \frac{V}{Z}\sin\phi\, \exp(-t/\tau), \text{ where : } \tau = \tfrac{X}{\omega R} \text{ is the time constant.}$$

This current is added to the symmetric current $i_r(t) = \frac{V}{Z}\sin(\omega t - \phi)$.

This results in a transient current illustrated in Figure 9.1, whose instantaneous maximum $i(t) = i_t(t) + i_r(t)$ is reached after $\omega t = \phi + \frac{\pi}{2}$ approximately.[2]

The passage of high currents in transformers causes several adverse consequences for the device:

1. increase in magnetic flux, with possible saturation of the core;
2. localized heatings due to increased thermal dissipation by the Joule effect and Foucault effect (i.e., eddy currents induced by leakage flux);
3. electrodynamic forces between current-carrying conductors (Ampère forces).

Core saturation results in the inability to faithfully transduce the voltage and current at the non-faulted side. However, since the voltage of the faulted side is

[2]This is the instant of maximum symmetric current; the maximum instantaneous current is reached just before this instant. However, the error that is made is not significant.

Figure 9.2 A Buchholz relay (left) and an overpressure valve (right)

already degraded, this is not a problem for the quality of service. Moreover, the reduced current transduction has a mitigating effect on the current at the non-faulted side, although it introduces harmonic distortions due to saturation.

Instead, it should be considered that in the case of a net short circuit, that is a fault near the transformer bushings, the faulted side current is limited only by the series impedance of the transformer, and the voltage at the terminals is practically zero. This case approaches the ideal short circuit, in which the magnetomotive forces of the two windings exactly compensate themselves without magnetizing the core.[3] In this case, the leakage flux is very high and can saturate the tank if the winding feeding the short circuit is the outer one, or the magnetic core if it is the inner one.[4]

9.2.2 Mechanical and electrical protections

ATRs are equipped with various protections, both mechanical and electrical.

The customarily installed mechanical protections are (Figure 9.2) as follows:

- the *Buchholz relay* (also called *gas and liquid actuated relay*) is a device that is installed on the oil conservator pipe connecting it to the tank's cover. It has a floating buoy that lowers due to gas accumulation at the top of the device and an impeller that measures oil flow. An electrical discharge or an excessive heating inside the transformer vaporizes the insulating oil, creating bubbles (also by the saturation of the gaseous solution in the oil in the case of

[3]This applies to transformers with concentric windings, such as power transformers. In contrast, it does not apply to current transformers where the windings are magnetically coupled by means of a toroidal core on which only the secondary winding is wound and where the primary winding bar passes through.
[4]See Figure 9.15 for the trend of the short circuit leakage flux.

slow-evolving defects) which tend to rise toward the conservator and get trapped in the relay, so the buoy lowers and causes the activation of two electrical contacts, related to the alarm and trip thresholds. In addition, a high-energy electrical breakdown creates an overpressure for the sudden generation of gas, which causes oil to flow toward the conservator. The impeller, upon reaching a threshold of instantaneous oil velocity, closes a contact that actuates the trip signal.

- the *overpressure valve* (also called *pressure relief device*) is a valve located on the cover of the tank. Through an opening, normally closed with a shutter held in place by a spring, it allows the release to the outside of an overpressure of oil due to a high-energy internal breakdown. The shutter is connected to an electrical contact that generates the trip signal when activated. In large ATRs, there are three valves, one at each phase column.

Calibration of mechanical protective devices is done by the transformer manufacturer, who, based on the oil mass and rated power of the machine, estimates the velocity and overpressure of oil in case of internal fault. The calibrations should be neither too low to prevent untimely intervention (in case of external fault or earthquake the transformer is subject to vibrations that may actuate the protective devices), nor too high to prevent failure to intervene. It should be emphasized that the overpressure device is in any case unable to dispose of high pressures in the case of internal faults with intense or persistent arcing, and therefore, it does not prevent a tank rupture in all possible faults.

The electrical protections normally installed are as follows:

- differential protection, which compares the currents of the circuit connections of the ATR, corrected according to the transformation ratio. The protection must be properly desensitized against untimely tripping in the presence of distorted currents due to core saturation (such as occurs at energization, in the presence of residual flux, and in the absence of pole synchronizer at the transformer's circuit-breaker). Desensitization is usually achieved by a block enabled by the second harmonic of the current.
- distance protection, which is often intended for power lines but can also be used for transformers given the high series impedance (Z_t). It should be calibrated to a sufficiently low value, e.g., first zone $\leq 50\%\ Z_t$ and subsequent zones used as backup of the interconnected line protections.
- overcurrent protection with an appropriately high calibration ($>1.5\ I_r$) and time delay (≥ 2 s) to ensure selectivity.

Because modern digital protection relays have very fast tripping times, even as little as 1 or 2 cycles for differential and distance protections, the issue of core saturation has little impact.

Moreover, although mechanical protections are faster than electrical ones, they intervene only for internal faults in the transformer, which is therefore already compromised at the time of their tripping. Therefore, tripping by the mechanical and differential protections always commands the blocking of the reclosure of the transformer's circuit-breakers, since an inspection is necessary before putting the device back into operation.

In the following, only faults external to the transformer but fed by it will be described. It should be remembered that the short circuit withstand test guarantees the behavior of the transformer only toward external faults, not for internal faults. However, a transformer that has passed the short circuit withstand test for external faults exhibits high reliability in not developing destructive internal faults.[5]

9.2.3 Thermal stresses

The heat produced by the Joule effect and the Foucault effect during the short circuit mainly causes the temperature rise of the conductors for the short duration of the fault event.

The standard only considers heat by Joule effect and an adiabatic heat exchange, so

$$Q = cm\Delta\theta = RI^2 t, \tag{9.3}$$

where

- Q is the heat produced (J);
- c is the specific heat of the conductor (J/kg·°C);
- m is the mass of the conductor (kg);
- $\Delta\theta = \theta_1 - \theta_0$ is the temperature difference before and after the event (°C);
- R is the resistance of the conductor (Ω);
- I is the RMS value of the symmetrical short circuit current (A);
- t is the duration of the short circuit (s), assumed to be equal to 2 s for verification.

Then, it is considered that:

$$m = \gamma \cdot \mathscr{V}, \tag{9.4}$$

where γ is the density (kg/m^3) and $\mathscr{V}$ the volume (m^3) of the conductor and further that:

$$RI^2 = \rho \frac{\ell}{S} I^2 t = \rho \ell S \left(\frac{I}{S}\right)^2 t = \rho \mathscr{V} J^2 t, \tag{9.5}$$

where

- ρ is the resistivity of the conductor (Ω m);
- ℓ is the length of the conductor (m);
- S is the cross-sectional area of the conductor (m^2), such that $\ell S = \mathscr{V}$;
- $J = \frac{I}{S}$ is the current density in the conductor (A/m^2).

[5] For example, a fault between two turns is less likely to evolve into a fault on other turns because of the deformation of the windings due to electrodynamic forces.

Compounding the different expressions, taking into account the dependence of resistivity on temperature through the coefficient α (°C^{-1}):

$$\rho = \rho_0[1 + \alpha\Delta\theta], \tag{9.6}$$

and solving the resulting equation gives the expression:

$$\theta_1 = \theta_0 + \frac{\alpha^{-1}}{\frac{c\gamma}{\alpha\rho_0 J^2 t} - 1}, \tag{9.7}$$

in which, by entering the material data (from the copper or aluminum datasheet), and the design values of J and θ_0, the final temperature θ_1 is obtained.

For large transformers, the temperature rise is normally within the range of 10–15°C/s [3], and considering that the standard allows final temperatures much higher than the thermal class of the conductor, verification to thermal stresses is usually demonstrated without problems and therefore no design arrangements dedicated to this requirement are needed.

9.2.4 Dielectric stresses

During a short circuit in the network, the voltage is greatly reduced for the phase(s) involved and may go to zero for nearby faults or weak networks.

The Grid Code of TSOs requires generators to be able to withstand voltage during short circuits, and thus, power plant transformers and autotransformers in connecting substations are in an abnormal working condition. For the latter:

- if the fault is on the lower voltage side (which is usually referred to as the secondary side), the voltage on the common winding of the ATRs is almost zero, so all the primary voltage is borne by the series winding;
- if the fault is on the higher voltage side (which is usually referred to as the primary), the voltage on the series winding is almost equal and the opposite in phase to that on the common winding.

Of the two cases, the former is definitely the more burdensome,[6] because it causes an increase in magnetic flux to which corresponds an increase in the voltage per turn of the series winding. Having said $\mathfrak{c} = \frac{V_{r1} - V_{r2}}{V_{r1}}$ the co-ratio, the voltage per turn becomes $\mathfrak{c}^{-1} = 1 + \frac{V_{r2}}{V_{r1} - V_{r2}}$ times its nominal value.

It is noted that the induced voltage test is considered useful [4] to test the dielectric strength of the ATR under the mentioned conditions that are characteristic of a short circuit. Although the test is performed, according to the standard [5], with the secondary open rather than short circuited, the phase-to-phase test voltage is set equal to $1.8U_{r1}$ for the duration of 120 s[7] and thus gives a stress between turns similar to that shown.

[6]It can be shown that in the second case, the voltage per turn becomes $\frac{1-\mathfrak{c}}{\mathfrak{c}} = \frac{V_{r2}}{V_{r1} - V_{r2}}$ times the nominal one, thus always lower than the first case.

[7]It is permissible to maintain the test voltage for a shorter duration, which is inversely proportional to the frequency of the applied test voltage, which may also be higher than the rated voltage in order not to saturate the core.

9.2.5 Mechanical stresses

It is well known that the Ampère force between two infinitesimal current-carrying conductors is given by the differential expression[8] ([6] p. 119):

$$d^2\boldsymbol{F}_{1\to 2} = I_2 d\boldsymbol{r}_2 \times d\boldsymbol{B}_1 = \frac{\mu_0}{4\pi} I_2 I_1 \frac{d\boldsymbol{r}_2 \times \left(d\boldsymbol{r}_1 \times \widehat{\boldsymbol{R}}_{12}\right)}{R_{12}^2} [\mathrm{N}], \tag{9.8}$$

where $\widehat{\boldsymbol{R}}_{12}$ is the unit vector of the direction from 1 to 2, and $d\boldsymbol{r}_i$ is the differential element of conductor length i.

In particular, considering transient currents such as those described in Section 9.2.1, and integrating over the conductor circuits Λ_i, the time dependence of the force is obtained:

$$\begin{aligned} \boldsymbol{F}_{1\to 2} &= \oint_{\Lambda_2} \oint_{\Lambda_1} d^2\boldsymbol{F}_{1\to 2} \\ &= F\left(\frac{1}{2} + \exp\frac{-2t}{\tau} - 2\exp\frac{-t}{\tau}\cos\omega t + \frac{1}{2}\cos 2\omega t\right)[\mathrm{N}], \end{aligned} \tag{9.9}$$

where F is the peak force at steady state, in which the behavior, given by the first and last terms in parentheses, has periodicity twice the fundamental frequency. However, in the first cycles of the short circuit event, the damped oscillating term at fundamental frequency prevails (Figure 9.3).

Since the damped oscillating term has the largest initial amplitude, the heaviest dynamic stresses are at the beginning of the transient. Consequently, the standard stipulates a duration of each short circuit test of 0.25 s for large power transformers.

Although formula (9.9) allows the overall force to be calculated, it is more useful for the design to know the mechanical stress for the infinitesimal section of

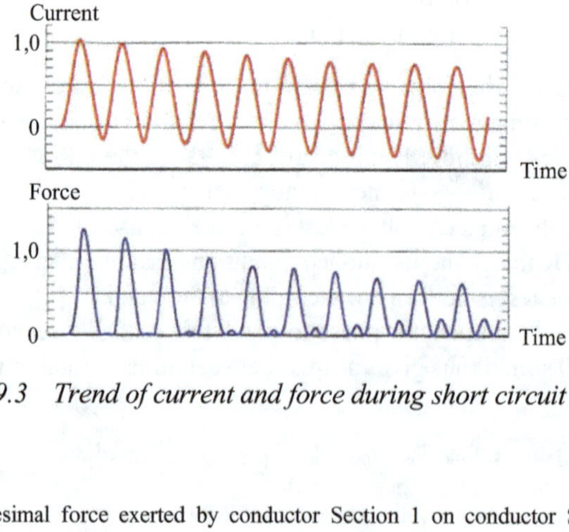

Figure 9.3 Trend of current and force during short circuit transient

[8]This is the infinitesimal force exerted by conductor Section 1 on conductor Section 2. Note that $d^2\boldsymbol{F}_{1\to 2} \neq -d^2\boldsymbol{F}_{2\to 1}$ because the vector product is not commutative; however, the total forces on two complete closed circuits are equal and opposite and therefore obey the third principle of dynamics.

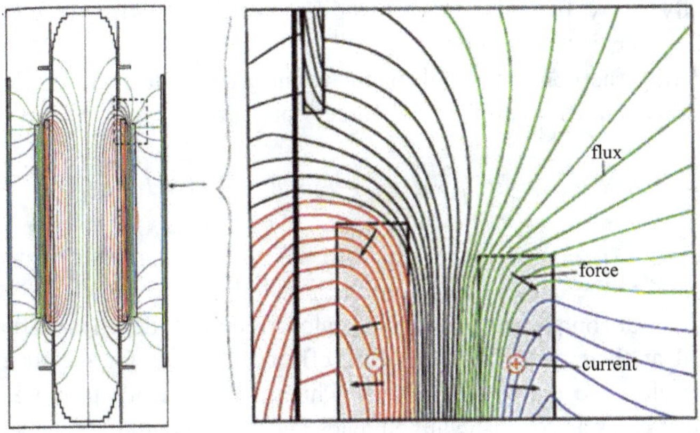

Figure 9.4 Qualitative trend of magnetic flux in a core-type transformer and consequent forces

conductor, i.e., the strain:

$$\sigma = \frac{\oint_{\Lambda_1} d^2\boldsymbol{F}_{1\to2}}{dS_2}[\mathrm{Pa}], \tag{9.10}$$

where dS_2 is the infinitesimal area of the conductor Section 9.2, over which the stress is distributed, according to De Saint-Venant's theory.

The calculation of the Ampère force requires the solution of the magnetic field equations, considering imposed currents circulating in the windings. Although solutions are commonly obtained with finite element simulation software, in the presence of particular symmetries, it is possible to arrive at analytical solutions.

Figure 9.4 represents the qualitative trend of the magnetic flux, associated with induction $\boldsymbol{B}$, and the consequent forces in the current-carrying windings.[9] The flux is almost exclusively axial for about two-thirds of the height of the windings, while it bends and frays at the upper and lower ends. Since the circulating currents are always longitudinal (in the figure, they enter and leave the plane of the page), to an axial induction corresponds, according to (9.8), a radial force. In contrast, in the presence of a radial component of magnetic induction (as in the ends), the force has an axial direction. Moreover, since the currents have opposite trends in the two windings, the forces have opposite directions. Thus, the outer winding tends to

[9]In the case of windings whose current is not determined by short circuit network calculations, such as in the case of tertiary delta winding connected in short circuit with stabilizing functions, the calculation of induced voltage by the Faraday effect is necessary in order to determine the current circulating in the same. The effect of this current is to counteract the magnetic induction that generates it. Therefore, the calculation of the overall induction must be done recursively, greatly complicating the simulation.

move radially away from the inner winding, which will instead compress the insulation layer against the core.

Considering the constant axial magnetic induction along the entire length of the winding, we can estimate the total force[10] on it to be equal to [7]:

$$F_{rad} = \pi D_m N_2 I_2 B_1 = \pi D_m N_2 I_2 \mu_0 \frac{N_1 I_1}{h_w} = \mu_0 \frac{\pi D_m}{h_w} (N_1 I_1)^2, \tag{9.11}$$

where we used $N_2 I_2 = N_1 I_1$, D_m is equal to the average diameter of the winding, and h_w is equal to the equivalent height of the windings.

The force that compresses the windings along their axis depends on how the flux is distributed at their ends. Normally, the effect is greater for the inner winding because it is closer to the core limb over which the flux tends to close. Since the compressive stresses are opposite at both ends and tend to compress the winding, the inner windings are the ones most subject to compressive stress. Figure 9.5 represents the qualitative trend of the compressive force along the axis of the windings.

Although an accurate evaluation of axial stresses requires finite element simulation, an estimate of the maximum axial force that is reached at the center of the winding is given by the following formula:

$$F_{ax} = \mu_0 \frac{\pi D_m}{h_w^2} (N_1 I_1)^2 \left(d + \tfrac{a_1 + a_2}{3}\right)(2k_R - 1), \tag{9.12}$$

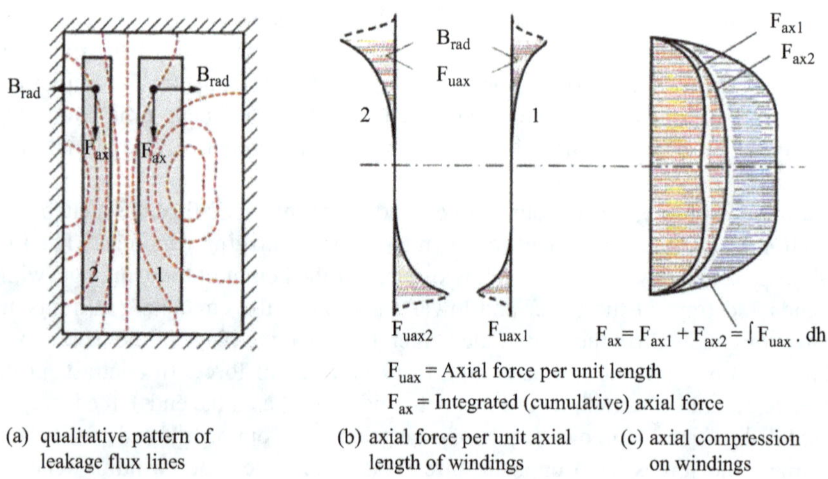

Figure 9.5 Qualitative trend of flow, axial stresses, and axial forces

[10]Using the RMS values of the currents, an average force equal to half the peak force is obtained. However, considering that the different conductors in each winding disc are subject to different forces depending on the different magnetomotive forces they concatenate, the average value is sufficiently representative of reality.

where d is the width of the channel between the two windings, a_i the thickness of the winding i, and the *Rogowski factor* k_R is given by

$$k_R = 1 - \frac{d + a_1 + a_2}{\pi h_w}. \tag{9.13}$$

Note that both types of forces are directly proportional to mean diameter and inversely proportional to height. Since the leakage inductance also has this dependence, taller and narrower transformers have a higher short circuit current, inversely proportional to the reactance, but experience lower forces. The reverse happens in lower and wider transformers. However, the size and short circuit impedance are often constrained by other design requirements (such as transport gauge limits and maximum voltage drop), so they cannot be fully optimized by the designer.

If the two windings do not have the same height or are not both centered with respect to the centerline, an equal and opposite axial force is established between the windings that tend to separate them in the direction with respect to which they are asymmetrical, roughly equal to:

$$F_s = 4\mu_0 \frac{D_m}{h_w^2} (N_1 I_1)^2 \upsilon, \tag{9.14}$$

where υ is the axial deviation between the windings (Figure 9.6).

The dynamic behavior of the active part, when the radial and axial forces have the periodic pattern of the type expressed in (9.9), is very complicated to model even in a finite element simulation software, because the mechanical response of the windings and the armature (on which the forces transmitted from the ends by the windings are applied) must be taken into account.[11]

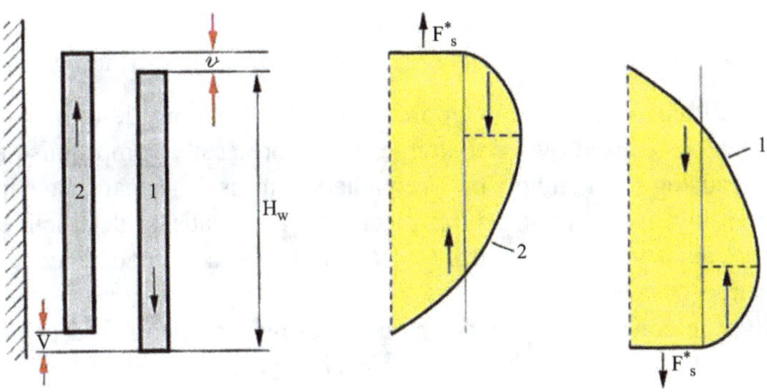

Figure 9.6 Axial displacement and separation force diagrams

[11]Since the vector sum of forces is always zero, by the third principle of dynamics, the core and armature are not subject to net forces but to internal stresses due to the different points of application of forces.

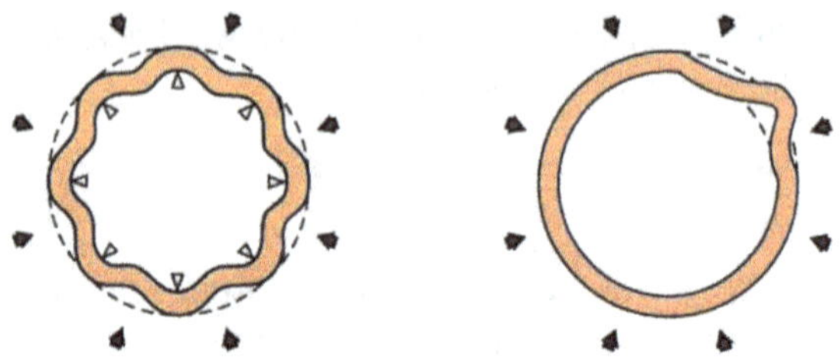

Figure 9.7 Diagram of deformations by forced or free bending

The winding coils are more prone to mechanical failure because the spaces required for oil circulation and insulation between discs do not allow structural tightness like that of the compact structure of core and armature.

Mechanical deformations on coils resulting from excessive radial stresses are as follows:

- inward bending, called *forced or free buckling* (Figure 9.7);
- longitudinal stretching, called *spiraling*.

These stresses, if small, do not compromise the dielectric strength of the winding, but they result in mechanical weakness that can lead to destructive failure in case of further short circuit events. These defects can be detected by leakage inductance and swept frequency response analysis (SFRA) measurements. Therefore, it is important to perform such measurements after each relevant short circuit event to monitor the condition of the machine.

Axial stresses can result in the following deformations (Figure 9.8):

- inclination (*tilting*), typically in disc windings;
- displacement (*tilting or telescoping*), typically in helical windings.

Figure 9.9 represents three examples of the deformations described above.

Deformations caused by axial stresses can more easily compromise the insulating tape, causing short circuits between adjacent turns. They can also damage the winding pressing, inducing structural instability that leads to destructive failure. Defects from axial stresses can also be detected by leakage inductance and SFRA measurements.

Finally, the connections inside the transformer are also subject to dynamic stresses. In fact, the leads of the connections have low stiffness, because they do not belong to a compact structure, and their tightness is ensured by fastening them to the wooden (or other insulating material) frame, with bands and bolts (Figure 9.10).

These connections, along with those of the bushings, are the most critical parts of the transformer because they cannot be modeled adequately in mechanical computer simulations.

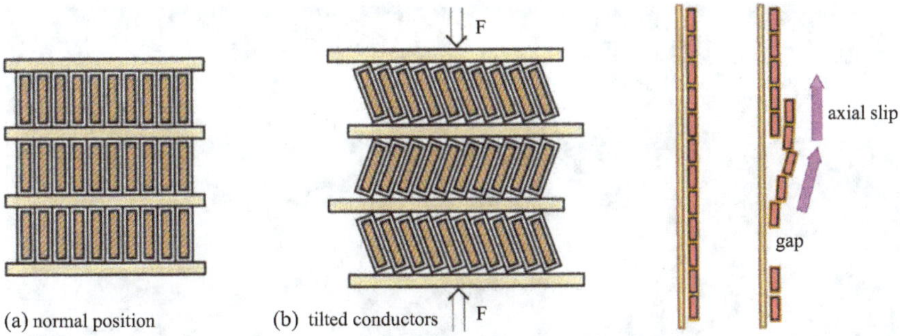

Figure 9.8 Tilting (left) and slipping (right) of wires

Figure 9.9 From left to right: free bucking, spiraling, and slipping

9.2.6 Manufacturing precautions

The construction measures to ensure the short circuit withstand of transformers are mainly as follows:

- mechanical stiffening of the active part, using bonded cables, pressing the windings through the core reinforcements, and properly fastening the active part to the tank;
- adaptive elasticity of the structures, with static pressing or with springs equipped with viscous dampers, which allow constant pressure to be maintained on the windings in the face of dynamic stresses during a short circuit;
- dimensional checks during all stages of construction to ensure compliance with design tolerances and to avoid weak points that may compromise the short circuit withstand.

Figure 9.10 Active part with the harp *of the connections to the on-load tap-changer*

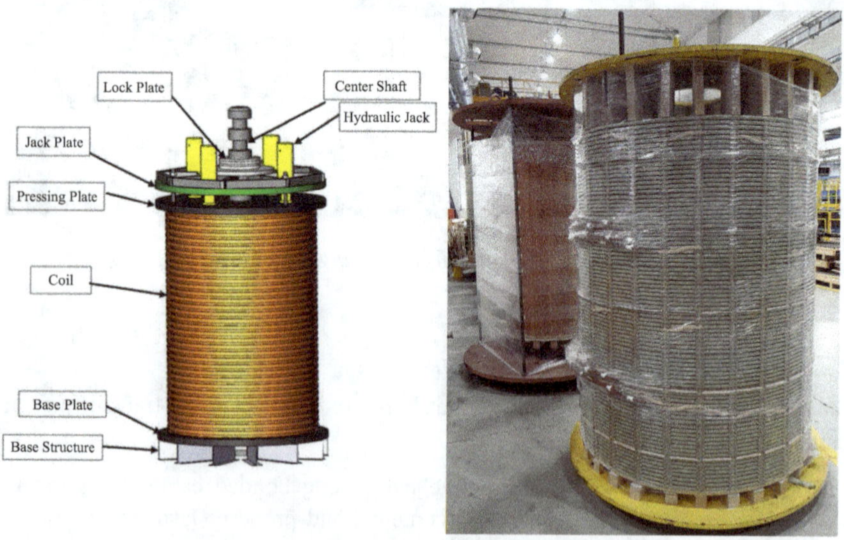

Figure 9.11 Pressing by hydraulic jacks

Pressing for medium-sized windings is done by hydraulic jacks (Figure 9.11), which compress the windings by means of metal plates and tie rods. In contrast, large windings are pressed with hydraulic presses (Figure 9.12), while the core is

Figure 9.12 Winding press and core assembly machine

assembled in special structures that ensure proper pressing until complete assembly with tie rods and armatures. This operation is sometimes carried out in a dedicated core-building facility, where sheet-metal cutting operations also take place.

The windings of the same phase after pressing are assembled (*coil assembling*) into a single cylinder, called a *phase winding*, which undergoes further pressing so that all windings of the same phase have the same height. The phase winding is then fitted onto the core limb, from which the top yoke has been removed (Figure 9.13).

Next, the upper yoke is put back in place, and pressing of the windings is achieved by a set of tie rods on the yoke clamps; this is a solution adopted mainly for distribution transformers. Alternatively, especially for large transformers, a cup spring system is used that implements a pressing of the windings, adaptive to the electrodynamic stresses, by means of viscous dissipators called *dashpots* (Figure 9.14).

The active part of the transformer obtained by the previous procedures, i.e., the core fitted with the windings, is subjected to drying in an oven to remove the moisture contained in the paper and in the pressboard. The oven, called *vapor phase oven*, exploits vacuum heating cycles with the circulation of kerosene vapor, which has a high water absorption capacity (Figure 9.15). Sometimes the windings undergo preliminary heat treatment in an autoclave to preemptively reduce most of the moisture.

At the end of the furnace treatment, the windings are subjected to a final dimensional check, because temperature excursion and moisture removal change the dimensions of the materials. Next, the phase windings are pressed (*registered*) to the design value for pressure between the core clamps, and the active part is placed in the tank for final filling with oil before factory testing.

It is widely believed that placing too many pressboard spacers and too many layers of Kraft paper on the conductors reduces the mechanical stiffness of the winding, as these insulations are more flexible than the bonded cable. However,

Figure 9.13 Fitting of phase windings in distribution transformers

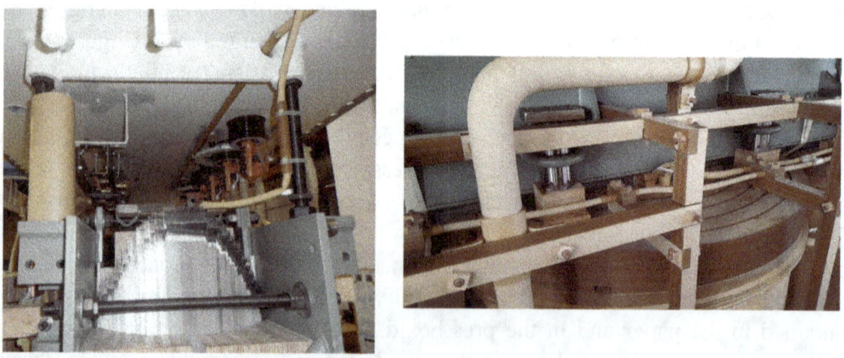

Figure 9.14 Fixing windings by static pressing (left) or adaptive pressing with dashpots (right)

pressing the winding allows to compensate for the winding's elasticity, and the correct pressure value depends on the manufacturer's experience, which can only be obtained after passing a large number of short circuit withstand tests.

The mechanical behavior of the active part is also affected by temperature excursions during transformer operation. In fact, the various metals that make up the transformer have a different thermal elongation coefficient that causes internal stresses in the turns when the temperature rises. In addition, copper has a yield stress

Figure 9.15 Schematic drawing of a vapor phase *furnace*

given by the work hardening process.[12] Therefore, if copper is brought to high temperatures, a recovery process[13] takes place, which reduces the yield stress[14] and the stiffness of the metal.[15] Thus, hot-spots inside the transformer can compromise both the solid insulation and the mechanical seal.

[12]*Work hardening* is the process of cold mechanical treatment to increase the stiffness of metal by plastic deformation, which for copper wire occurs during drawing. Plastic deformation creates structural defects in the crystal lattice (*dislocations*), which make the material stiffer but increase its brittleness. The corresponding hot process is *quenching*, in which the metal is abruptly brought from recrystallization temperature to room temperature, blocking the diffusive process that eliminates dislocations and resulting in a very stiff but potentially brittle material.

[13]At high temperatures, the phenomenon of *recovery* allows the crystal lattice to recombine, eliminating dislocations. Ductility increases but material stiffness is reduced. Recovery can be an undesirable phenomenon, when it occurs during the useful life of the material, or desired through special treatments called:

- *annealing*: is a slow heating and cooling process that allows complete recrystallization of the metal;
- *tempering*: is a process of slow heating and natural cooling, which occurs at a temperature below that of complete recrystallization and is intended to eliminate dislocations that make the material excessively brittle, preventing its cold processing.

[14]*Yield stress* is the stress at which the material ceases to deform elastically and begins to deform plastically, that is, with permanent deformations. Normally, it is equal to the stress that results in 0.2% permanent deformation in the tensile direction of the specimen ($R_{p0.2\%}$).

[15]*Stiffness* is the slope of the stress–strain curve. Greater stiffness implies less elastic deformation for the same force (tension) applied to the material. Similar but different concepts are as follows:

- *hardness*, which represents resistance to surface plastic deformation and is measured by the Vickers indentation test;
- *resilience*, which represents the energy absorbed by the material during elastic deformation and is equal to the integral of the stress–strain curve to the yield point;
- *toughness*, which represents the energy absorbed by the material during elastic and plastic deformation to failure and is measured by the Charpy pendulum test.

Where the mechanical strength of copper is to be increased, special alloys with silver content can be adopted. However, because of the greater difficulties in processing copper–silver alloy, this expedient is rarely adopted.

9.3 Short circuit testing

9.3.1 Test sequences and configurations

The short circuit withstand test or, more briefly, the short circuit test of transformers consists of applying the rated voltage to one winding and rigidly connecting the terminals of the other winding together to form the short circuit. The standard allows two test sequences:

1. The *pre-set* sequence, in which the unenergized winding is short circuited before the energization of the other winding; core magnetization occurs simultaneously with the test.
2. The *post-set* sequence, in which the unenergized winding is shorted after the other winding is energized; the core magnetization transient is already exhausted when the test begins.

Sequence 1 is less representative of a short circuit in service because it sums two transient events that do not occur simultaneously (except for the infrequent case of energization with shorted terminals). The effect of this sequence depends strongly on the residual induction in the core and on the position of the energized winding (Figure 9.16). In fact, if the energized winding is the inner winding, the leakage flux tends to close preferentially in the core limb, with possible saturation of the core and higher short circuit current (see Section 9.2.1). However, in tests on ATRs, power is normally supplied from the HV side with the LV side[16] shorted, and since the series winding is external, the risk of saturation is reduced.

Sequence 2 is more representative of an actual fault and allows greater control over the test conditions. It requires an additional operating device, the *circuit-closer*, which is a switch with only current-making functions.

The energizing switch in sequence 1, or the circuit-closer in sequence 2, must be equipped with a pole synchronizer having the function of establishing the short circuit in the instant that determines the peak current, corresponding to the passage through zero of the voltage.[17]

Figure 9.17 depicts the three-wire circuit diagram of the test bay built at the Rondissone substation, which allows power supply at 400, 230, and 135 kV.[18]

[16]Recall that the standards require that the sides of a transformer be identified as HV and LV regardless of their actual rated voltage.

[17]The voltage to be considered is the voltage between the terminals of the fed winding, which corresponds to the phase-to-ground voltage for star windings (including those of ATRs) and the phase-to-phase voltage for delta windings.

[18]The scheme, taken from the preliminary report for the construction of the test bay, called for the test substation to be built at an electrical substation where a 150 kV section was present. Later, the choice turned to Rondissone, which has considerable short circuit power, where only the 135 kV section is present instead. Testing of transformers with 150 kV primary rated voltage is always possible but must be done with the equivalent single-phase configuration (Section 10.3.4).

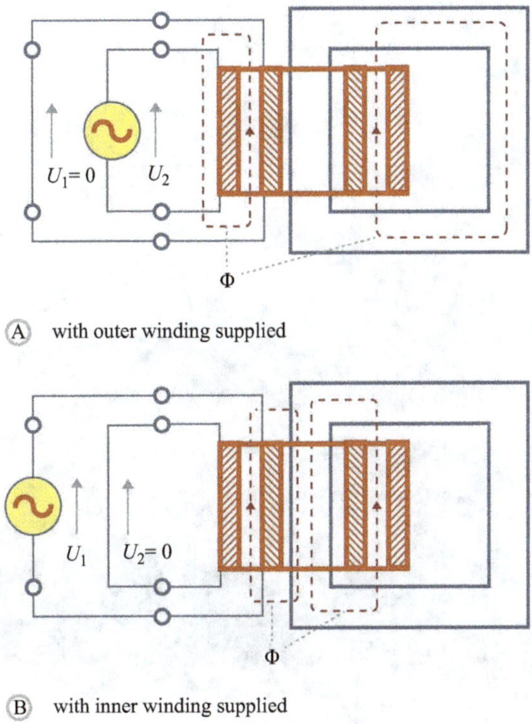

Figure 9.16 Leakage flux trend for the two short circuit winding cases

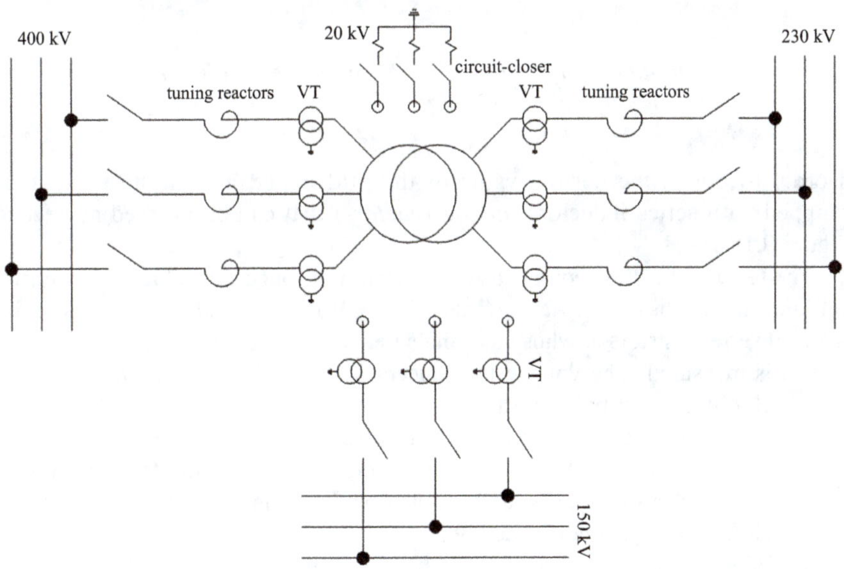

Figure 9.17 Three-wire diagram of the short circuit test stall at the Rondissone electrical substation

Figure 9.18 EDF laboratory test bay "Les Renardières" (FR); tuning reactors are highlighted

In order to obtain the desired value of the grid impedance, the test circuit is equipped with series inductors (*tuning reactors*) that can be inserted as needed (Figure 9.18).

The test bay must be equipped with current transducers capable of measuring high and highly dissymmetric currents (Figure 9.19). Normally, this is done by means of series resistors at whose terminals the voltage, proportional to the passing current, is measured. The voltage transducers used are RC dividers, to which the pole synchronizer must be connected.

The transformer is placed in an area protected by walls (Figure 9.20), to avoid damage to the system in the event of destructive failure of the test, and on an isolated platform in order to be able to also perform the equivalent single-phase test, which involves the ungrounded neutral.

The alternative to feeding the test bay through the power grid is *synthetic* feeding (Figure 9.21) through an alternator that has been previously energized and

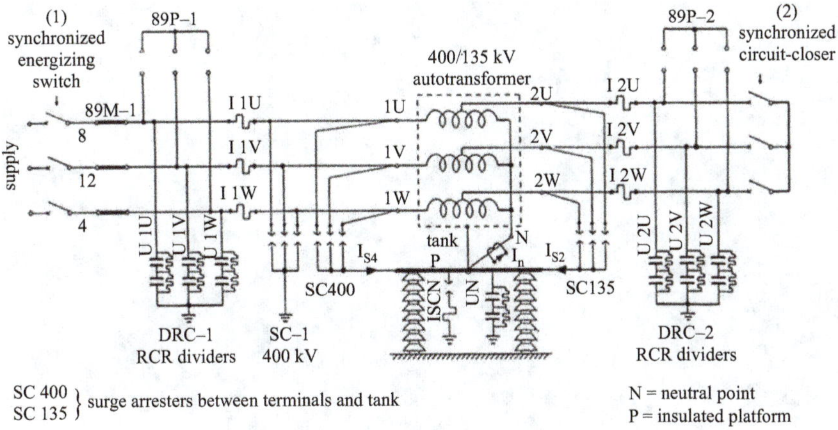

Figure 9.19 Detailed diagram of the Rondissone test bay (three-phase test configuration)

Figure 9.20 Two ATRs assembled in the Rondissone test bay

brought to rated condition by an external grid, from which it will be disconnected before performing the test. Next, a switch with pole synchronizer will connect the alternator to the transformer under test, possibly through a transformer,[19] in case it is necessary to obtain a test voltage different from the nominal voltage of the alternator.

[19]The connection transformer must be designed to withstand repeated short circuit events, so it has to be appropriately oversized. Normally, a calibration stroke, by closing the test switch, leads the connection transformer to saturate the core; the next test stroke is by closing the switch at the transition with the opposite slope through zero (e.g., increasing vs. decreasing voltage) so that the transient leads to the demagnetization of the core at saturation. In this way, the entire linear stretch of the core magnetization characteristic of the connecting transformer (from $-B_s$ and B_s) is exploited. It should be considered that, as a consequence, also the transformer under test undergoes additional strokes for calibration.

Figure 9.21 Aerial view of the KEMA (CESI) test bay (Arnhem, NL). The transformer under test is positioned on a barge moored in front of the bay; the alternators are located in the central building.

This method has advantages: The energy associated with the short circuit is limited to the kinetic energy of the alternator's rotor (since it is disconnected from the grid), and it is also possible to adapt the test voltage and also the frequency, bringing the alternator to different rotational speeds, within the limit of its nameplate data.

The short circuit test is not classified as a destructive test; however, a failure of the transformer's internal insulation during the test can cause[20] a high-energy breakdown resulting in tank rupture and oil release or fire.

The synthetic test has limited available energy, so the probability of destructive failure is very small. However, a transformer that has undergone modest mechanical deformation during a synthetic test creates the uncertainty of possible failure if tested with grid's power.

9.3.2 Testing requirements

The standard allows two modes of feeding the three-phase transformer:

- the three-phase test, in which the transformer is fed with a symmetrical system of voltages at the terminals;
- the equivalent single-phase test, in which the transformer is fed in two-phase mode between the terminal[21] of the phase under test and the terminals of the other two phases connected together.

[20]The test bay is not equipped with rapid protections such as differential protection, but only by overcurrent protection. The power supply is automatically cut off at the end of the rated duration of the test even if no such protection intervenes. A short circuit of the same duration as in the test is not conceivable in operating reality with a properly functioning protection system.

[21]The equivalent single-phase test is applicable only to a star-connected winding, because delta-connected windings have a common terminal between two phases.

A third mode, which is not envisaged in the standard but is of interest, is a pure single-phase supply between the line and neutral terminals of the phase under test, with the other two phases fed at no-load but not short circuited during the test.

Each of the above modes has advantages and disadvantages that are described below.

The series impedance of the test circuit causes a voltage drop during the passage of the short circuit current, so the supply voltage must be appropriately increased from the rated value of the supplied winding (U_r)[22] to compensate for the voltage drop and obtain the rated short circuit current calculated with reference to the voltage U_r measured at the transformer terminals.

The standard requires that, in the case of a post-set short circuit, the open-circuit voltage should not exceed 115% of U_r, in order to avoid overexcitation and possible saturation of the core, as well as the risks of dielectric breakdown of the transformer insulation.

Said U_s the no-load source voltage, the relationship between the short circuit impedance of the transformer and the series impedance of the test circuit is related to the relationship between the voltages by the relationship:

$$\frac{Z_t}{Z_s} = \frac{U_s^2}{U_r(U_s - U_r)}, \tag{9.15}$$

having to be $U_s \leq 1.15U_r$ it follows that it is necessary that $Z_t \geq 8.82Z_s$, which relationship gives an indication of the short circuit power of the source, which must be at least nine times that of the transformer under test.

The standard allows an indeterminate number of calibration strokes (also called *calibration tests*) to be performed, as long as the initial peak current does not exceed 70% of the prescribed test value. Since the electrodynamic forces depend on the square of the current, the corresponding stresses will be at most 49% of the expected ones, which is still a relevant value. It is therefore essential to limit the calibration strokes as much as possible, ideally only one stroke per phase, or to perform them at a current much lower than the prescribed limit.[23]

The standard requires that each phase of the transformer under test be subjected to three short circuit shocks, each with the prescribed initial peak current and duration. If an on-load tap-changer is present, it must be placed in the positions corresponding to the voltages: minimum, rated,[24] and maximum, depending on the phase under test, so as to test the entire range of regulation.

[22]For simplicity of discussion, the rated winding voltage is that at the winding terminals, i.e., the phase-to-phase voltage for delta windings and the phase-to-ground voltage for star windings. This description mainly refers to the tests for ATRs, and U therefore means the phase-to-ground voltage, unless otherwise stated.

[23]With a current equal to 50% of the test current, which can be obtained by inserting an additional series reactance, the stresses are limited to 25% of those expected, a value that is not a cause for concern, and the short circuit power of the source can be estimated sufficiently accurately. Higher values of the calibration current are needed only if there is a connecting transformer to be saturated prior to the test.

[24]It is noteworthy that, for transformers with variable-flux regulation such as ATRs, the rated voltage tap is not necessarily the middle tap but must be checked on the nameplate.

9.3.3 Test with three-phase power supply

The three-phase power supply test is the preferred solution in the standard because it is most representative of the occurrence of a fault in the power grid with the transformer in service. However, it is also the most onerous test, because the windings not under test are subjected to the symmetrical short circuit current, although the peak asymmetrical current is only required for the phase under test.[25]

9.3.4 Single-phase equivalent test

In the equivalent single-phase test, the two terminals of the phases not under test in the star windings (the fed winding must necessarily have such a connection) are connected together (Figure 9.22).

This configuration is realized at the test bay with special disconnectors, denoted 89P-1, 89P-2, and 89M-1 in the scheme of Figure 9.23.

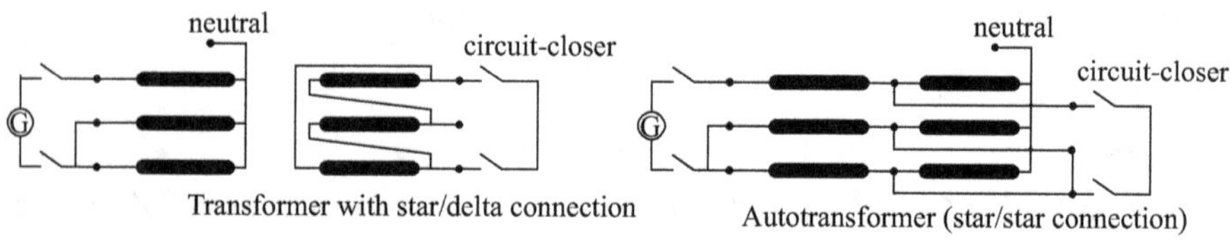

Figure 9.22 Single-phase equivalent test's connection diagrams

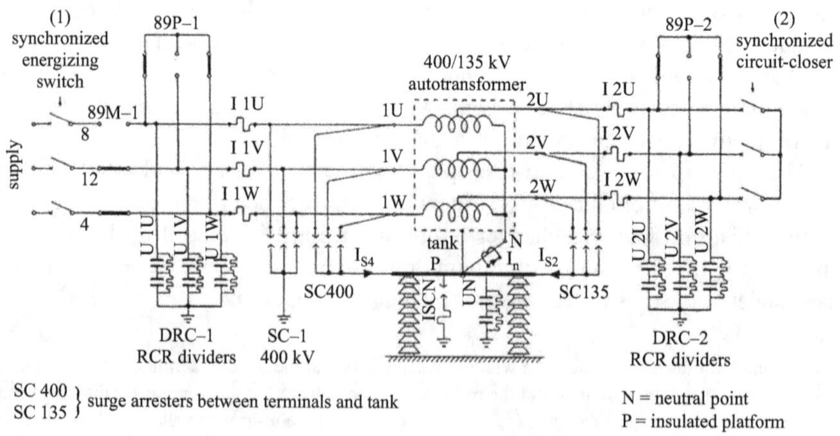

Figure 9.23 Test bay in equivalent single-phase configuration (phase under test: V)

[25]The peak asymmetry is estimated to be as high as 75% on non-test phases ([6] p. 81), compared with 50% in the equivalent single-phase configuration.

The terminal of the phase under test, and that of the other two connected to each other, are connected to two different poles of the energizing switch, to obtain a supply with phase-to-phase voltage. In a 3-limb transformer, since this voltage is distributed ⅔ on the winding under test and ⅓ on the other two non-under test windings,[26] the voltage at the ends of the winding under test results:

$$U_t = \frac{2}{3}\sqrt{3}U_s \simeq 1.155U_s, \tag{9.16}$$

which is permissible according to the standard. In this way, an increased voltage can be obtained to compensate for any insufficient short circuit power from the grid.

In a 5-limb, or shell-type transformer, the distribution of voltages depends on the specific magnetic core, but still the resulting voltage for the winding under test is greater than ⅔ of the applied voltage.

This test mode has the disadvantage of leaving the neutral floating. In a 3-limb transformer, the neutral is brought to $U_s/\sqrt{3}$ (Figure 9.24), that is, at ⅓ of the phase-to-phase voltage. Therefore, for this configuration to be possible, the neutral insulation must be sized for this value.[27]

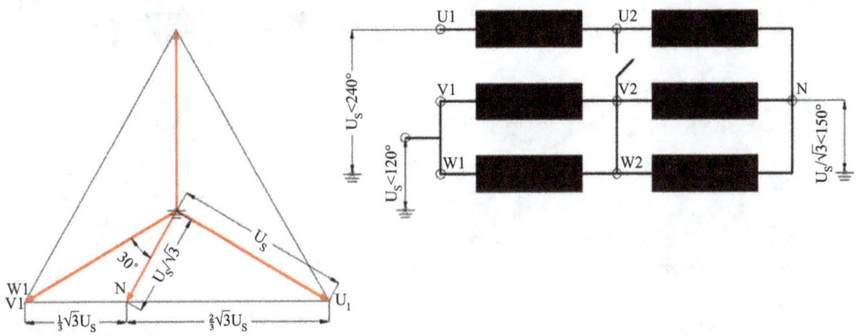

Figure 9.24 Vector diagram and phase voltages on the windings of a 3-limb transformer

[26]In a 3-limb transformer, the magnetic circuit is given by the core phase columns that have identical reluctance $\mathcal{R}_f$ and by the yokes. Since the two non-test windings are in parallel, their reactance, inversely proportional to the reluctance, is half that of the other winding. In a 5-limb transformer, the magnetic circuit also includes the side columns without windings with reluctance $\mathcal{R}_l$, so the magnetic flux circulating in the columns of the windings not under test is less.

[27]The insulation value to be considered for the suitability of neutral insulation is that of the short duration voltage test (1 min), because the transformer is kept energized for a limited period of time during short circuit tests. For example, for ATRs with rated voltage of 400 kV in the Italian grid, the maximum voltage of neutral is 72.5 kV, which corresponds to a short time withstand voltage of 140 kV, thus enough to sustain ⅓ of the primary phase-to-phase rated voltage (133 kV).

In contrast, in a 5-limb transformer, the magnetic flux is free to circulate in the side columns and thus the value to which the floating neutral terminal is brought is greater than ⅓ of the phase-to-phase voltage (Figure 9.25).

The experience of evidence ([3] pp. 82–83) and dynamic electromagnetic transient (EMTP) simulations confirm the unconstrained neutral voltage trend at transformer energization in the equivalent single-phase configuration (Figure 9.26). The problem is mitigated, but not completely eliminated, by running the test with preset sequence.

In addition, this test configuration creates a zero sequence flux in the core that can saturate it and invest the tank and other metal parts of the transformer. Therefore, it is necessary to have an isolated platform to which the neutral bushing and the tank can be connected, in order to avoid possible flashover to the ground.

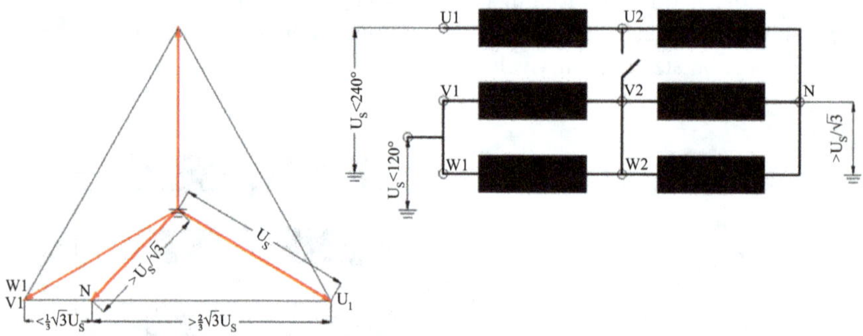

Figure 9.25 Vector diagram and phase voltages on the windings of a 5-limb transformer

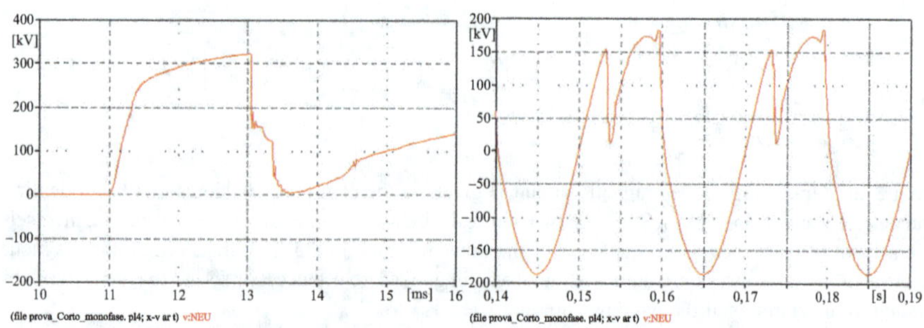

Figure 9.26 Neutral voltage trend simulation for a 5-limb transformer in equivalent single-phase configuration: on the left at the time of energization and on the right at no-load steady state

9.3.5 Single-phase test

The test configuration for pure single-phase short circuit is not covered by the standard. However, it is a representative test of the most frequent fault in electric transmission networks.

This configuration requires feeding a star-connected winding with the phase voltage between the line and neutral terminals. This can be done in several ways:

- connecting the neutral terminal to ground: feasible if the ground mesh can effectively dissipate the test current;
- connecting the phase and neutral terminal to the corresponding terminals of a connecting transformer, with fully insulated neutral: it requires an isolated platform such as for the equivalent single-phase test;
- by connecting the phase and neutral terminal to two terminals of a connecting transformer, with appropriate transformation ratio $\sqrt{3}$ to obtain the phase-to-ground voltage from the phase-to-phase voltage: it requires an isolated platform such as for the equivalent single-phase test.

This test does not stress the non-short circuited phases, reducing collateral stresses on the windings not under test.

In addition, it allows testing of the possible delta stabilization winding, in which current circulates only in the presence of zero sequence component; to obtain this, the line terminals of the fed winding can be connected together, reducing the test current on it (Figure 9.27).

The possibility of a pure single-phase test on a 5-limb ATR can be explored, in case the test laboratory does not allow the single-phase equivalent test, although it requires a special agreement between the manufacturer and the purchaser since it is not envisaged in the standard.

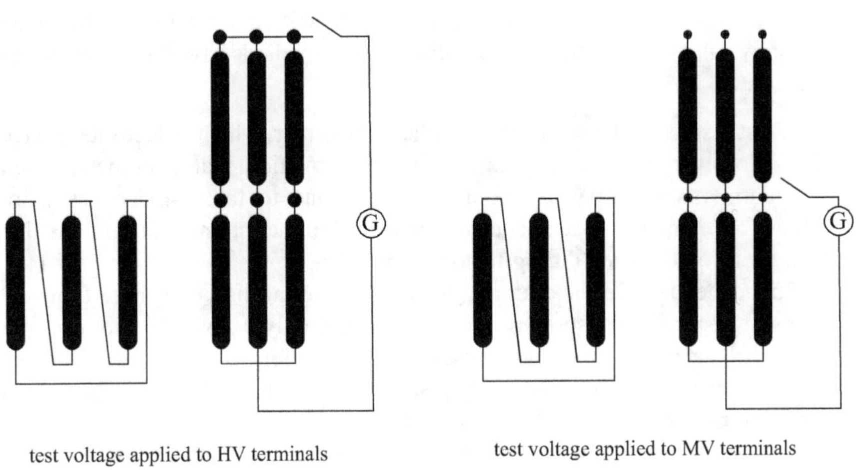

Figure 9.27 Zero sequence test setup for stabilization winding

9.3.6 Conditions for passing the test

Test voltages and currents, including at the neutral (if connected to ground or to the isolated platform), should be recorded so to obtain oscillograms, allowing to analyze transient trends. It may also be useful to connect the Buchholz relay and overpressure valve contacts to a data acquisition system to observe any untimely tripping.

At the end of each stroke, the leakage inductance is measured, which must remain within the normative limit (±1% for large power transformers) of the value measured before the start of the test. Recently, SFRA measurements have also come into use, but these do not allow for a quantitative comparison; it is suggested to carry out these only at the end of all the strokes on each phase, so as not to prolong the duration of the test excessively.

Upon completion of the short circuit tests, the device has to be returned to the factory for removal from the tank and inspection of the active part. The outcome of the inspection is favorable if there are no signs of discharges, deformations, cracks, or other internal defects that could affect the reliability of the transformer. Normally, modest displacements of the internal connections are acceptable, and these can be repositioned and refixed before proceeding with factory testing.

Short circuit withstand capability is considered conclusively proven if the transformer passes the factory dielectric tests and especially the lightning impulse test, which is the most severe test for insulation.

9.4 Verification by calculation

9.4.1 Advantages and disadvantages of calculation versus actual testing

Although the standard considers demonstrations of short circuit withstand by actual testing or by calculation to be equally valid, the two methods have some significant differences:

- The test results in abnormal stress to the transformer, which reduces its service life. However, there is no engineering confirmation that a properly sized transformer suffers permanent deterioration due to the test. The standard allows for "small displacements and changes" in the internal connections that can be accommodated during factory inspection.
- Testing is expensive, laboratories have long waiting lists, for large transformers, exceptional transportation is needed, and only a few laboratories large enough exist. Economic and logistical considerations clearly depend on the user's needs. In general, testing is justified for series transformers, for which only the prototype is tested (see next point).
- The prototype undergoing the test is built with special attention and arrangements that will not be adopted normally in series production. Thus, the test does not guarantee the same reliability for other devices, even though they

have the same design and will be built with the same techniques. Instead, verification by calculation can account for construction uncertainties by adopting appropriate safety coefficients. However, the rules of verification by design calculation, and by comparison with similar devices, contain conventional and simplifying assumptions that cannot be compensated for by simply increasing the safety coefficients.

Due to the above, verification by calculation is recommended for non-series devices, such as phase shifting transformers, ATRs, and TRs with uncommon transformation ratios, or for series devices for which there are similar specimens on which a meaningful comparison can be made, such as ATRs and TRs with similar nameplate data.

In any case, since verification involves rules of calculation and comparison that are not completely established by the standard, but left to the experience of the manufacturer, it is important that the purchasers of the machine have confidence in the same manufacturer built up through the examination of the list of tests already passed or through *design review*. To this end, it is essential to have a verification report that is sufficiently detailed to allow review by an independent third party.

9.4.2 Verification conditions

The methods of verification by calculation and design comparison are governed by the standard, which can be complemented by the Cigré technical brochure 209 [11].

Verification can be conducted in two ways:

1. The comparison of the design of the transformer under consideration with that of a similar transformer that has already successfully undergone the short circuit test, using similarity criteria described in the standard. The verification is considered passed if the expected stresses in the transformer under consideration are not greater than those expected in the reference transformer that passed the test;
2. Verification of compliance with design rules for short circuit withstand. The verification is considered passed if the expected stresses are not greater than the maximum allowable stresses of the materials and structures.

The first way offers greater assurance for the purchaser, but it is rare for the manufacturer to have available a verified transformer having similar design and construction techniques to the transformer under consideration. However, the standard allows the transformer to be compared with multiple reference transformers with different designs, in case the similarity is not applicable for all the components with a single reference. For example, a comparison is made with a first design for the series and common windings and a second design for the tertiary winding, considering two different ATRs that have passed the short circuit test.

The second way allows verification even in the absence of transformers already subjected to the short circuit test but implies a comparison with theoretical limits, which are not completely defined in the standard and therefore left to the discretion of the manufacturer.

In addition, it should be considered that the calculation of stresses is based on design rules that are defined by the manufacturer and thus are subject to a wide margin of discretion.

In the presence of axial flux, the stresses for the conductors are of two types (Figure 9.28):

- an outward shear force, which results in a tensile stress;
- an inward shear force, which results in a compressive stress.

Given the radial shear force F_{rad} from (9.11), or rather from a finite element simulation, the stress is obtained with the expression:

$$\sigma_t = \frac{F_{rad}}{2\pi a h_w}, \tag{9.17}$$

where the thickness a of the winding is related to the current density by the following formula:

$$J = \frac{N_1 I_1}{a h_w}, \tag{9.18}$$

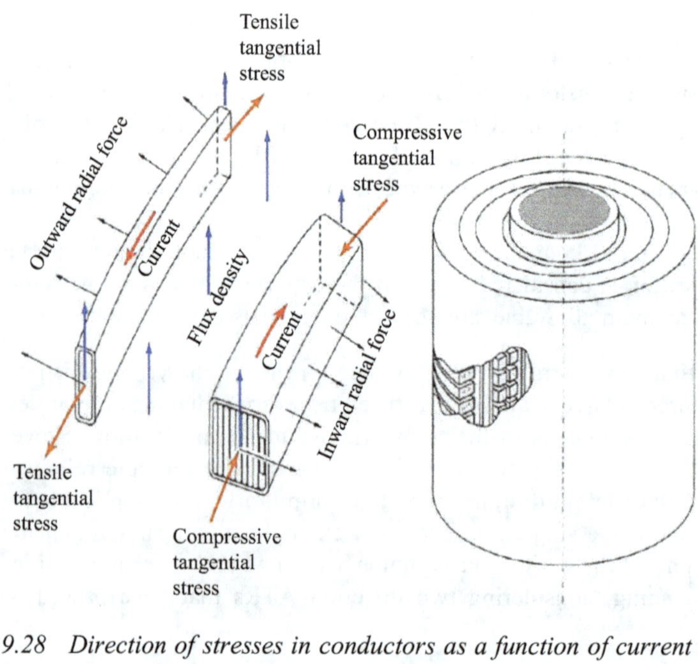

Figure 9.28 Direction of stresses in conductors as a function of current and flux

so the expression of the stress can be written as

$$\sigma_t = \frac{\mu_0}{2} N_1 I_1 J \frac{D_m}{h_w}. \tag{9.19}$$

In a disc winding, each layer is subject to a different stress, with the maximum for the outermost layer. In such a case, one can use the correction coefficient $1 + \frac{n-1}{n}$, where n is the number of layers.

The calculated stresses should be compared with the structural limits of the material used, adopting appropriate safety coefficients. Table 9.1 summarizes the relationships to be verified according to the standard. The verification must be carried out for each coil.[28]

In many cases, the standard refers to the yield stress ($R_{p0.2}$) of the material. Therefore, it is essential to have accurate mechanical properties for the conductors used. The new edition of IEC 60317-0-2 [12] for enameled conductors introduced the possibility of requesting the tensile test according to ISO 6892-1. It is recommended to request the test to be performed in this manner in order to obtain the correct yield stress.[29]

In some cases, where the standard does not provide a verification criterion, it makes reference to the Cigré brochure 209 [11], considering in particular Chapter 3. The formulas adopted in the mentioned document are given in the following for completeness:

$$\text{Buckiling } \sigma_{bu} = \frac{F_{rad}\ell^2}{2\pi R \cdot 2bt^2}, \tag{9.20}$$

$$\text{Buckiling, limit without supports } \sigma_{bu,crit} = \frac{Et^2}{4R^2}, \tag{9.21}$$

$$\text{Buckiling, limit with supports } \sigma_{bu,crit} = \frac{Et^2 N^2}{48R^2}, \tag{9.22}$$

$$\text{Slipping } \sigma_{te} = \frac{F_{ax}L^2}{2\pi R \cdot 2tb^2}, \tag{9.23}$$

$$\text{Slipping, limit } \sigma_{te,\ crit} = \frac{Eb^2}{14R^2} + \frac{msE_c t^2}{12\pi Rb^2}, \tag{9.24}$$

where

- ℓ is the distance between two adjacent radial supports;
- L is the distance between two consecutive axial supports (spacers);

[28] Please note that some windings may consist of several coils (e.g., fine and coarse regulation).

[29] The previous edition of IEC 60317-0-2 included a test mode that did not conform to ISO 6892-1; therefore, the value resulting from the tensile test was not the yield stress.

Table 9.1 Design verification criteria for core–core transformers

Type of force/effort	Relationship
Mean hoop tensile stress on outer coils (MPa)	$\sigma_t \leq 0.9\ R_{p0.2}$
Mean hoop compressive stress on disc-, helical-, single-layer-type inner coils (MPa)	• Unbonded $\sigma_c \leq 0.35\ R_{p0.2}$ • Bonded $\sigma_c \leq 0.6\ R_{p0.2}$
Equivalent mean hoop compressive stress on layer-type inner coils (MPa)	• Unbonded $\sigma_{c,eq} \leq 0.35\ R_{p0.2}$ • Bonded $\sigma_{c,eq} \leq 0.6\ R_{p0.2}$
Stress due to radial bending on conductors in the span between axial sticks and between spacers used to build any axial cooling ducts within the coil radial width (MPa)	$\sigma_{br} \leq 0.9\ R_{p0.2}$
Stress due to axial bending on conductors in the span between radial spacers with disc- and helical-type coils (MPa)	$\sigma_{ba} \leq 0.9\ R_{p0.2}$
Thrust force acting on the coil leads (kN)	v. TB 209 ch. 3
Axial compressive force on each physical coil (kN)	$F_{ax} \leq F_{ax,crit}$
Axial compressive force on coil compared to critical force for torsional failure of conductors (kN)	$F_c \leq 0.8\ F_{tilt}$
Thrust force on physical coil heads: • Upward (kN) • Downward (kN)	$F_t \leq 0.8\ F_{press}$
Compressive stress on radial spacers with disc- and helical-type coils (MPa)	• Paper-coated $\sigma_{sp} \leq 80$ MPa • Enameled $\sigma_{sp} \leq 120$ MPa
Compressive stresses on end stack insulation structures (MPa)	• Wrapped type $\sigma_{er} \leq 40$ MPa • Stacked type $\sigma_{er} \leq 80$ MPa
Compressive stress on conductor paper insulation with layer-type coils	$\sigma_{pi} \leq 35$ MPa
Compressive stresses on end stack insulation structures end rings (MPa)	• Wrapped type $\sigma_{es} \leq 40$ MPa • Stacked type $\sigma_{es} \leq 80$ MPa
Compressive stress on common press rings (or plates) (MPa)	$\sigma_{pr} \leq 80$ MPa
Tensile stress on tie rods (flitch plates) of the clamping structure (MPa)	$\sigma_{rod} \leq R_{eL} \approx R_{p0.2}$
Clamping force per column (kN)	$F_{clamp} \leq F_{clamp,\ crit}$
Free or forced buckling (MPa)	$\sigma_{bu} \leq \sigma_{bu,\ crit}$ (see TB 209 ch. 3)
Slipping (MPa)	$\sigma_{te} \leq \sigma_{te,\ crit}$ (see TB 209 ch. 3)
Spiraling (MPa)	$\sigma_{ti} \leq \sigma_{ti,\ crit}$

- b is the axial height of the conductor;
- t is the radial thickness of the conductor;
- R is the average radius of the coil;
- N is the number of radial supports;
- E is the modulus of elasticity of the conductor and E_c that of the paper/pressboard (Pa);
- m is the number and s thickness of the spacers.

Recent studies [13] have shown that the formula for $\sigma_{te,crit}$ in many cases underestimates the actual value of the stress. In cases where the standard and the Cigré brochure do not provide sufficient detail, the manufacturer must define the critical limit value (F_{crit}, F_{tilt}, F_{press}). Normally, the maximum allowable compressive force is calculated using *Euler's formula* (also called Timoshenko's formula):

$$F_{ax,crit} = \frac{\pi^2 EI}{H_w^2}, \tag{9.25}$$

where the moment of inertia is worth $I = \frac{bt^3}{12}$.

The method of calculating stresses in complex cases is necessarily left to the manufacturer, who depending on his or her experience will adopt different approaches with the use of theoretical formulas (e.g., De Saint-Venant's cases), or semiempirical ones including safety coefficients, or through the systematic use of finite element simulation (Figure 9.29).

In more complex cases, some manufacturers resort to coupled magnetic-mechanical simulations, in which the magnetic induction is more realistically calculated as a function of the mechanical strain and thus of the actual geometry of the windings (Figure 9.30).

Simulations do not allow for counter-verification in design review. Therefore, they should be supported by sufficient experience in performing successful short circuit tests. Safety coefficients, which can be discussed during the review, should also be adopted after gaining sufficient experience with actual tests and not only on the basis of theoretical considerations.

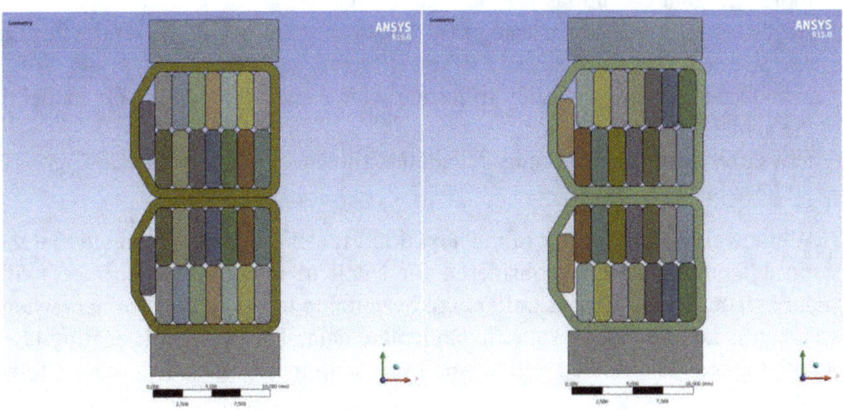

Figure 9.29 Simulation of a continuously transposed cable, without and with offset between strands (right)

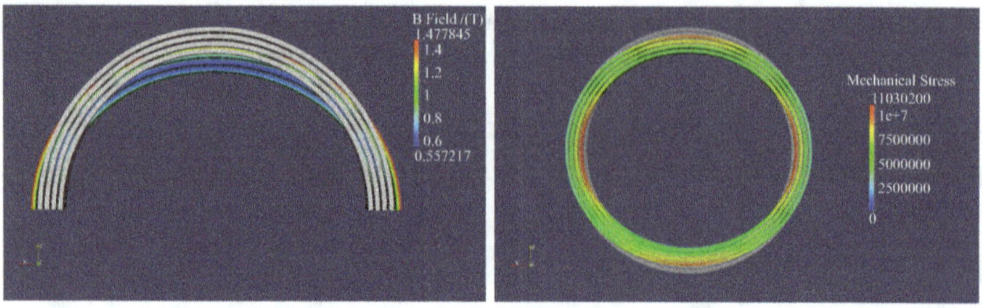

Figure 9.30 Magnetic induction and mechanical strain calculated considering conductor strains

In conclusion, the choices of design criteria remain largely the responsibility of the manufacturer, and the design review is an opportunity to focus attention on the most significant design and construction criticalities, leaving to each actor in the process (manufacturer, purchaser, and the possible consultant), the role that belongs to each, without wanting to impose alternative construction solutions in the absence of a justified reason.

9.5 Prospects for developments in the standardization

IEC 60076-5:2006 has been under revision for a few years. The revision work is difficult because of the need to meet the opposing requirements of making the test meaningful without unnecessarily stressing the transformer. New features included in the first draft of new edition are as follows:

- more stringent conditions for thermal verification;
- review of the transformer categories based on rated power, new rules for the calculation of the reactances in the peak factor, new tolerances;
- introduction of the possibility of testing with a *mock-up*, i.e., a scale model of the actual transformer;
- new criteria for verification by calculation (however, the reference to Cigré TB 209 has been removed).

While waiting for the text of the next draft, it can only be pointed out that the test on a mock-up can be considered for small to medium transformers with standard characteristics but is unlikely to be suitable for large transformers where the design is adapted to the specific nameplate data. The applicable scaling laws to make the results obtained with a small mock-up meaningful also remain to be demonstrated.

It would be desirable for the following topics to be examined as well:

- differentiation between three-phase test and single-phase equivalent test;

- possibility of testing in pure single-phase configuration,[30] at least for 5-limb transformers.

In particular, it would be desirable to reduce the number of strokes provided for the three-phase test, since this involves higher stresses. A compromise might be one stroke per phase, but with a test duration twice that of the equivalent single-phase configuration, for example, 0.5 s instead of 0.25 s for large power transformers.

Finally, the experience gained by manufacturers should make it possible to arrive at more defined and precise rules of verification by calculation that would avoid discussions during design review. However, this goal is hindered by the reluctance of manufacturers to share their internal knowledge about transformer design.

References

[1] IEC 60076-5: *Power Transformers – Part 5: Short Circuit Withstand Capability*, 4th ed., 2006.
[2] IEC 60909-0: *Short Circuit Currents in Three-Phase Alternating Current Systems – Part 0: Calculation of Currents*, 3rd ed., 2016.
[3] G. Bertagnolli, G. Caprio, A. Inesi, and G. Zafferani: *Short Circuit Withstand Capability of Large Power Transformers*, Cigré paper, 1990.
[4] T. Aragona, A. Babare, G. Doglio, U. Di Marco, G. Redaelli, and M. Ubaldini: *Importance of short circuit tests for large-power autotransformers*, The Electric Energy, no. 74-5, September-October 1997.
[5] IEC 60076-3: *Power Transformers – Part 3: Insulation Levels, Dielectric Tests and External Clearances in Air*, 3rd ed., 2013.
[6] A. Shadowitz: *The Electromagnetic Field*, McGraw-Hill, 1975 (reprint Dover, 1988).
[7] G. Bertagnolli: *Short Circuit Duty of Power Transformers*, ABB, 3rd ed., 2006.
[8] T. Fogelberg: *Surviving a Short Circuit*, ABB Review 1/2008.
[9] G. Caprio, and G. Cannavale: *Tests of Tightness to Electrodynamic Stresses on Large Power Transformers*, ENEL Report, 1986.
[10] Terna Technical Note INS MA N 04: *Short Circuit Test of Three- and Five-Limb Autotransformers Without Tertiary with Equivalent Single-Phase Scheme*, 05/13/2011.
[11] Technical brochure (TB) 209: *The Short Circuit Performance of Power Transformers*, Cigré, 2002.
[12] IEC 60317-0-2: *Specifications for Particular Types of Winding Wires – Part 0-2: General Requirements – Enamelled Rectangular Copper Wire*, 4th ed., 2020.

[30]In the current edition, this possibility is mentioned only in a footnote (footnote 3 to Section 3.2.5.4), so such a test does not allow to declare compliance with the standard.

[13] B. Bošnjak, and M. Stuhne: *Analysis and experimental verification of static tilting behavior of CTC in transformer windings*, 3rd International Colloquium Transformer Research and Asset Management, Split (Croatia), October 2014.

[14] B. Bošnjak, A. Hauck, and H. Landes: *Coupled magneto-mechanical finite element analysis of a power transformer in short circuit conditions*, 3rd International Colloquium Transformer Research and Asset Management, Split (Croatia), October 2014.

Chapter 10

Acoustic noise emitted by transformers and reactors

10.1 Introduction

Electrical devices such as autotransformers, transformers, and reactors are a source of acoustic noise in and around electrical substations.

Electrical substations are normally located outside the urban perimeter; however, the increasing demand for land urbanization implies both bringing housing closer to existing substations and making more difficult to find areas without human presence to build new substations.

The social sensitivity to noise has introduced a regulatory risk for transmission system operators (TSOs) with respect to noise legislation, which can vary over time, and the introduction of new corporate environmental goals regarding public acceptability toward electrical installations. Therefore, TSOs must find and adopt technological and layout solutions to address these issues without penalizing the operational needs of the electrical system.

This chapter describes different possible solutions for the mitigation of noise emitted by electrical devices. The results of the experiments allow some evaluations, with the goal to identify the most promising solutions and direct future study activities.

10.2 Principles of acoustics

10.2.1 Propagation of sound

Sound is a change in pressure within an elastic medium (such as air) that propagates longitudinally. The propagation of the pressure changes causes the transmission of sound. According to the Boyle–Mariotte gas law, a change in pressure corresponds to a proportional change in density in the gas (Figure 10.1).

Pressure changes with a periodic pattern generate sound in the strict sense of the term, while pressure changes with an aperiodic pattern generate what is perceived as noise.

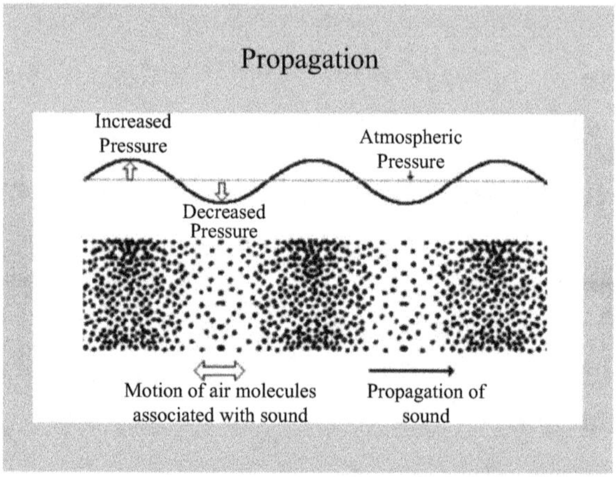

Figure 10.1 Propagation of sound

10.2.2 Physical quantities

Sound power W is the power emitted by the sound source and is measured in watts (W).

Sound pressure p is the force of the acoustic wave incident perpendicularly on a surface and is measured in pascals (Pa).

Sound intensity I is the sound power impinging perpendicularly on a surface of area A and is measured in watts per square meters (W/m^2)

$$I = \frac{P}{A}. \tag{10.1}$$

Customarily, the previous quantities are expressed in decibels and related to a reference value to obtain power, pressure, and sound intensity levels:

$$L_W = 10log\frac{W}{W_0}\ [\text{dB}],\ W_0 = 10^{-12}\text{W}, \tag{10.2}$$

$$L_p = 10log\frac{p^2}{p_0^2}\ [\text{dB}], p_0 = 20 \cdot 10^{-6}\text{Pa}, \tag{10.3}$$

$$L_I = 10log\frac{I}{I_0}\ [\text{dB}], I_0 = 10^{-12}\text{W}/\text{m}^2. \tag{10.4}$$

The sensitivity of the human ear to sound intensity is logarithmic, and thus the use of a logarithmic scale makes easier to express the high range of variability of measured values in acoustics (Figure 10.2).

In addition, the sensitivity of the human ear varies with the frequency of the perceived sound. Using experimental measurements, weighting curves have been defined to weigh the values measured at different frequencies (Fletcher and

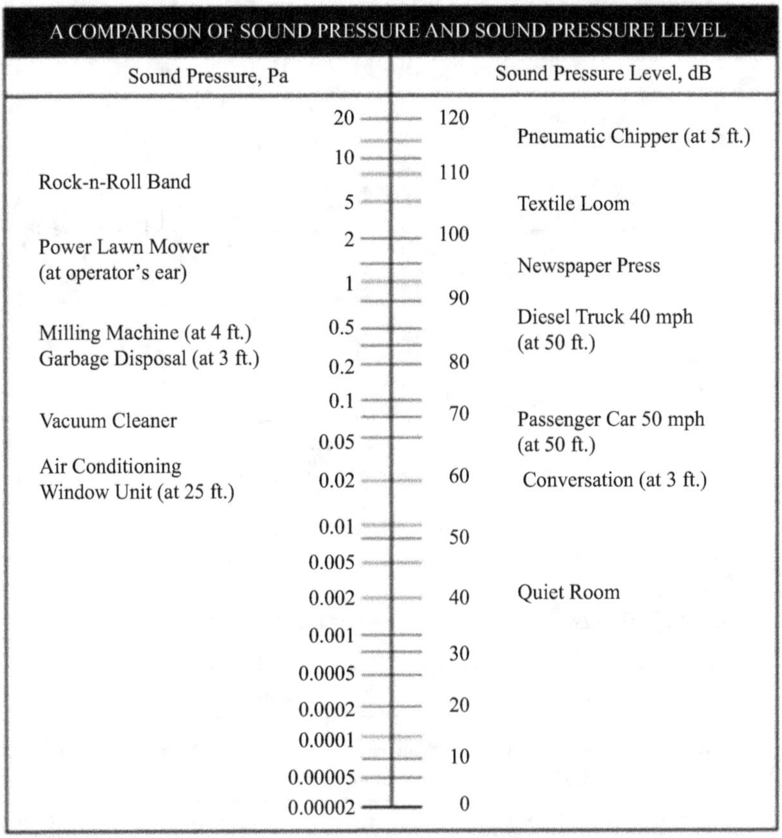

Figure 10.2 Examples of sound pressure levels

Munson studies). The commonly used curve is the "A" curve, on which the legal limits for sound emissions are based (Figure 10.3).

Weighting curves introduce corrections at different frequencies. The frequency spectrum of a measured sound can be divided into *octave* or *one-third octave bands*, which correspond to bands of amplitude frequencies in geometric proportion, as shown in Table 10.1.

Each third-octave band can be associated with a weighting value C_i, as shown in Table 10.2.

Therefore, the weighted sound pressure level in dB(A) is obtained by logarithmically summing the measured values in dB at the different one-third octave bands L_{p_i} with the weighting values:

$$L_{p_A} = 10log\left[\sum_{i=1}^{N} 10^{\frac{L_{p_i}+C_i}{10}}\right] \,[\mathrm{dB(A)}]. \tag{10.5}$$

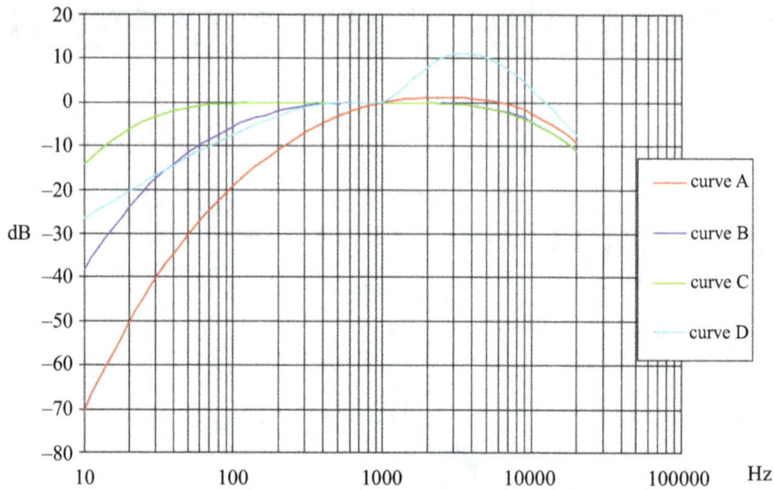

Figure 10.3 Acoustic weighting curves

Table 10.1 Bands of octave and one-third octave

One-third octave		Octave	
Central frequency (Hz)	**Band limits (Hz)**	**Central frequency (Hz)**	**Band limits (Hz)**
25	22.4–28.2	31.5	22.4–44.7
31.5	28.2–35.5		
40	35.5–44.7		
50	44.7–56.2	63	44.7–89.1
63	56.2–70.8		
80	70.8–89.1		
100	89.1–112	125	89.1–178
125	112–141		
160	141–178		
200	178–224	250	178–355
250	224–282		
315	282–355		
400	355–447	500	355–708
500	447–562		
630	562–708		
800	708–891	1,000	708–1,410
1,000	891–1,120		
1,250	1,120–1,410		
1,600	1,410–1,780	2,000	1,410–2,820
2,000	1,780–2,240		
2,500	2,240–2,820		
3,150	2,820–3,550	4,000	2,820–5,620

(Continues)

Table 10.1 (Continued)

One-third octave		Octave	
Central frequency (Hz)	**Band limits (Hz)**	**Central frequency (Hz)**	**Band limits (Hz)**
4,000	3,550–4,470		
5,000	4,470–5,620		
6,300	5,620–7,080	8,000	5,620–11,200
8,000	7,080–8,910		
10,000	8,910–11,200		
12,500	11,200–14,100	16,000	11,200–22,400
16,000	14,100–17,800		
20,000	17,800–22,400		

Table 10.2 Weighing values for one-third octaves

Central frequency (Hz)	A-weighting (dB)
25	−44.7
31.5	−39.4
40	−34.6
50	−30.2
63	−26.2
80	−22.5
100	−19.1
125	−16.1
160	−13.4
200	−10.9
250	−8.6
315	−6.6
400	−4.8
500	−3.2
630	−1.9
800	−0.8
1,000	0
1,250	+0.6
1,600	+1.0
2,000	+1.2
2,500	+1.3
3,150	+1.2
4,000	+1.0
5,000	+0.5
6,300	−0.1
8,000	−1.1
10,000	−2.5
12,500	−4.3
16,000	−6.6
20,000	−9.3

Figure 10.4 Sound level meter

10.2.3 Measuring instruments

Sound pressure is measured using an instrument called a *sound level meter*, which converts sound pressure (averaged over a specified time interval), into an electrical signal that can be viewed on the instrument's display (Figure 10.4). The instrument is capable of applying the acoustic weighting filter, directly returning a value in dB(A). In addition, the sound level meter can be equipped with a bank of filters, in octave or one-third octave bands, which enable the measurement of the sound spectrum of a noise signal and the detection of any tonal components, i.e., frequencies at which the noise is particularly intense compared with adjacent frequencies.

10.3 The noise regulations

As an example of a representative case, the Italian regulations [1] stipulate that the land is divided into areas according to its class of use. Each area is associated with a limit value of noise pressure level, according to Table 10.3.

Each TSO must evaluate the acoustic characteristics of electric substations to verify compliance with emission limits. However, the acoustic classification may change over time (e.g., when the use of areas changes), and this may cause substations initially compliant with the limits to violate the revised limits. As a result, it may be necessary to adopt technical and layout solutions to reduce the noise perceived outside the facility.

In case of installation of new devices in existing substations, it is necessary to verify compliance with the limits of overall disturbance.

It is pointed out that the nighttime limits are particularly stringent. This affects the installation of shunt reactors that are very noisy and are mainly used during the nighttime hours of low electrical demand.

Table 10.3 Limits of noise levels in Italy

Land use classes	Daytime limit L_{pA} [dB(A)]	Nighttime limit L_{pA} [dB(A)]
I—Especially protected areas	45	35
II—Predominantly residential areas	50	40
III—Mixed-type areas	55	45
IV—Areas of intense human activity	60	50
V—Predominantly industrial areas	65	55
VI—Exclusively industrial areas	65	65

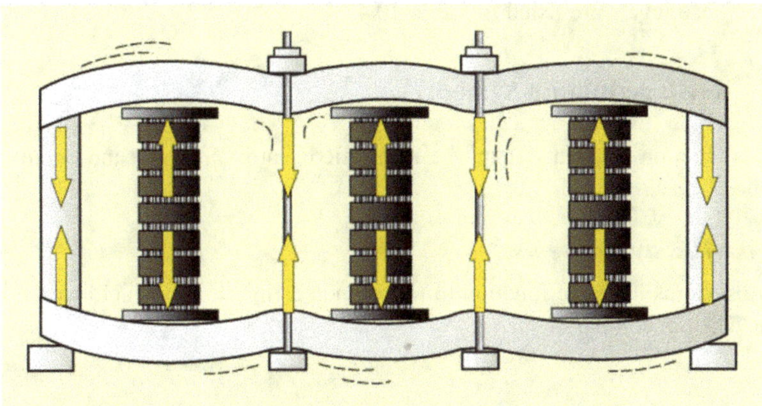

Figure 10.5 Amplified representation of vibrations on a reactor

10.4 Noise emission by electric devices

10.4.1 Characteristics of the emitted noise

The noise emitted by the electric devices is mainly caused by two factors:

- magnetostriction of the core;
- noise from the cooling system (pumps and fans).

Magnetostriction is a change in the configuration of the magnetizable material when it is immersed in a magnetic field. If the magnetic field changes periodically, the material also deforms in an oscillatory manner, causing vibrations of the transformer and, consequently, a sound emission (Figure 10.5). The magnetostriction force has twice the frequency of the magnetic field, so oscillations, and therefore noise, occur at the fundamental frequency of 100 Hz and higher harmonics.

In gapped-core reactors, besides the magnetostriction problem there are the electrodynamic stresses at the various air gaps in the magnetic core's gapped limbs, which are adopted to prevent saturation of the ferromagnetic material.

A study [2] confirmed that for the transformer sound source, the main emissions are at the frequencies 100, 200, and 300 Hz. Instead, for the reactor sound source, the main emission frequency is more concentrated at 100 Hz and, to a lesser extent, at 200 Hz.

Conversely, the noise produced by the cooling system can be likened to white noise, that is, it has no dominant tonal frequency.

10.5 Examples of technical requirements

The technical specifications issued by TERNA, the Italian TSO, for the purchase of electric devices prescribe noise emission limits expressed as sound power levels in dB(A). These levels are listed in Table 10.4.

10.6 Noise reduction systems

Noise reduction systems can be classified into *passive* and *active*, as described below.

10.6.1 Passive systems

Passive systems have the function to reduce noise emission by shielding the noise source. They encompass:

Table 10.4 Limits of sound power emissions for electric devices purchased by TERNA

Device	Cooling system	Year of issue	Sound power limit [dB(A)]
ATR 400/230 kV 600–400 MVA ATR 400/155–135 kV 250 MVA	OFAF	2014–	90
ATR 230/155–135 kV 250–400 MVA	OFAF	2014–	88
ATR (all types)	OFAF	2010–2014	92
ATR (all types)	OFAF/ONAF	1993–2010	95
ATR (all types)	OFAF	1986–1993	110
Single-phase 400 kV 86 Mvar reactors	ONAN	2007–	88
Single-phase 230 kV 60 Mvar reactors	ONAN	2007–	85
HV/MV transformers 63 MVA	ONAN	2011–	75
HV/MV transformers 40 MVA	ONAN	2011–	70
HV/MV transformers 25 MVA	ONAN	2011–	67
HV/MV transformers 16 MVA	ONAN	2015–	66

In the case of reactors, TERNA's bay configuration includes three single-phase units, so a factor 4.8dB = 10*log*3 must be added to the value of the single-phase unit to get the overall value of the three-phase bank.

- on-tank paneling;
- sound-absorbing walls;
- vibration damping systems.

Panels have a small footprint but can limit accessibility to accessories installed on the tank of the electric device. In addition, the possibility of retrofitting with paneling is limited, so they are usually included early in the design stage.

Sound-absorbing walls can limit cooling efficiency by reducing air circulation and thus heat-disposal capacity, if not properly spaced from the device for reasons of limited space. They have the advantage of leaving the possibility to install them after the device is assembled, and this is what usually happens when more stringent acoustic constraints have to be met (Figure 10.6).

Vibration damping systems consist of materials with anti-vibration properties, such as natural or synthetic rubber, which are mounted on accessory supports at the edge of the tank or between the tank and the supporting feet. They mainly have the function of damping vibrations transmitted to the foundation (Figure 10.7).

According to ISO 9613-1, the noise perceived by a receptor can be expressed as follows:

$$L_p = L_W + D_I - A, \tag{10.6}$$

where

- L_p is the sound pressure level perceived by the receptor;
- L_W is the average sound pressure level produced by the sound source at the distance of the receptor;
- D_I is the directivity index, which describes how much the source sound pressure level at the receptor point differs from the mean value at the same distance;
- A is the attenuation.

The attenuation of noise by barriers depends mainly on the height and absorption capacity of the material from which they are made. In particular, attenuation is a function of the difference between the distance of the receptor from the source with and without a barrier.

According to Maekawa's formula, the resulting attenuation is given by

$$\Delta L = 10log\frac{20m}{\lambda}(a + b - c), \tag{10.7}$$

where λ is the wavelength of sound (given by the speed of sound divided by its frequency), and the parameters a, b, and c are defined as shown in Figure 10.8.

Therefore, according to Maekawa's theory, the greater the height of the barrier in comparison with the distance to the receptor, the greater the attenuation, as can be seen from the graph shown in Figure 10.9. Noise reductions on the order of 4–10 dB(A) can be achieved with this method ([3] p. 456).

Figure 10.6 Sound-absorbing paneling around a reactor

10.6.1.1 Experimentation with on-tank dampers

TERNA, the Italian TSO, has experimented the installation of dampers fixed on the tank of a reactor to achieve a distributed vibration damping. In Figure 10.10, the experimental installation is shown.

The observed noise reduction was limited to 0.5–1 dB(A). This is because the dampers cover only a small part of the tank's surface. The dampers were effective in reducing vibration at 100 Hz because they were designed in accordance with that

Figure 10.7 Foot with anti-vibration baseplate

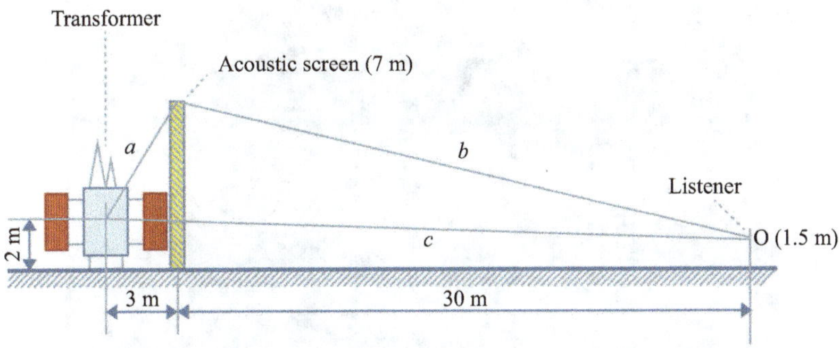

Figure 10.8 Diagram of the relevant distances in Maekawa's theory

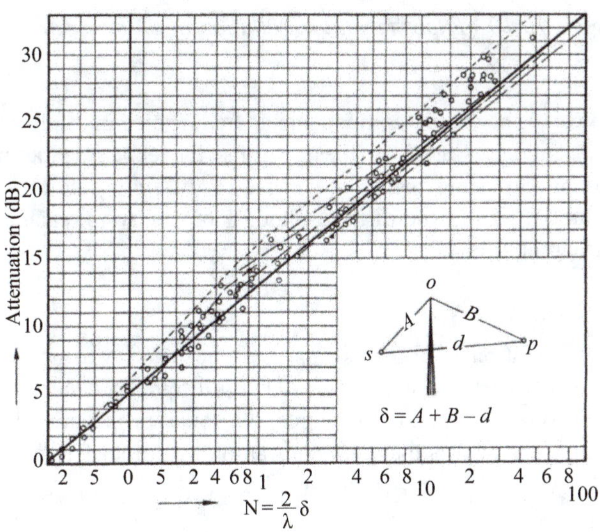

Figure 10.9 Maekawa attenuation graph [4]

Figure 10.10 Experimental installation of on-tank dampers

frequency. However, harmonic components at frequencies multiple of 100 Hz are not effectively damped. Based on these results, the experimentation should be further refined by exploring the effectiveness of different solutions in terms of number, installation, sizing, and distribution of dampers, so to reach a target value of a sufficient damping level.

10.6.2 Active systems

Active noise cancellation (ANC) systems use loudspeakers to generate sound in counterphase to that of the noise source. This is useful to cancel the noise at a specific point or to reduce, in a defined area, the sound propagation by means of a superposition of sound waves that has a substationary point of zero at the receptor (Figure 10.11).

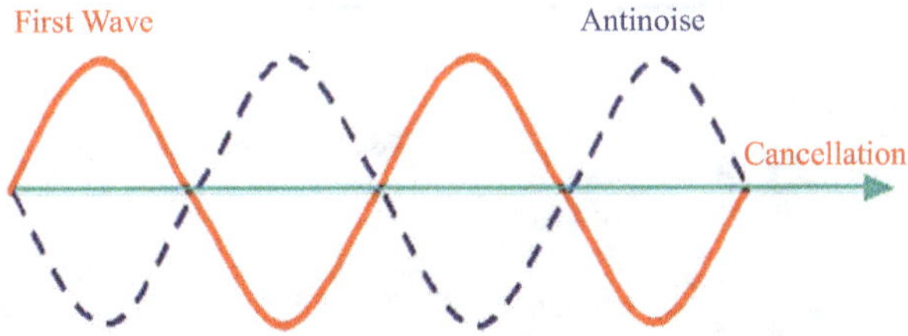

Figure 10.11 Principle of active noise cancellation

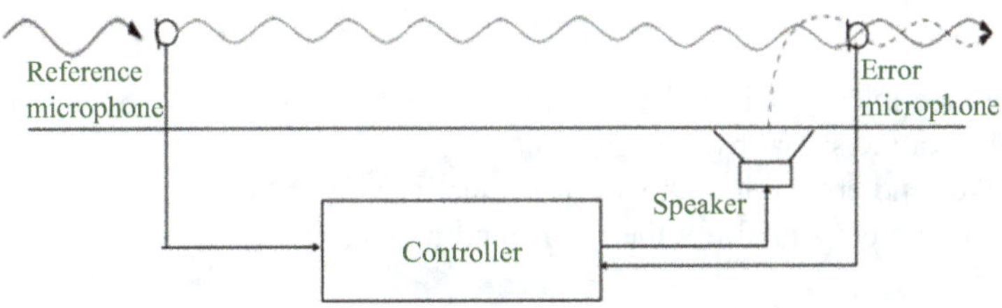

Figure 10.12 Schematic diagram of the active compensation system

The system consists of the following components (Figure 10.12):

- one or more loudspeakers placed near the receptor;
- one or more microphones at the sound source (*reference microphone*);
- a *controller* that, based on the measured noise, generates the signal for the loudspeakers;
- one or more microphones at the receptor (*error microphones*) to be used for calibration and feedback signal to the controller.

Depending on the number of microphones present, several solutions are possible:

- *point-to-point*: point sound source (one reference microphone) and point receptor (one error microphone);
- *zone-point*: distributed sound source (multiple reference microphones) and point receptor;
- *point-zone*: point sound source and distributed receptor (multiple error microphones);
- *zone-zone*: distributed sound source and receptor.

For the system to be effective, it is essential that the signal processing time by the controller is shorter than the transmission time of the sound wave. In addition, the system must be able to distinguish the noise emitted by the source from the echo emitted by the loudspeaker (feedback effect).

Calibration of the system consists of determining the transfer functions (Figure 10.13) between

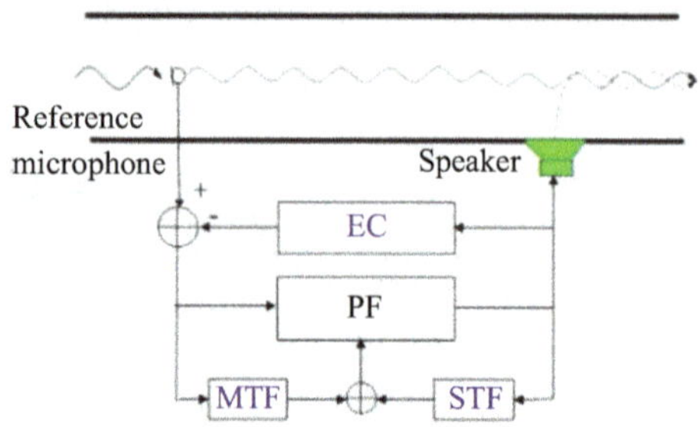

Figure 10.13 Schematic diagram of the transfer functions

- reference microphone and error microphone (MTF function), by measurement of the source signal by the error microphone;
- speaker and error microphone (STF function), by measurement of the speaker signal (sample signal) by the error microphone.

The system calculates the echo cancellation (EC) transfer function and the predictive speaker filter (PF).

The estimated noise reduction with this system is not predictable because it is dependent on the type of calibration performed, the location of the error microphones, and the location of the reduction measurement. It is also dependent on the frequency spectrum of the source, with greater effectiveness at low frequencies. Therefore, it is best to calibrate the system for a specific frequency band. The system tested by TERNA contains filters for different frequency bands, up to 7812 Hz (the maximum sampling frequency of the instrument).

10.6.2.1 Experimentation with an active system

An ANC system was initially experimented with sample sound sources generated from loudspeakers emitting a reactor-like noise and arranging the loudspeaker close to the source (Figure 10.14).

The observed reduction reached a maximum of 10 dB(A) around the error microphone, placed 1.5 m above the ground and 3 m from the source. The effectiveness of the system is drastically reduced at distances greater than 13 m, with attenuation <5 dB(A), because a single loudspeaker cannot compensate for the diffuse noise at these distances. Even with a microphone 3 m above the ground, the effectiveness of the system is less than 5 dB(A).

The described experimental system was installed for noise mitigation of a 258 Mvar, 400 kV single-phase reactor bank (Figure 10.15).

For this facility, the following solutions were planned to reduce noise disturbance:

- installation of walls with sound-absorbing barriers (Figure 10.16);

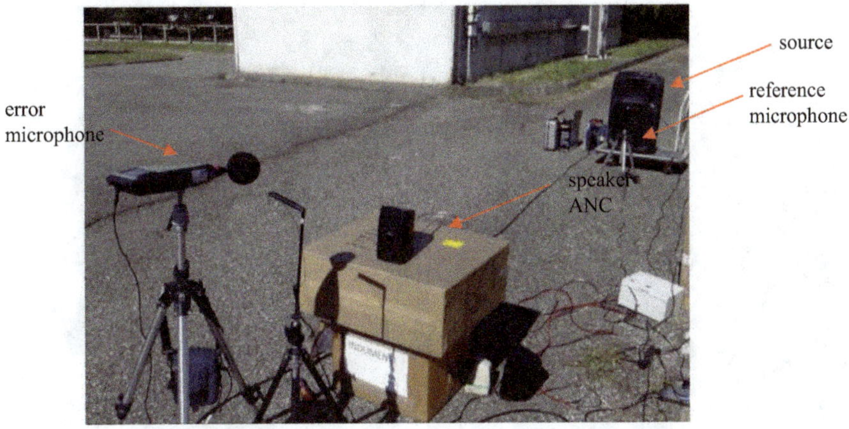

Figure 10.14 Arrangement of the test setup of the ANC system

Figure 10.15 Overall view of a shunt reactor bay in 400 kV substation

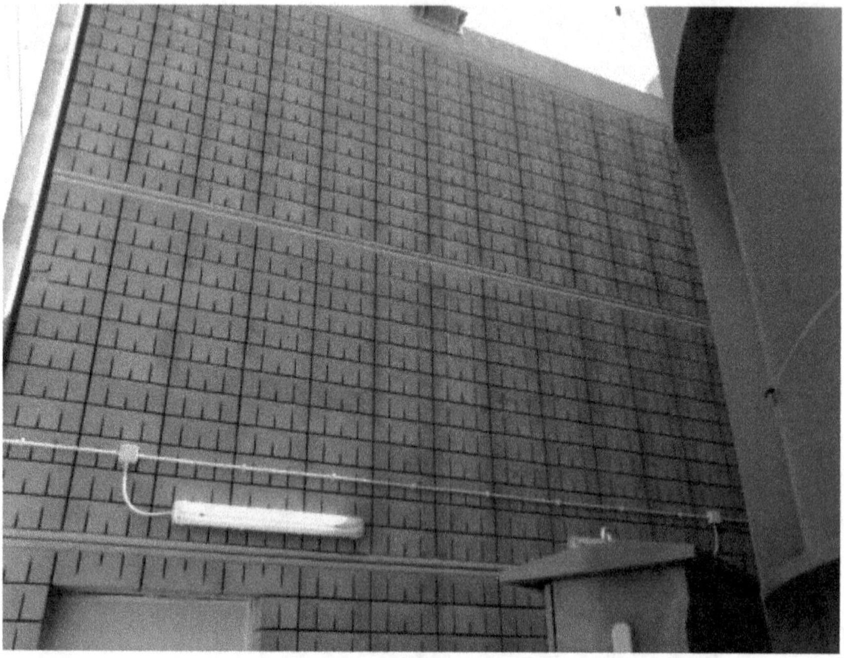

Figure 10.16 Interior view of walls with sound-absorbing panels

- installation of feet with anti-vibration baseplate;
- installation of active compensation system with loudspeakers on the top of the walls (Figure 10.17).

Further experimental steps consist of installing speakers on the outer sides of the walls and reference microphones on the inner sides. To supplement this, there are accelerometers installed directly on the reactor tank, whose signal will be converted so to make it comparable with the acoustic signal of the microphones. Error microphones for calibration are placed at the perimeter of the substation.

In this present experiment, the noise source is large, and noise reduction is carried out for the large area of the substation perimeter. Therefore, the expected reduction is less than that obtained in the experiment with sample sources as described above.

10.6.3 Sound-absorbing panels on the tank

A 150 kV submarine cable connection in Southern Italy required the installation of a 37 Mvar shunt reactor in a distribution substation placed in a heavily urbanized environment. Therefore, the reactor manufacturer was asked to investigate a solution to reduce the noise impact without compromising the cooling capacity of the device. The result is a sound-absorbing system of panels enclosing the tank (Figure 10.18), which does not compromise accessibility to accessories on the tank (such as thermostats). The reactor was placed inside a building for added sound

Figure 10.17 Exterior view of walls with setup for active compensation system

Figure 10.18 The substation reactor with sound-absorbing panels

Figure 10.19 External battery of radiators

mitigation, while the battery of radiators was placed outside to allow effective heat exchange with the environment (Figure 10.19).

10.7 Conclusions regarding the experiments

Noise emission by transformers and reactors at electrical substations is a major problem because it can hinder or limit the operation of the devices if the noise exceeds the regulatory limits in the installation zone.

There are some possible technical solutions to reduce noise, whose different effectiveness has been tested by the Italian TSO as follows:

- tank paneling: estimated reduction 5–6 dB(A);
- walls with sound-absorbing panels: estimated reduction 4–10 dB(A), depending on receptor location;
- anti-vibration feet: limited estimated reduction, but useful for reducing vibrations transmitted to the foundation;
- on-tank dampers: estimated reduction 1 dB(A), as they do not cover the entire radiating surface;
- ANC systems: still experimental with variable effectiveness, to be assessed in the specific zonal reduction.

In addition, a low-noise transformer or reactor can be specified from the outset. Some design expedients are as follows: the adoption of low-magnetostriction steel sheets, reduced working induction, and improved core clamping; all of which can reduce the sound power by 10–15 dB(A) ([3] p. 455), but are more expensive than external noise reduction structures.

All the previously described actions have advantages and disadvantages. Therefore, the most effective solution, for a certain system and for a certain level of

Figure 10.20 Example of an encapsulated autotransformer with the exchangers outside the containment

attenuation, must be identified on a case-by-case basis, as a combination of different possibilities. If this is not sufficient, changes in the design of electrical devices should be implemented for the part concerning the cooling system, which most contrasts with the noise reduction solutions.

The adoption of panels, walls, and encapsulation is always the preferred solution because it is static, more effective, and has a precisely calculable a priori effect. However, it always implies verification of the efficiency of the cooling system, which must not be reduced. In unfavorable cases, such as those of encapsulation required to meet small space constraints (Figure 10.15), the modification of the cooling system is necessary, such as

- moving the cooling system outside the containment on the side not penalized by the noise limits, in cases where the receptive area is localized in a specific direction and the noise of the cooling system itself does not exceed the limits (Figure 10.20);
- use of radiators and heat exchangers with vertical airflow, instead of the usual horizontal flow, in cases where the machine is fully encapsulated, when the cooling system is by itself also a source of noise to be limited;
- use of the ONAN or OFAN/ODAN system, always to be preferred for new installations and where space permits, because of the added value of energy sustainability and greater reliability.

In any case, the installation of panels or barriers must not reduce the functionality of the transformer; in particular, air exchange to the cooling system must be guaranteed. It is recommended that walls be installed at least 3 m from the cooling system. In addition, reflective surfaces placed at multiple wavelength distances (1.7 m at 50 Hz) should be avoided to prevent echo and standing wave phenomena. Alternatively, holes on the panels should be left for air exchange, but these are also pathways for noise to pass through and thus decrease the effectiveness of damping.

References

[1] Italian Government's decree (DPCM) dated 14 November 1997, *Determination of Limit Values of Sound Sources*.
[2] F. Borchi, S. Luzzi, M. Goretti, *et al.*: *Mitigating Electrical Substation Noise*, Technical Journal, "ProgettandoIng", 2015, p. 15–23.
[3] Alstom, Power transformers, vol. 2 Expertise, 2011.
[4] G. Alfaro Degan: *Acoustic Barriers: Performance and Experimental Measurements*, Tecniacústica Conference, Bilbao, 2003.

Chapter 11
DC core saturation of converter transformers

11.1 Introduction

Renewable energy plants and HVDC links are connected to the HVAC electric transmission grid with AC/DC conversion systems, which generate and feed into the AC grid harmonic current components that disturb the operation of electrical devices, including interconnection power transformers. The direct current component appears to be very dangerous because it can cause core saturation with worsening of voltage quality and possible localized heating inside the active part.

This chapter describes the problem of possible saturation of converter transformers in HVDC substation which, at present, are the most affected electric devices.

11.2 Description of an HVDC connection

TERNA, the Italian TSO, has planned to build an HVDC link on an international interconnection with another grid, at a voltage of ±320 kV to transmit a rated power of 600 MW.

The route chosen for the connection includes running the cable through tunnels and under highway viaducts, so it was decided to use a cable with a low specific weight such as extruded insulation (XLPE) cable (Figure 11.1), rather than using the heavier traditional cable with paper-oil insulation.

However, unlike paper-oil cables, XLPE cables do not withstand the rapid voltage reversal that is adopted in the traditional HVDC system, namely the LCC (*line commutated converter*) type, to change the direction of the power transmitted over the link. This weakness results from the high depolarization time constant of the XLPE cable screen, which, during the rapid transient of voltage reversal, causes a high overvoltage in the insulation between the outer screen and the inner conductor.

This problem prevents the use of XLPE cables in HVDC systems that adopt LCC-type conversion systems, which use power thyristors to rectify the voltage. Therefore, the choice of cable directed toward the type of the conversion system with the technology called VSC (*voltage source converter*) [1], which uses IGBT (*insulated gate bipolar transistor*), to reverse the current flow without changing

Figure 11.1 Pair of XLPE cables

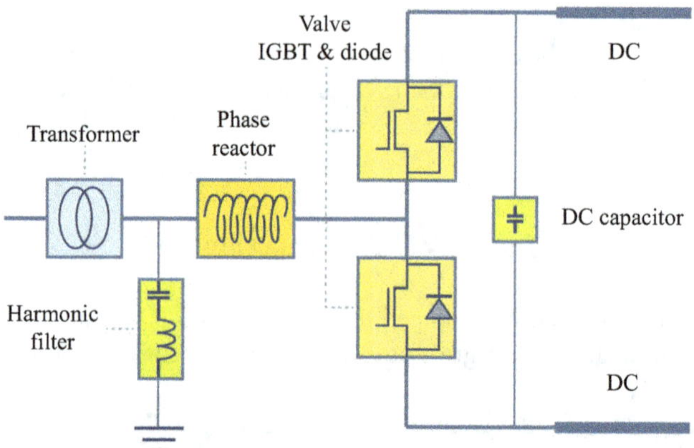

Figure 11.2 Principle diagram of a VSC

the direction of the voltage in case the power flow has to be reversed (Figure 11.2).

The VSC technology has the additional advantages of producing lower voltage and current harmonics and the ability to generate an alternating voltage from the direct voltage in the absence of voltage in the receiving grid, which would allow the use of a HVDC-VSC system for black start after a black out in the electrical system.

Because of the characteristics indicated, the conversion facilities of the HVDC link do not include harmonic filters, but only a "phase reactor" on the alternate side, and the converter transformer are not subjected to the DC separate source-voltage withstand test as for HVDC-LCC plants.

11.3 The problem of core saturation in VSC converter transformers

In the ideal operation of the VSC conversion bridge, all the direct current circulates in the conversion branches and the DC link, and no direct current component is transferred to the AC link.

In contrast, due to the imperfect symmetry of the conversion branches and the imperfect synchrony of the transistor turn-on system, it is possible for a small proportion of direct current to be injected into the AC link. To reduce this drawback, a feedback control corrects the on and off instants of the valves according to the value of the DC component measured by special transducers at the output of the conversion bridge.

Unfortunately, this correction is not complete for inherent errors in the measurement chain (transducer+meter), and thus a residual DC component may remain in the AC link. Apart from voltage quality, this component circulates in the conversion transformer and can cause saturation of the ferromagnetic core.

The technical specifications of the converter substation do not describe the problem of the DC component in the AC link. However, the converter transformer specification document requires that: "The flux density of the magnetic core and any other magnetic parts shall be ≤ 1.8 T under the worst operating conditions (maximum voltage and maximum current)."

According to the data from the manufacturer, the "residual long term direct current in valve winding for rating" is 10 A. This was assumed as the maximum DC component circulating inside the converter transformer, called *DC bias*. However, it is not known whether this is a conservative assumption and which is the actual maximum DC bias in service.

11.4 Study of the problem

The circulation of a direct current in one winding of a transformer participates entirely in the magnetization of the core and is not transferred by electromagnetic induction to the other winding.

The magnetic core of the converter transformer under study has the dimensions shown in Figure 11.3.

The law of magnetic circulation states that the line integral applied to a circuit with magnetizing force H is equal to the algebraic sum of the currents flowing through the surface bounded by the circuit ([2] eq. 5.25):

$$\oint \vec{\boldsymbol{H}} \cdot d\vec{l} = \sum I_j \tag{11.1}$$

Considering the circuit highlighted in red in Figure 11.3, and linearizing the integral, with the simplification of $\vec{\boldsymbol{H}}$ almost constant and parallel to the circuit, we obtain:

$$H \cdot p = NI \tag{11.2}$$

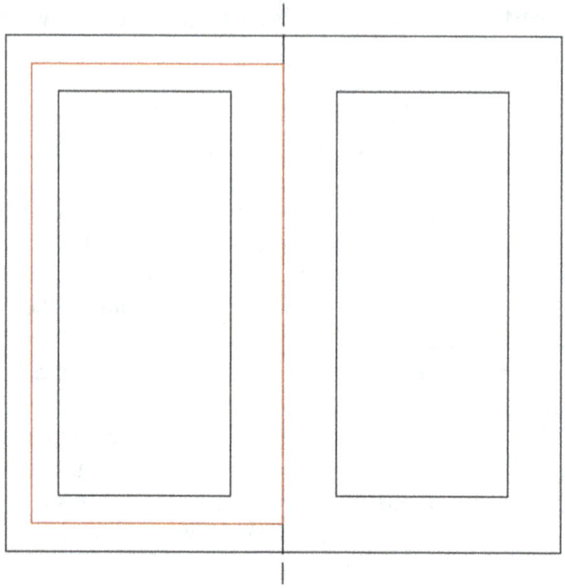

Figure 11.3 Schematic drawing of the magnetic core

where

$p = 10,27$ m is the perimeter of the rectangle calculated as $p = 2(l_1 + l_2)$, being: $l_1 = 1855$ mm and $l_2 = 3280$ mm;
$I = 10$ A is the assumed magnetization current;
$N = 634$ is the number of turns of the valve winding.

Thus the magnetization current in DC generates a magnetizing force in the core equal to $H = 617.3$ A/m.

The material of the magnetic core is steel sheet from Nippon Steel (product code 23ZH85). Figure 11.4 shows its magnetization curve.

A nominal working induction for the transformer equal to 1.61 T was considered. This value guarantees a limited production of current harmonics and corresponds to an AC magnetizing force of about 20–22 A/m. The direct component calculated above, when added to the alternating component, would lead to a magnetizing force value of about 640 A/m to which corresponds an induction value of about 1.9 T, too high for the ferromagnetic steel employed.

The manufacturer also estimates that the superimposition of a DC component (*DC bias*) of 10 A causes an offset of the flux pattern reaching 1.92 T.

To verify the effect of core saturation, the magnetization characteristic of the core was simulated with a calculation in which the nominal AC voltage and DC bias are superimposed.

The result of the calculation indicates the presence of saturation and a strongly biased current at the primary (Figure 11.5).

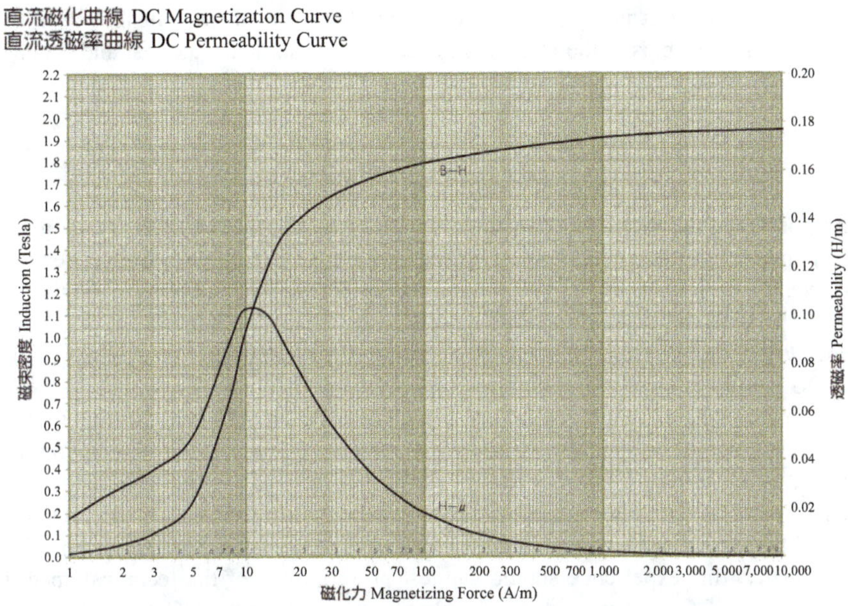

Figure 11.4 Magnetization curve of the steel sheet used in the converter transformer

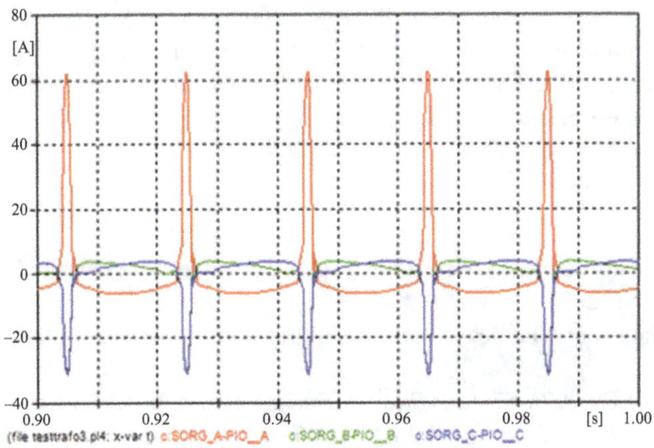

Figure 11.5 Primary current calculated from the power simulation of the conversion transformer with the addition of DC bias

This current distortion, if confirmed in actual operation, would result in unacceptable harmonic content being fed into the grid that is not preventable because the VSC system is not equipped with harmonic filters.

In addition, the saturation of the core would also cause increased losses in the core and more noise produced by higher magnetostriction.

Unfortunately, it is not possible to test the behavior of the transformer in the presence of a DC bias at the factory because there is no dedicated test and it is not possible to have a DC supply in the test room.

11.5 Conclusions

The possible saturation of a converter transformer, due to the presence of a DC component of the current drawn or emitted by the VSC converter bridge, reduces the quality of service of the connecting AC grid. It is not possible to estimate a priori the probability of occurrence of this event, nor its magnitude, because it depends on the measurement errors of the control chain and the ability of the HVDC converter's control system to correct the anomaly.

In cases where the DC component is unavoidable, an online monitoring system of core temperatures should be installed because an increase in core temperatures caused by hysteresis losses is certain. This solution is of low cost, since modern transformers are already equipped with core temperature sensors, which, however, should be properly placed where the specific overtemperatures are expected.

Finally, this experience should suggest prescriptions of the technical specifications of HVDC systems with VSC technology, in which the DC bias calculation method should be stated in detail in the reference conditions and possibly require a dedicated DC test of the converter transformer.

Unfortunately, the current international standard does not support this requirement. In fact, the standard IEEE/IEC 60076-57-129 [3] indicates that no special test should be done for conversion transformers for HVDC systems with VSC technology. If the TSO deems it necessary, this requirement needs to be added on a case-by-case basis.

References

[1] Technical description of HVDC Light® technology, ABB AB Grid Systems – HVDC, 2008.

[2] J.D. Jackson: Classical Electrodynamics, Wiley & Sons Publisher, 3rd ed., 1998.

[3] IEC/IEEE 60076-57-129: *Power Transformers – Part 57-129: Transformers for HVDC Applications*, 1st ed., 2017.

Chapter 12
Shunt reactors

12.1 Generalities

The *shunt reactors* are large inductors used in electrical grids to store reactive power (hence the name). Reactors can be used also for other purposes, such as limiting fault currents, earthing, and filtering, although those applications are less common in HV grids.

For the circuital point of view, a reactor is identical to a star-connected and earthed winding of a power transformer, but without another winding that balances the magnetic flux on the core. This means that all the current flowing in the reactor's winding is a magnetizing current.

However, from a constructive point of view, a transformer's core cannot sustain a large magnetizing current without saturation, but a large magnetizing force H is necessary to store the magnetic energy[1]:

$$\mathcal{E} = \frac{1}{2}\int_{\mathcal{V}} \boldsymbol{B} \cdot \boldsymbol{H}\, dV. \tag{12.1}$$

Therefore, it is necessary to allow the magnetic flux to pass into a region without ferromagnetic material, in order to obtain a large magnetizing force. This is accomplished by inserting air gaps in the wounded limb; these gaps must be not too long, otherwise the magnetic flux would spread out, thus reducing the magnetic induction B. The space inside the gaps is filled with spacers, usually cylindrical pieces made of nonmagnetic material (like soapstone or ceramic), glued to the surfaces of the gap (Figure 12.1).

The drawback of air gaps is the increase of electromagnetic stress on the limb, because the electromagnetic field tends to squeeze to length of the gap, in order to minimize the magnetic energy, according to the formula of the surface pressure [1]:

$$\boldsymbol{p} = \frac{1}{2}\left[\left(\tfrac{1}{\mu_0} - \tfrac{1}{\mu_{\mathrm{Fe}}}\right)B_n^2 - (\mu_0 - \mu_{\mathrm{Fe}})H_t^2\right]\boldsymbol{n}, \tag{12.2}$$

where $\boldsymbol{n}$ is the outward pointing unit vector perpendicular to the ferromagnetic material's surface, B_n is the normal component of the magnetic induction, and H_t is the tangential component of the magnetizing force, which are both conserved at the

[1]This is (2.30) with the assumption $\boldsymbol{B} = \mu\boldsymbol{H}$.

Figure 12.1 Gapped section of a reactor's limb with spacers

interface between two materials with different magnetic properties (although H_t is negligible in the present case).

This stress, that like the short circuit forces has a frequency twice the frequency of the alternating current, requires a particular mechanical reinforcement on the core, usually by means of one or more tie-beams that compress the limb and stabilize the yokes of the magnetic core (Figure 12.2).

In long high voltage transmission lines and cable networks, it is the most compact device for reactive power adjustment.

Shunt reactors have usually a fixed reactance, but to obtain a higher flexibility, it is possible to add a tapped winding connected to an on-load tap-changer (Figure 12.3). However, since the range of variation of the magnetic induction must be maintained within reasonable limits, usually the range of regulation does not exceed 30% of the maximum reactance.

12.2 Reactor's constructive types

The constructive type of shunt reactors can be distinguished by their design characteristics, identified by the presence or absence of a wounded magnetic limb [2].

Shunt reactors are therefore identified as *gapped* or *coreless* reactors. Figures 12.4 and 12.5 show these basic constructive types.

The reactors can be manufactured as single or three-phase units. The single-phase solution has the advantage of reduced transportation's size and weight and a more effective management of the spare unit, but the overall size and cost of the three-phase bank are higher. Therefore, the three-phase solution is preferred when feasible, also considering that the magnetic fluxes partially compensate in the outer

Figure 12.2 Tie-beam for reactor's core

frame of the magnetic core, thus reducing the amount of ferromagnetic material required.

The constructive type is chosen by the manufacturer according to the user's specific requirements, such as reactive power, reactance to be guaranteed, and its limit of linearity considering the possible saturation of the core.

The main difference between the two main categories is the way they provide their reactance.

Coreless reactors, not having magnetic steel gapped limbs inside the windings, behave like an air reactance, only providing their reactance by means of the winding turns, with only a magnetic steel frame around the windings to collect the flux. In this way, they provide a high level of linearity (limited only by the saturation of the outer frame), but with a low value of reactance, since the reluctance in air is greater than inside ferromagnetic materials.

On the other hand, gapped reactors can avail of the ferromagnetic properties of the wounded limb in order to obtain a higher value of reactance. This comes at the expense of a lower limit of linearity (since the cross section of the limb is less than equivalent section of the outer frame, especially in the 5-limb solution), and with a higher price given the additional material needed for the central limb.

According to the peculiarities of the two types, coreless reactors are normally used as series reactors for industrial applications (usually requiring lower values of

Figure 12.3 Active part of a gapped reactor with OLTC

power and reactance), while gapped reactors are the favorite solution for shunt applications in grids.

Several dispositions of the limbs are possible in gapped reactors, identifying different constructive solutions. The standard solution is placing the limbs side by side, just like an ordinary transformer.

However, when there are space constraints, a different disposition of the limbs is more beneficial, where they are placed at the vertices of a triangle with a single circular upper and lower yoke, in the so-called *temple style* core. This kind of disposition allows to save space and to reduce the dimension of the yokes, obtaining a significant reduction of materials and consequently reducing the cost. However, this solution is available only for units with small power rating, because a circular yoke cannot be made too wide. Figure 12.6 shows an example of a temple style reactor.

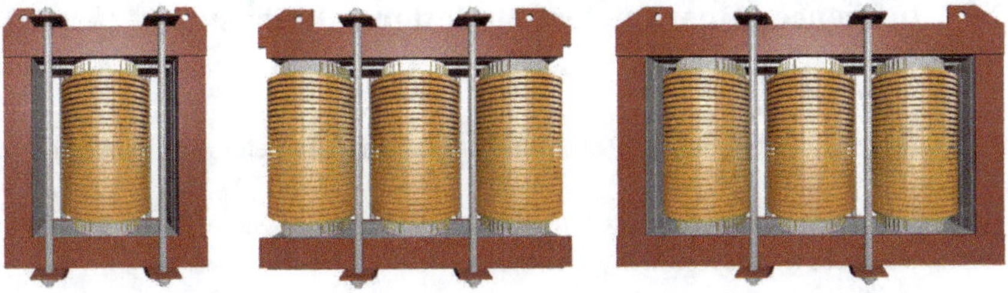

Figure 12.4 Gapped reactors, single-phase and three-phase (with 3 and 5 limbs)

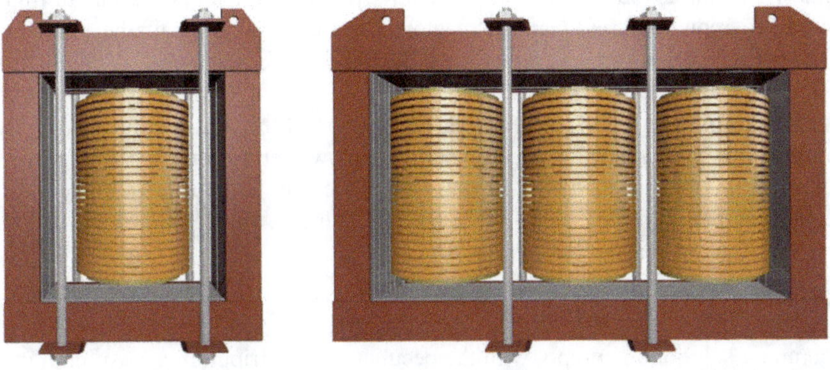

Figure 12.5 Coreless reactors, single-phase and three-phase

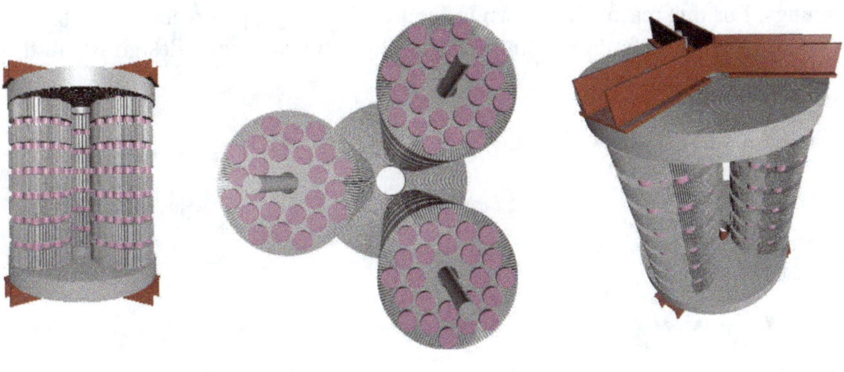

Figure 12.6 Gapped reactors, three-phase temple style

12.3 Influence of parameters on design solutions

12.3.1 Losses

In addition to power, reactance, and linearity, other parameters influence reactor design.

An important parameter to consider when designing a reactor is the amount of acceptable losses the device will generate.

Due to the inherent nature of the reactors, it is not possible to split the core losses, the ohmic losses and the additional losses when measuring. It is, however, possible to evaluate the impact of different solutions capable of limiting the magnetic and Joule losses when necessary.

As in the transformers, controlling how the stray flux distributes along the core window's height is the best method to identify possible critical points and to find a solution to reduce losses. Gapped reactors have a distribution of the stray flux different from transformers, due to the peculiar conformation of the limb that affects the ways the flux propagates.

As seen in Figure 12.7, the peculiar distribution of the flux is influenced by the length of the gaps: a narrower gap results in a lower stray flux (but also a lower limit of saturation), instead a wider gap allows the flux to repeatedly fringe out of the limb and then forces it to find a way back in the limb with an nonoptimal direction (thus increasing the losses).

12.3.2 Windings eddy losses

A common solution to comply with the peculiar flux distribution is positioning the windings accordingly. As previously shown, reactors have a very intense stray flux zone in proximity of the limb.

Positioning the windings on the limb without considering this phenomenon would place the windings in a zone where the stray flux direction is not regular but continuously changing, this increases dramatically the eddy current losses in the windings. For this reason, as shown in Figure 12.8, in gapped reactor's design, the windings are placed farther from the limb than in transformer's design where they can be placed much closer.

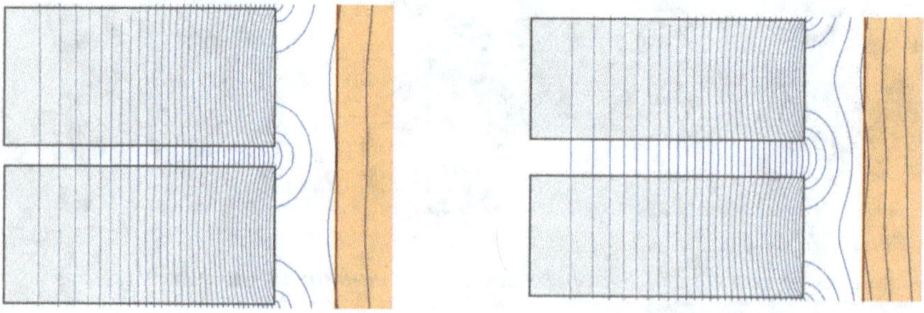

Figure 12.7 Gapped reactors, examples of flux lines: smaller gap (left) vs. larger gaps (right)

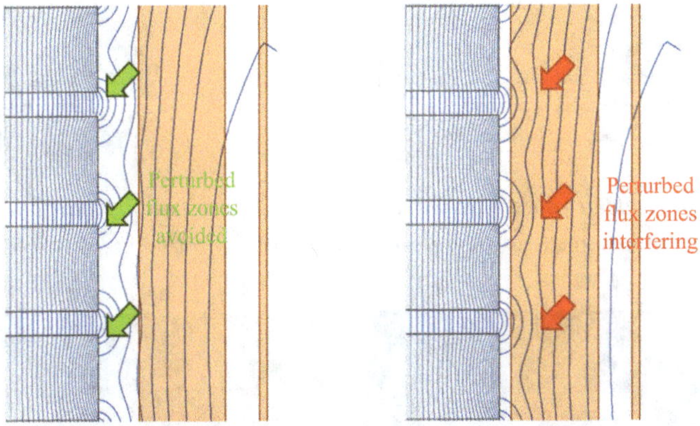

Figure 12.8 Effect of the distance to the limb on the stray flux on the winding

In this way, the windings are less affected by the stray flux, which is unable to cause local increments of the losses and consequently hot-spots. However, the total reactance is slightly reduced due to the wider distance between limb and winding.

12.3.3 Core losses

The whirly direction of the flux in gapped reactors also affects the core losses, due to the fact that it has to pass through the magnetic steel sheets against their preferential direction.

Considering a core traditionally interwoven, this is strongly noticeable, as the flux crosses the sheets along multiple directions, not only the preferential one.

Because of this phenomenon, traditionally interwoven limbs have a very limited gap's length, and also the distance between gaps cannot be reduced below a certain limit. This hinders the linearity of a reactor with a traditionally interwoven limb. Furthermore, compensating the limited height of the gaps with an increase in their number makes the core's manufacturing more expensive.

Therefore, the layout of steel sheets in the limb is adapted to the orientation of the magnetic flux: the magnetic sections of limbs are built with the steel sheets disposed radially.

With this shrewdness, even though the flux still keeps going out and into the core, the preferential direction of the steel sheets is radial; therefore, the losses are lower than in the traditionally interwoven design; such a solution allows for a more distorted flux to be tolerated, essentially meaning higher gap heights are possible.

The re-entering of the flux into the core can also be furtherly aided beveling the edges of the core sections between gaps. An example of this building type, along with the others previously discussed, is shown in Figure 12.9.

However, a radial disposition of the steel sheets is more expensive to manufacture than the interwoven disposition, and the bevelment of the sheets comes with an additional cost.

Figure 12.9 *Gapped reactors, limb section's building types: traditionally interwoven (left), radial (center), and beveled edge radial (right)*

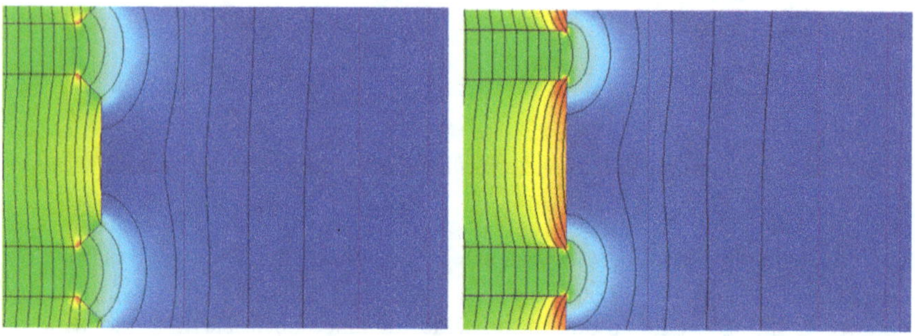

Figure 12.10 *Gapped reactors, induction distribution comparison: Beveled edge (left) and non-beveled edge radial limb (right)*

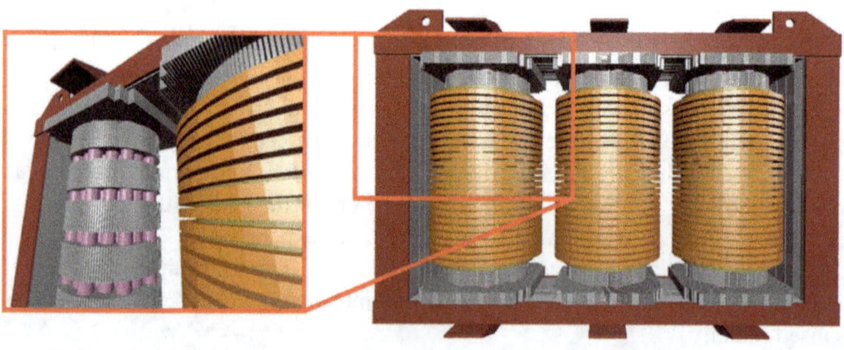

Figure 12.11 *Gapped reactors, flux collecting plates*

Although the advantage in loss reduction of the bevelment is definitely noticeable, as shown in Figure 12.10, its use is limited to the cases where the value of the losses makes worth adopting it.

Another case where the stray flux encounters the same kind of issues is when approaching the yokes, where lines of flux leaving the windings radially do not enter along the yoke's preferential magnetic direction. This effect can be avoided placing a magnetic steel plate between the limb and the yoke, arranging the steel sheets inside the plates transversally, as shown in Figure 12.11.

These devices are able to collect the radial stray flux and direct it into the yokes, following better the preferential direction. This solution has also a positive impact on the eddy loss affecting the tank since a fair amount of the undirected radial flux would end up in the tank.

It should be noted that while all of these solutions make the designer able to control the amount of the losses, they also increase the manufacturing price and therefore are used only for large units, especially when the losses are priced separately.

12.3.4 Noise

Due to the high level of electromagnetic stress and hence vibrations in the core of gapped reactors, the noise is an important parameter to consider in the design, more than in power transformers.

The vibrations of the core are induced by magnetostriction too, and it becomes particularly severe when approaching the material saturation. Magnetostriction is essentially due to the material properties, such as chemical composition, crystal structure, material thickness.

The magnetic force causes the attraction of the two steel surfaces on each side of a gap, tending to compress the gap and so posing a significant stress also to the spacers in the gaps; this induces noisy vibrations as well.

Since many of these parameters are material-dependent and as such are not variable by design, the only design measures that can be taken to limit this problem are as follows:

- a careful assembly of the core:
  - correct tightening of the tie-beams, in order to control the compressive force on the gapped limb (a loose limb vibrates due to the transversal component of the field on the sheets, while in a too tight limb the magnetostriction increases);
  - precise alignment of the magnetic sections of the gapped limb;
- a design aimed at reducing as much as possible leakage flux and local peaks of induction:
  - the height of the gaps is a flux-control factor;
  - beveling the edge of the core section between gaps can ease the stress due to local induction peaks.

Other solutions not affecting the active parts design are as follows:

- employ of anti-vibration devices connecting the active parts to the rest of the structure and the tank;
- thickening of the tank floor;
- employ of sound-proof structures (although extremely expensive);
- when possible, a properly considered positioning of the device in the installation site.

Figure 12.12 Examples of enclosed and paneled solutions to reduce the acoustic emission

For example, an integrated solution of noise suppression could be developed using sound panels (Figure 12.12), vertical and horizontal stiffeners and increasing the thickness of tank; additionally, anti-vibration pads could be interposed between the tank and the laying supports in order to reduce the vibrations transmitted to the foundation.

The achieved sound's reduction depends on the sound panel's technology; this is mainly related to the thickness and to the material of the panel.

References

[1] Z. Ren: *Comparison of different force calculation methods in 3D finite element modelling*, IEEE Transactions on Magnetics, Vol. 30, issue 5, pp. 3471–3474, September 1994.

[2] S. Zunino, P. Portoghese, S. Badini, G. Cannavale, M. Giussani, and F. Ferrari: *Use of shunt reactors for smart voltage control in EHV/HV grid*, paper 4-05, First South East European Regional CIGRÉ Conference, Portoroz, 2016.

Chapter 13
Industrial transformers

13.1 Introduction

In this chapter, we will briefly discuss some basic concepts useful in defining industrial transformers, mainly represented by two broad categories: *furnace transformers* and *rectifier transformers*.

The industrial transformers have fundamental importance in the operation of an industrial plant, and all possible efforts should be made in order to avoid any outages. A specific design, manufacturing process, and suitable sets of external protections for preventing overvoltages are very important factors for a well-functioning, reliable, and safe machine.

Furnace transformers are used to feed electric arc furnaces that use the heating of the electric arc to melt the materials contained in the furnace.

Rectifier transformers are used to feed rectifiers, composed by electronic valves (diodes or thyristors), so to produce the direct current needed in several industrial processes (e.g., electrolytic processes).

These devices can transform electric power while operating under notably stressful conditions, significantly different than the usual conditions of grid power transformers.

Also differently from grid transformers, there is no redundancy in the installation at the industrial plant, so the failure of an industrial transformer leads to a long interruption of the production process. For this reason, sometimes a spare identical transformer is stored at the plant, although this requires additional space and costs.

Due to the peculiarity and prominence of industrial transformers application, optimized and dedicated designs are necessary.

13.2 Furnace transformer's design

Furnace transformers are usually tailored to the plant design. The transformer main parameters can be adapted according to the plant designers' studies, in order to comply with the operational conditions needed.

Notably, wide regulation ranges are commonly used to meet the needs of the oven, according to the particular kind of duty cycle required. An example of a typical power locus diagram of an arc furnace can be seen in Figure 13.1. To achieve such

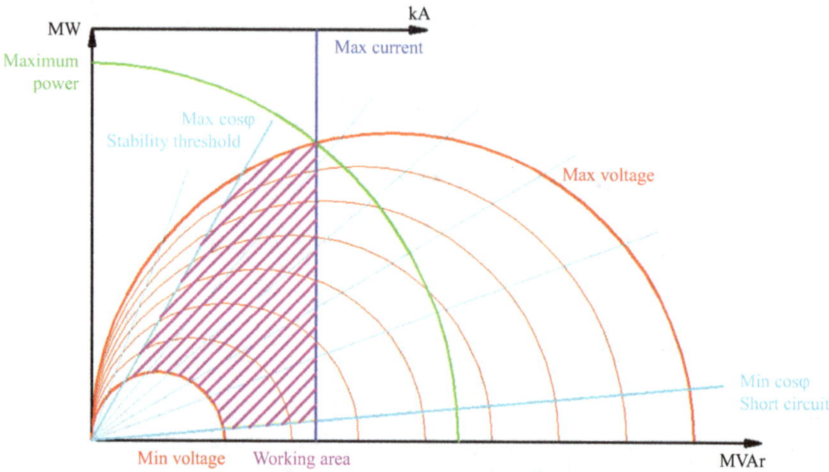

Figure 13.1 Typical diagram of the working locus of a furnace transformer

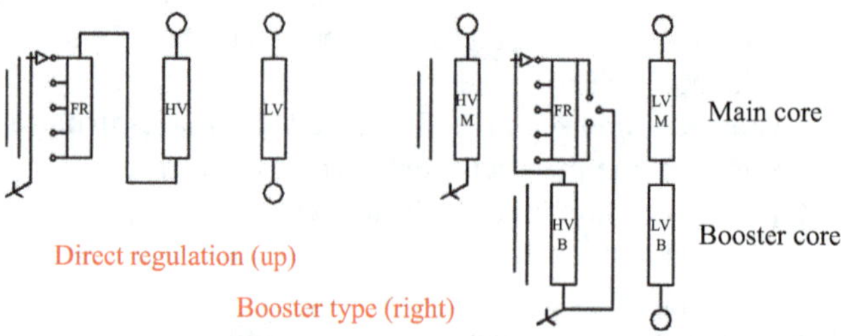

Figure 13.2 Furnace transformers typical design models

performances, the principal design models for the transformer are the "direct regulation" model with single core and the "booster type" model with double core.

The single-phase diagrams of these transformer models can be seen in Figure 13.2.

The choice of the most suitable model for the application depends on the desired regulation voltage steps and range and should be evaluated according to the specific plant's data.

13.3 Rectifier transformer's design

Rectifier transformers, just like furnace transformers, are adapted to the plant needs according to the desired values of the general duty parameters, but unlike a furnace transformer, a rectifier transformer must also provide the correct number of terminal and phase displacement so as to accommodate the number of pulses (6 to 72) of the rectifier.

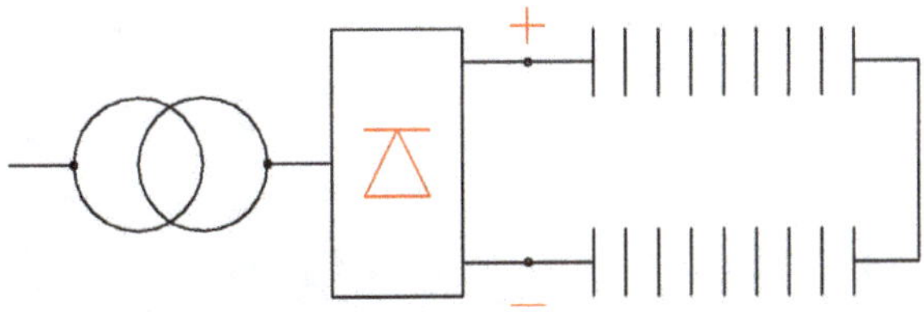

Figure 13.3 Typical arrangement for an electrolytic plant

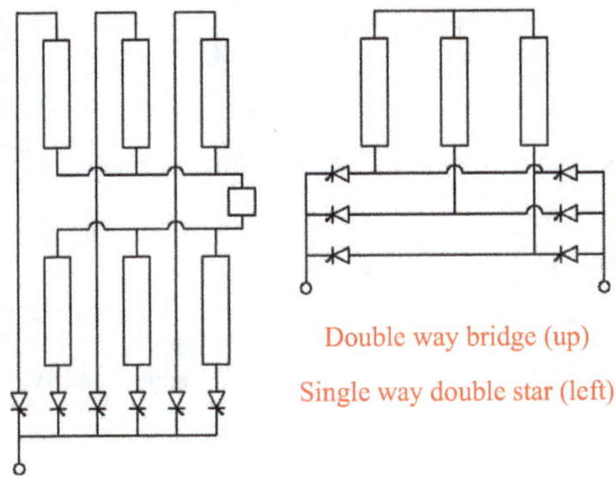

Figure 13.4 Most common connection types for rectifier transformers

The choice of the parameters is strongly dependent on the type of rectifier used and, most importantly, on the type of process to be fed. A typical application is an electrolytic plant that can be schematically represented as seen in Figure 13.3.

According to the customer specifications, different connection schemes are possible, the most common of which being "single way double-star" connection and "double way bridge" connection. The diagrams referring to these kinds of connection can be seen in Figure 13.4.

The aim of these transformers is to feed an equally phase displaced set of AC voltages to the rectifier, which will provide an almost DC voltage output to the electrodes inside the furnace, across which the DC current will pass through an electric arc.

An example of a double-star transformer behavior is shown in Figure 13.5, where each phase of a star-connected winding is displaced by 180° with respect of the same phase in the other winding. Thanks to this configuration six equally displaced pulses are available, from which an almost DC voltage output is obtained by the rectifier, and so an almost direct current is allowed to flow into the load.

However, this implies that the waveshapes (Figure 13.6) of the currents in the rectifier transformer are strongly distorted, far from the normal sinusoidal wave shape, and are intermittent in the secondary windings.

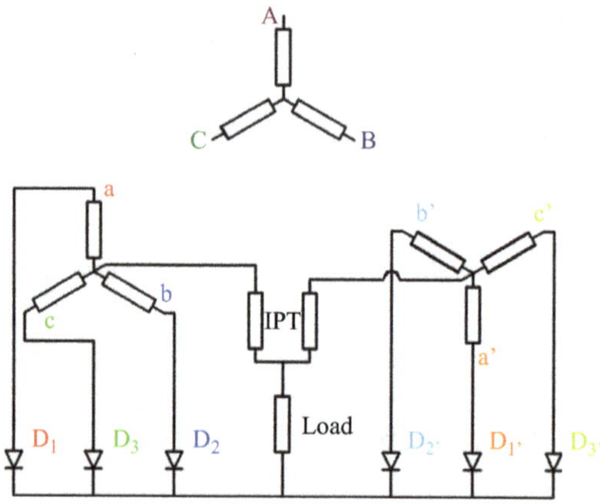

Figure 13.5 Connection diagram of the winding in double-star rectifying system

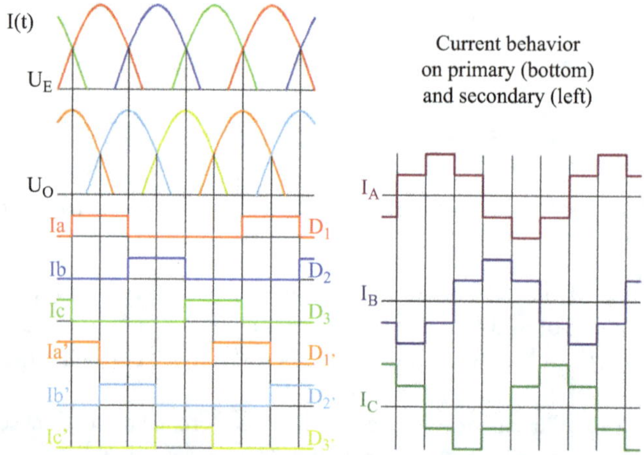

Figure 13.6 Voltage and current trends in a double-star rectifier

13.4 Operational conditions of industrial transformers

Industrial transformers are heavily stressed machines, due to several peculiarities of their usage.

Furnace transformers suffer a large number of daily tap-changing operations, causing severe contacts wear in the OLTC and transients with inrush currents and high-frequency switching overvoltages, leading to dielectric and mechanical stresses.

Both types of transformers endure extremely high-rated currents, especially furnace transformers that operate near short-circuit conditions, with significant harmonic distortion, mostly prominent in rectifier transformers, leading to thermal stresses and potentially dangerous overtemperatures.

13.5 Dielectric design

The maximum primary voltage for industrial transformers is usually within 52 kV. Thus, the dielectric insulation of the primary side at power frequency is not an issue. Nonetheless, the usually wide regulating range implies a large variable-flux operation of the primary winding.

Instead, during high-frequency overvoltages, the voltage distribution among the windings is strongly influenced by the coupling with the stray capacitances.

Studying this capacitance network is necessary to evaluate voltage gradients in oil ducts and in solid insulation, with consequent evaluation of the risk of creepage and partial discharges. A sample simulation can be seen in Figure 13.7.

13.5.1 Lightning overvoltages

In industrial transformers, lightning overvoltages are quite a rare occurrence, and therefore no specific design adjustment is usually envisaged for them; the risk can be mitigated, just like any other kind of transformer, installing surge arresters close to the transformers.

13.5.2 Switching overvoltages

As stated earlier, frequent switching operations are a common occurrence in furnace transformers.

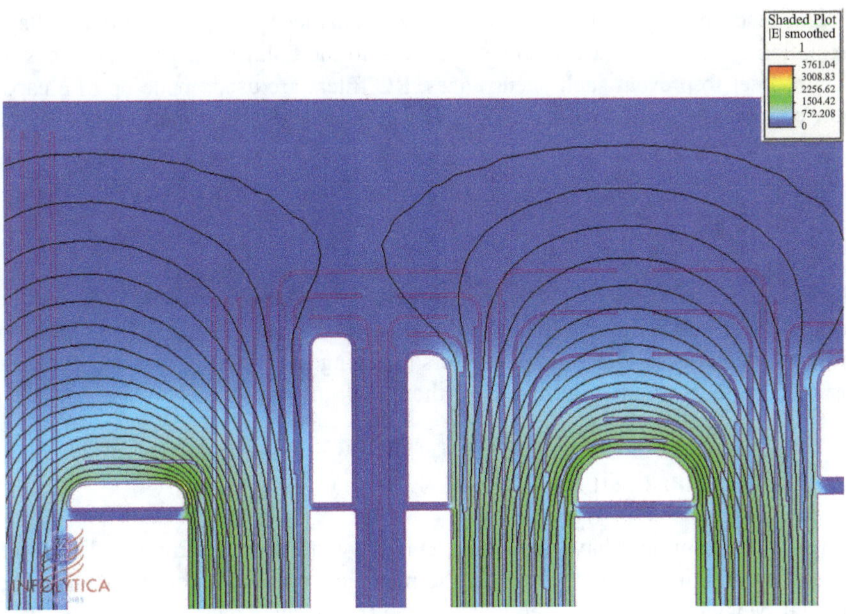

Figure 13.7 Electric field distribution simulation

Considering switching-on, the amplitude and the frequency of the overvoltages depend on the equivalent circuital parameters of the transformer, like coil inductances and turn-to-turn capacitances, and contribution from the circuit-breaker's duty cycle.

Usually magnitude of these overvoltages, caused by the switching-on operation, is not dangerous for the transformer; the most significant feature to be taken in consideration is the frequency of the oscillations at the transformer's terminals, which should not approach the transformer's resonant frequencies, because otherwise this could cause high dielectric stress on the windings.

Thus, it is suggested to install the plant circuit-breaker as close as possible to the transformer to reduce as much as possible the resonant coupling with the capacitances in the connection cables.

Considering switching-off, a typical stress peculiar to this kind of application is induced by the usage of vacuum-type circuit-breakers. This kind of circuit-breakers is gaining widespread use in the medium voltage applications due to their lower cost, reduced maintenance, safety, and suitability to the application. However, vacuum-type circuit-breakers can be affected by reignition of the arc inside the interruption chamber during the breaking operation, which causes additional overvoltages on the transformer.

The overvoltages are caused by the current chopping phenomenon, occurring when the current is approaching its natural zero crossing but is interrupted before reaching it. The transient recovery voltage caused by the interruption (and possible reignition) causes an electrostatic voltage accumulation in the stray capacitances of the transformer.

In the worst case when the resonance causes excessive transient overvoltages, several restrikes occur inside the interruption chamber, and a dangerous voltage escalation is possible that can sometimes lead to the failure of the circuit-breaker.

In order to prevent such occurrences, RC filters are used, made up of a capacitor connected in series with a resistor. This device must be properly tuned on the specific plant arrangement and operative conditions so as to avoid resonances and mitigate switching overvoltages.

13.5.3 Secondary side overvoltages

Hazardous overvoltages can be produced by the irregular nature of the load and the possibility of sudden current interruptions in the furnace (Figure 13.8). To limit these overvoltages, a system of capacitors behaving as a filtering system should be installed on the secondary terminals of the transformer.

13.6 Thermal design

Industrial transformers have to endure currents with strong harmonic distortion, generating additional eddy current in the magnetic materials. For this reason, the thermal behavior of the affected parts of the transformer must be carefully evaluated during design stage.

Figure 13.8 Connections on the secondary side of a furnace transformer

The parts mainly affected that need to be considered are the followings:

- high voltage and regulating coils, low voltage coils;
- tank and low voltage terminals.

13.6.1 Thermal calculation in HV coils

A correct analysis of the leakage magnetic flux is fundamental to determine the worst positions inside the HV and regulating coils, where eddy current produce hot-spots. Using finite elements simulation programs (FEM), it is possible to check in the most accurate way local flux distribution, according to the specific geometrical features of the wires, thus obtaining local losses values for each part of the coils. An example of this kind of analysis can be seen in Figure 13.9.

Conversely, to properly design the low voltage coils, a totally different approach is employed, due to the mechanical nature of this kind of windings. In fact, LV coils need a considerably large cross section in their wires, because of the need to carry large currents (up to one hundred kA) and have a very small number of turns.

The LV coils in rectifier transformers are placed in the outermost position of the phase winding, so to allow an easier connection of the several leads to the LV terminals, instead of the usual innermost position in grid transformers.

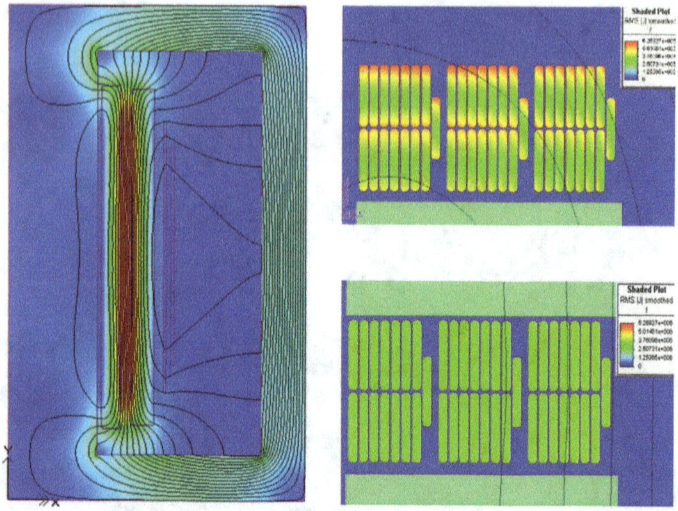

Figure 13.9 Stray flux lines simulated by FEM and localized losses inside the conductors

Moreover, it is not practical to employ helical coils with a large cross section and a small number of turns: The helix pitch would be unbearably large, preventing the achievement of a good electromagnetic balance and wasting a notable amount of space at each end of the coil.

For these reasons, LV coils usually consist of a number of parallel disc-wound sections axially arranged with their leads connected directly to vertical copper busbars.

It would be desirable that these parallel discs carry evenly distributed currents among them. However, the different geometrical locations, characterized by nonuniform magnetic flux, cause an uneven distribution of currents among the parallel sections.

This problem is particularly noticeable in furnace transformer rather than in rectifier transformer, because the former ones usually have a higher number of parallel sections and so controlling current unbalance is critical.

Figure 13.10 shows an example of the current distribution among parallel sections of a low voltage coil, evaluated by a FEM program. Top sections of the coil are located on the left of the graphs, bottom sections are located on the right.

The top and bottom sections, located in the regions of the coil where the magnetic flux deviates the most from vertical direction, experience a current content higher than the central ones, causing an uneven losses and temperature distribution.

This is more evident in the picture on the left, representing a furnace transformer with a large number of parallel sections; while the picture on the right, representing a rectifier transformer with few parallel sections, shows a less pronounced uneven distribution.

Nonetheless, it is worth noticing that rectifier transformers usually work with a much higher harmonic content, leading to an increase in the eddy losses.

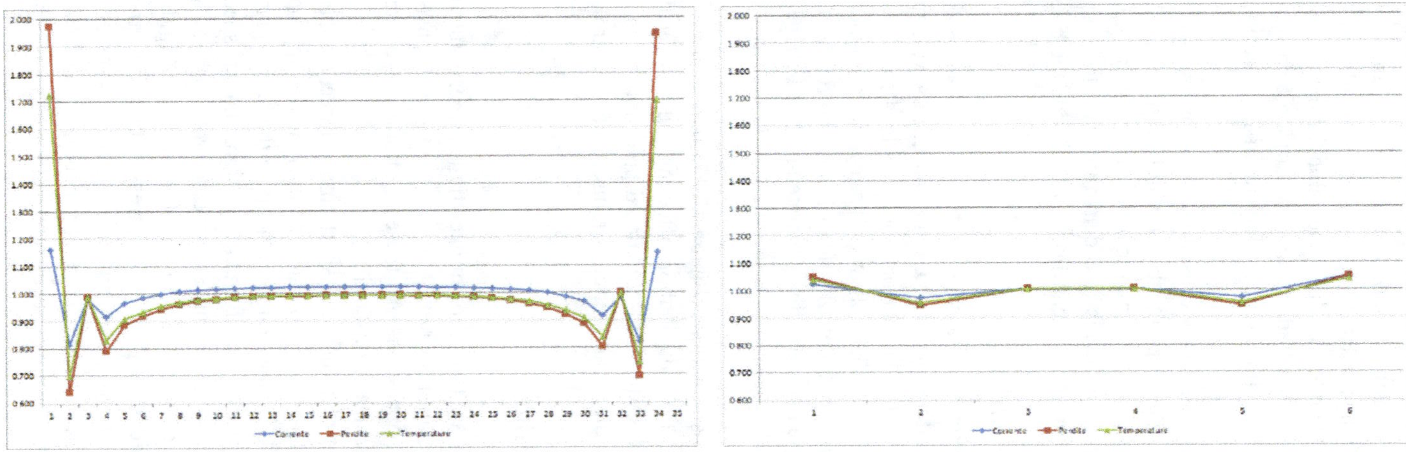

Figure 13.10 *Current (blue), losses (red), and temperature (green) distribution in LV windings in a furnace transformer (left) and in a rectifier transformer (right)*

Table 13.1 shows a typical harmonic spectrum for both kinds of industrial transformers. A remarkably strong harmonic content is present in the spectrum of rectifier transformers, showing a harmonic order up to the 49th harmonic; this leads to an increase in eddy losses up to five times the value the device would experience in pure sinusoidal condition. On the other hand, the harmonic content in the furnace transformer reaches only the 14th order, leading to an increase in eddy losses of about 50%.

To cope with the thermal stresses caused by the abovementioned issues, special design solutions have to be implemented. Possible choices are a local increase of radial cooling ducts to allow a better current distribution, employ of high-temperature material for coil insulation in order to increase the thermal class, and employ of CTC to reduce uneven current distribution inside the coil.

13.6.2 Thermal calculation in LV coils

The high intensity of the current flowing in the low voltage side represents a challenge also for layout and definition of LV terminals. Low voltage terminals for industrial transformers consist in copper bars or tubes located either on the top or on the side of the tank. Figure 13.11 shows a typical delta connection of a secondary voltage system, made of a combination of bars and tubes.

The study of the intense magnetic flux generated by low voltage connections is crucial to prevent dangerous local overheating due to leakage flux in the tank or other metallic parts.

Also in this case FEM simulations (Figure 13.12) represent the most accurate instrument to evaluate the flux distribution affecting the tank walls, in order to check where the expected heating needs additional precaution, like the addition of cooling circuits so to keep the local overtemperatures under control.

13.7 Mechanical design

Severe mechanical stresses are typical of furnace transformers, due to their heavy-duty cycles, including operation at nearly short-circuit conditions, and very frequent switching on and off operations. All these conditions have a relevant impact on the transformer's mechanical withstand.

The correct evaluation of inrush and short-circuit forces in the windings are key factors for an appropriate design, including the choice of appropriately robust materials, such as bonded CTC and reinforcements in the active part.

Moreover, some industrial transformer's designs adopt special clamping devices in low voltage windings, to assure an increased robustness. These consist of iron rings, fixed to the structure of the frame by four tie rods (Figure 13.13).

The iron ring is sized considering the stresses affecting the transformer through its life, performing deformation analysis and nodal stresses analysis. This guarantees a proper tightness of the winding's geometry and assures the capability to withstand the most severe mechanical stresses in all the operational conditions.

Table 13.1 Typical harmonic spectrums for industrial transformers

Typical harmonic spectrum for rectifier transformers

Harmonic order	2nd	3rd	4th	5th	6th	7th	8th	10th	11th	12th	13th	14th	16th	17th	18th	19th	20th	22nd	23rd
%	2.50	4.00	2.50	3.50	1.00	2.00	2.00	1.50	8.10	0.10	6.60	1.50	1.50	1.20	0.10	0.90	1.00	1.00	2.70
Harmonic order	25th	26th	28th	29th	31st	32nd	34th	35th	37th	41st	43rd	47th	49th						
%	2.20	0.50	0.50	1.80	1.20	0.50	0.50	0.70	0.50	0.50	0.50	0.30	0.30						

Typical harmonic spectrum for furnace transformers

Harmonic order	2nd	3rd	4th	5th	6th	7th	8th	9th	10th	11th	12th	13th	14th
%	7.50	7.00	3.50	4.50	2.00	2.50	1.30	1.20	1.00	1.00	0.80	0.80	0.70

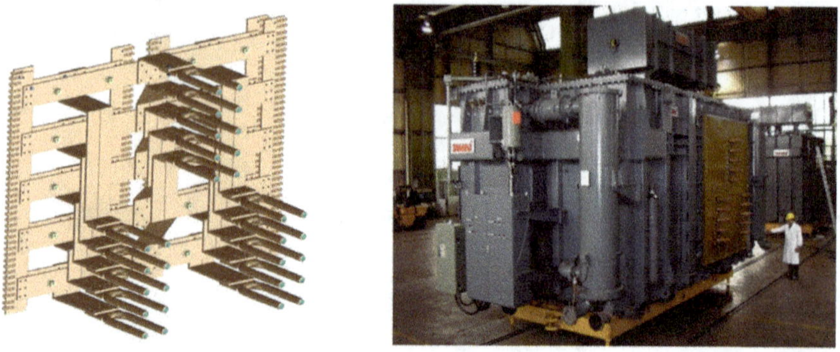

Figure 13.11 Typical layout of low voltage terminals

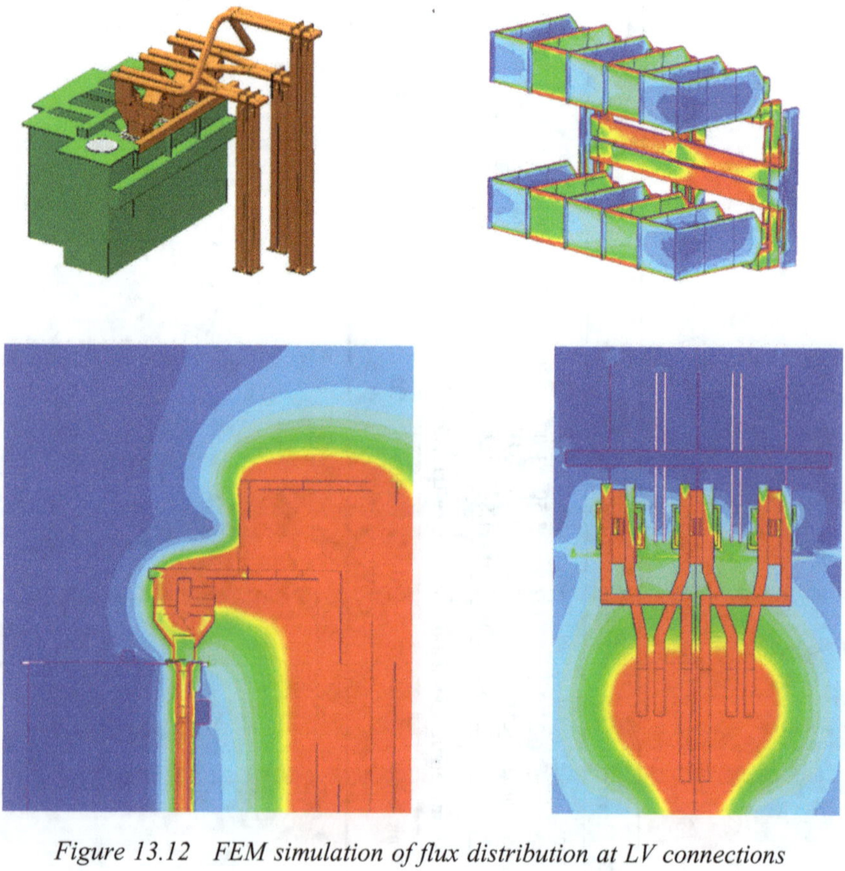

Figure 13.12 FEM simulation of flux distribution at LV connections

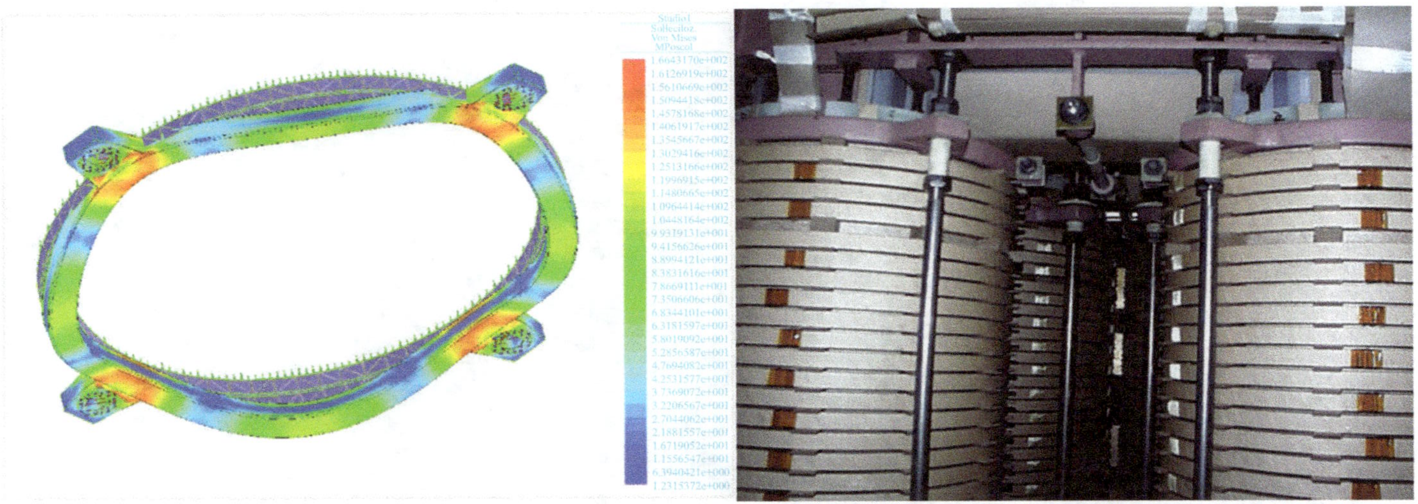

Figure 13.13 Dimensioning and study of stress on winding pressure ring during short circuit

13.8 Conclusion

Industrial transformers have peculiar working conditions, specific to their intended application. This peculiarity requires the employ of detailed simulation tools and specific design solutions, different from grid's interconnection transformers, so to guarantee the transformer's reliability and durability under all the working conditions of the industrial plant.

References

[1] L. Lombini, R. Gamba, and M. Giussani: Industrial transformers features, Doble "Life of a Transformer Seminar", 2015.

[2] D. L. Pelfrey: Specifying arc furnace transformers, Electrical Furnace Proceeding – Pittsburgh, Vol. 38, 1980.

[3] R. Pisila: Transformer response to transients in three-phase vacuum switched circuits, IEEE Summer Power Meeting, 1974.

[4] W. J. McNutt, T. J. Blalock and R. A. Hinton: Response of Transformer Windings to System Transient Voltages, IEEE Transactions on Power Apparatus and Systems, vol. PAS-93, no. 2, pp. 457–467, 1974.

[5] M. Ubaldini: *Trasformatori per elettrolisi*, Programma di istruzione permanente – Politecnico di Milano, Facoltà di Ingegneria, 1986.

Credits of the pictures

The pictures not mentioned in this list of credits are by the authors or in public domain.

Figure 1.1, left:	Edison Tech Center website (courtesy of Edison Tech Center)
Figure 1.2, bottom:	Tamini archive
Figure 1.3:	Technical Brochure 642: *Transformer reliability survey*, Cigrè, 2015 (courtesy of Cigrè)
Figure 2.1:	Gábor Gurszky: *Comparison of two technologies*, JST Presentation, January 2011 (courtesy of JST transformateurs SAS)
Figure 2.2:	Nippon steel corporation: *Orientcore® grain-oriented electrical steel sheets*, 2015 (courtesy of Nippon steel)
Figure 2.3:	JFE steel corporation: *SuperCore™ magnetic property curves*, 2014 (courtesy of JFE steel)
Figure 2.4:	William H. Yeadon, Alan W. Yeadon, *Handbook of Small Electric Motors*, McGraw-Hill, 2001 (courtesy of McGraw-Hill)
Figure 2.5:	Brockhaus website (courtesy of Dr. Brockhaus Messtechnik GmbH)
Figure 2.6:	JFE steel corporation: *SuperCore™ magnetic property curves*, 2014 (courtesy of JFE steel)
Figure 2.7:	AK steel: *Selection of electrical steels for magnetic cores*, 2007 (courtesy of AK steel)
Figure 2.8:	Gábor Gurszky: *Comparison of two technologies*, JST Presentation, January 2011 (courtesy of JST transformateurs SAS)
Figure 2.9:	Tamini archive
Figure 2.10:	LTC archive (courtesy of Legnano Teknoelectric Company)
Figure 2.11:	LTC archive (courtesy of Legnano Teknoelectric Company)
Figure 2.12:	LTC archive (courtesy of Legnano Teknoelectric Company)
Figure 2.13:	LTC archive (courtesy of Legnano Teknoelectric Company)
Figure 2.14:	Christian Freitag: *Magnetic properties of electrical steel, power transformer core losses and core design concepts*, Thesis at the Karlsruher Institutes für Technologie, February 2017 (courtesy of C. Freitag)
Figure 2.15:	Kent Olsson: *First class shunt reactors*, ABB presentation, May 2007 (courtesy of Hitachi energy, formerly ABB)
Figure 2.16:	Kent Olsson: *First class shunt reactors*, ABB presentation, May 2007 (courtesy of Hitachi energy, formerly ABB)
Figure 2.17:	https://www.britannica.com/science/Brillouin-function (courtesy of Encyclopædia Britannica, Inc.)
Figure 2.18:	https://www.britannica.com/science/molar-susceptibility (courtesy of Encyclopædia Britannica, Inc.)

(Continues)

(*Continued*)

Figure 2.21, left:	F. J. Himpsel, J. E. Ortega, G. J. Mankey & R. F. Willis: *Magnetic nanostructures, Advances in Physics*, Taylor & Francis, 1998 (courtesy of Taylor & Francis)
Figure 2.21, right:	https://commons.wikimedia.org/wiki/File:Stoner_model_of_ferromagnetism.svg (courtesy of Wikimedia)
Figure 2.22:	Massimo Solzi: *Basic magnetic properties*, Presentation at the Italian School of Magnetism, Rome, 3-7 February 2020 (courtesy of M. Solzi)
Figure 2.29:	https://www.periodni.com/crystal-systems-and-bravais-lattices.html (courtesy of Eni Generalic)
Figure 2.30:	https://www.periodni.com/crystal-systems-and-bravais-lattices.html (courtesy of Eni Generalic)
Figure 2.31:	https://commons.wikimedia.org/wiki/File:Hexagonal_close_packed.svg (courtesy of Wikimedia)
Figure 2.32:	https://commons.wikimedia.org/wiki/File:Pure_iron_phase_diagram_(EN).svg (courtesy of Wikimedia)
Figure 2.34:	https://www.doitpoms.ac.uk/miclib/full_record.php?id=617 (courtesy of Wilifried Kurz, University of Cambridge)
Figure 2.36:	Walter Nicodemi: *Metallurgia*, Zanichelli editore, 2nd ed. 2007 (courtesy of Zanichelli)
Figure 2.41:	Magneti permanenti, http://web.mclink.it/MC5690/index.html (courtesy of Giancarlo Franceschetti)
Figure 2.42:	Magneti permanenti, http://web.mclink.it/MC5690/index.html (courtesy of Giancarlo Franceschetti)
Figure 2.44:	Vladimír Roha *et al.*: *Basic classification and processing of electrical steels*, The International Journal of Engineering and Science (IJES), 9(7), 2020 (courtesy of IJES)
Figure 2.46:	POSCO: *Electrical steel*, Brochure, 2014 (courtesy of POSCO)
Figure 2.47:	I. Petryshynets, V. Puchý, F. Kovác, M. Šebek: *Effect of laser scribing on soft magnetic properties of conventional grain-oriented silicon steel*, Proceedings of the 16th Czech and Slovak Conference on Magnetism, Košice, Slovakia, 13-17 June 2016 (courtesy of CSMAG)
Figure 2.48:	Ludger Lahn, Chaoyong Wang, Andreas Allwardt, Thierry Belgrand, Jonathan Blaszkowski: *Improved transformer noise behavior by optimized laser domain refinement at ThyssenKrupp Electrical Steel*, IEEE Transactions on Magnetics, vol. 48, no. 4, April 2012 (courtesy of ThyssenKrupp)
Figure 2.49:	Rüdiger Koshorst: *How grain oriented electrical steel can help on continuous improvement on transformer efficiency*, ThyssenKrupp Electrical Steel presentation at Coiltech, 27-28 September 2017, Pordenone (courtesy of ThyssenKrupp)
Figure 2.50:	Rüdiger Koshorst: *How grain oriented electrical steel can help on continuous improvement on transformer efficiency*, ThyssenKrupp Electrical Steel presentation at Coiltech, 27-28 September 2017, Pordenone (courtesy of ThyssenKrupp)
Figure 2.51:	Rüdiger Koshorst: *How grain oriented electrical steel can help on continuous improvement on transformer efficiency*, ThyssenKrupp Electrical Steel presentation at Coiltech, 27-28 September 2017, Pordenone (courtesy of ThyssenKrupp)

(*Continued*)

Figure 2.52:	Metglas, Inc.: *Amorphous metal distribution transformers*, Brochure, 2016 (courtesy of Metglas)
Figure 2.53:	Alex Hubert, Rudolf Schäfer: *Magnetic Domains*, 3ª ed. 2009 (courtesy of Springer)
Figure 2.54:	Metglas, Inc.: *Amorphous alloys for transformer cores*, Datasheet, 2011 (courtesy of Metglas)
Figure 2.55:	A.J. Moses: *Electrical steels: past, present and future developments*, IEE Proceedings A, vol. 137, no. 5, September 1990 (courtesy of A.J. Moses)
Figure 2.56:	MK Magnetics, http://www.mkmagnetics.com/image-gallery.php (courtesy of MK Magnetics)
Figure 2.57:	Arnold Magnetic Technologies: *Soft magnetic applications guide*, 2015 - Steve Constantinides: *Energy savings with thin gage silicon-iron*, Arnold Magnetic Technologies, 2013 (courtesy of Arnold Magnetic Technologies)
Figure 2.58:	Paola Tiberto: *Electric motors*, Presentation at the Italian School of Magnetism, Milan, 18-22 April 2016 (courtesy of P. Tiberto)
Figure 2.59:	Paola Tiberto: *Electric motors*, Presentation at the Italian School of Magnetism, Milan, 18-22 April 2016 (courtesy of P. Tiberto)
Figure 2.60:	Paola Tiberto: *Electric motors*, Presentation at the Italian School of Magnetism, Milan, 18-22 April 2016 (courtesy of P. Tiberto)
Figure 2.60:	Hitachi Metals: *FINEMET® EMC components*, Catalogue, 2016 (courtesy of Hitachi Metals, Ltd.)
Figure 3.1:	https://commons.wikimedia.org/wiki/File:Solid_state_electronic_band_structure.svg (courtesy of Wikimedia)
Figure 3.2:	https://commons.wikimedia.org/wiki/File:Band_filling_diagram.svg (courtesy of Wikimedia)
Figure 3.3:	Charles Kittel: *Introduction to Solid State Physics*, Wiley, 7th ed. 1996 (courtesy of John Wiley & Sons, Inc.)
Figure 3.5:	https://copperalliance.org/wp-content/uploads/2022/03/Copper-Cathode.png (courtesy of Copper Development Association)
Figure 3.6:	CTP archive (courtesy of CTP s.r.l.)
Figure 3.7:	Copper Development Association: *High conductivity copper for electrical engineering*, Publication No. 122 (courtesy of Copper Development Association)
Figure 3.8:	Copper Development Association: *High conductivity copper for electrical engineering*, Publication No. 122 (courtesy of Copper Development Association)
Figure 3.9, left:	https://www.coptechwire.com/copper-profile-section-rods/ (courtesy of Coptech Wire & Cable Private Ltd.)
Figure 3.9, right:	Tamini archive
Figure 3.10:	CTP archive (courtesy of CTP s.r.l.)
Figure 3.11:	Erikoğlu Emaye: Enamelled Copper Wires (courtesy of Erikoğlu Emaye A.S.)
Figure 3.14:	Essex archive (courtesy of Essex Italy S.r.l., a Superior Essex company)
Figure 3.18:	Essex archive (courtesy of Essex Italy S.r.l., a Superior Essex company)
Figure 3.23:	Essex archive (courtesy of Essex Italy S.r.l., a Superior Essex company)

(Continues)

(Continued)

Figure 3.24:	Essex archive (courtesy of Essex Italy S.r.l., a Superior Essex company)
Figure 3.25:	Nynas: *Transformer oil handbook*, 2nd ed. 2009 (courtesy of Nynas)
Figure 3.26:	Radosław Szewczyk: *High temperature solid insulation, Presentation at Doble conference*, 26-28 October 2016, Dublin (courtesy of DuPont de Nemours, Inc.)
Figure 3.27:	Radosław Szewczyk *et al.*: *Diagnostic tools (DGA) for resilient transformers with aramid-based insulation systems*, Paper n° 11052 for CIRED Rome, 12-15 June 2023 (courtesy of DuPont de Nemours, Inc.)
Figure 3.28:	DuPont: *Nomex® 910 engineered cellulose*, data sheet, 2020 (courtesy of DuPont de Nemours, Inc.)
Figure 3.35:	G. E. Sauber: *Interlaced disc winding with improved impulse voltage gradient*, U.S. Patent 3528046, Filed Nov. 22, 1966
Figure 3.36:	R. C. Degeneff, W. N. Kennedy: *Esternally shielded disc windings for transformers*, U.S. Patent 4318066, Filed Mar. 2, 1982
Figure 3.39:	F. Coppadoro: *Variation of short circuit voltage in autotransformers for interconnection on different regulating sockets as a function of patterns and winding arrangement*, L'elettrotecnica, vol. LX no. 3, March 1973 (courtesy of AEIT)
Figure 3.40:	Tamini archive
Figure 3.42, left:	Copper Development Association: *High conductivity copper for electrical engineering*, Publication No. 122 (courtesy of Copper Development Association)
Figure 3.42, right:	J. P. Dubey: *Role of copper in continuously transposed conductors*, Transformers Magazine (courtesy of Merit Media Int.)
Figure 4.1:	Tamini archive
Figure 4.3, left:	https://www.freepik.com/free-vector/wood-paper-manufacturing-flat-infographic-schema-from-cut-logs-lumber-chips-pulp-converted-paperboard_6804314.htm
Figure 4.3, right:	https://www.graymont.com/en/markets/pulp-paper/pulp-and-paper-mills (courtesy of Graymont Limited)
Figure 4.4:	Lars E. Schmidt: *Transformer insulation materials – an introduction*, ABB presentation at the "Life of a Transformer" Seminar, 19-23 May 2014, Padua (Italy) (courtesy of Hitachi energy, formerly ABB)
Figure 4.5:	Tobias Wolfinger: *Lamination of Transformerboard using the structural elements of cellulosic fibers*, Transformer Magazine vol. 2 issue 4, October 2015 (courtesy of Weidmann)
Figure 4.9:	*DuPont Nomex Pressboard – Introductory data sheet*, K-29470 (02/17) (courtesy of DuPont de Nemours, Inc.)
Figure 4.10:	https://www.roechling-industrial.com/products/composites/laminated-densified-wood/fasteners - https://www.roechling-industrial.com/products/composites/gfrp-cfrp/fasteners (courtesy of Roechling)
Figure 4.13:	Lars E. Schmidt: *Transformer insulation materials – an introduction*, ABB presentation at the "Life of a Transformer" Seminar, 19-23 May 2014, Padua (Italy) (courtesy of Hitachi energy, formerly ABB)
Figure 4.14:	Christoph Krause: *Selection & processing of pressboard for prolonged transformer service life*, Weidmann presentation at the "MyTransfo" Seminar, 15-16 November 2016, Turin (Italy) (courtesy of Weidmann)

(*Continued*)

Figure 4.15:	Christoph Krause: *Selection & processing of pressboard for prolonged transformer service life*, Weidmann presentation at the "MyTransfo" Seminar, 15-16 November 2016, Turin (Italy) (courtesy of Weidmann)
Figure 4.16:	Tesar archive (courtesy of Tesar)
Figure 4.17:	Tamini archive
Figure 4.18:	MESSKO® archive (courtesy of Maschinenfabrik Reinhausen GmbH)
Figure 4.20:	Brizio Basi website (courtesy of Brizio Basi)
Figure 4.21:	Nynas: *Transformer oil handbook*, 2nd ed. 2009 (courtesy of Nynas)
Figure 4.22:	Tamini archive
Figure 4.23:	Technical Brochure 349: *Moisture equilibrium and moisture migration within transformer insulation systems*, Cigrè, 2008 (courtesy of Cigrè)
Figure 4.24:	Brian D. Sparling, Jacques Aubin: *Assessing water content in insulating paper of power transformers*, Electric Energy T&D Magazine, n° 4-11, July/August 2007 (courtesy of Electric Energy T&D Magazine)
Figure 4.25:	Steven Jiroutek: *Moisture in transformer oil behavior*, Vaisala presentation, May 2016 (courtesy of Vaisala)
Figure 4.27:	S.V. Kulkarni, S.A. Khaparde: *Transformer engineering*, Marcel Dekker, 2004 (courtesy of Marcel Dekker Inc.)
Figure 4.28:	Nynas: *Transformer Oil Handbook*, 2nd ed. 2009 (courtesy of Nynas)
Figure 4.29:	Tamini archive
Figure 4.30:	Bo Gao, Rui Yu, Guangcai Hu, Cheng Liu, Xin Zhuang, Peng Zhou: *Development processes of surface trucking and partial discharge of pressboards immersed in mineral oil: effect of tip curvatures*, Energies 2019, 12, 554 (courtesy of Energies)
Figure 4.31:	Tamini archive
Figure 4.32:	Martin J. Heathcote: *The J&P Transformer Book*, Newnes, 12th ed. 1998 (courtesy of Newnes)
Figure 4.33:	Technical standard IEC 60060-1:1989
Figure 4.34:	https://www.highvolt.com/en/productdetail/impulse-voltage-and-current-test-systems/ip (courtesy of Highvolt Prüftechnik Dresden GmbH)
Figure 4.38:	Tamini archive
Figure 4.39:	Tamini archive
Figure 4.40:	http://www.rancan.com/rancan/ita/industria_elettromeccanica.html (courtesy of Rancan)
Figure 4.41:	Tamini archive
Figure 4.42:	MESSKO® archive (courtesy of Maschinenfabrik Reinhausen GmbH)
Figure 4.43:	Brochure L2673: *Testing and diagnostics of power transformers*, December 2016 (courtesy of Omicron electronics GmbH)
Figure 4.44:	Qualitrol presentation (courtesy of Qualitrol)
Figure 4.45:	Maik Koch, Michael Krueger, Markus Puetter: *Advanced insulation diagnostic by dielectric spectroscopy*, paper Omicron, 2011 (courtesy of Omicron electronics GmbH)
Figure 4.46:	Power Diagnostix archive (courtesy of Megger Group Limited)

(Continues)

(*Continued*)

Figure 5.5:	R. Frotscher *et al.*, *Behaviour of ester liquids under dielectric and thermal stress - from laboratory testing to practical use*, CIGRE Session 2012, paper D1-105 (courtesy of Cigrè)
Figure 5.6:	Brizio Basi archive (courtesy of Brizio Basi)
Figure 5.9:	Tamini archive
Figure 6.1:	Gábor Gurszky: *Comparison of two technologies*, JST Presentation, January 2011 (courtesy of JST transformateurs SAS)
Figure 6.2:	Gábor Gurszky: *Comparison of two technologies*, JST Presentation, January 2011 (courtesy of JST transformateurs SAS)
Figure 6.4:	Gábor Gurszky: *Comparison of two technologies*, JST Presentation, January 2011 (courtesy of JST transformateurs SAS)
Figure 6.5:	Gábor Gurszky: *Comparison of two technologies*, JST Presentation, January 2011 (courtesy of JST transformateurs SAS)
Figure 6.6:	Gábor Gurszky: *Comparison of two technologies*, JST Presentation, January 2011 (courtesy of JST transformateurs SAS)
Figure 6.7:	Gábor Gurszky: *Comparison of two technologies*, JST Presentation, January 2011 (courtesy of JST transformateurs SAS)
Figure 6.8:	Gábor Gurszky: *Comparison of two technologies*, JST Presentation, January 2011 (courtesy of JST transformateurs SAS)
Figure 6.9:	Gábor Gurszky: *Comparison of two technologies*, JST Presentation, January 2011 (courtesy of JST transformateurs SAS)
Figure 6.10:	Gábor Gurszky: *Comparison of two technologies*, JST Presentation, January 2011 (courtesy of JST transformateurs SAS)
Figure 6.11:	Gábor Gurszky: *Comparison of two technologies*, JST Presentation, January 2011 (courtesy of JST transformateurs SAS)
Figure 6.12:	Gábor Gurszky: *Comparison of two technologies*, JST Presentation, January 2011 (courtesy of JST transformateurs SAS)
Figure 6.13:	Gábor Gurszky: *Comparison of two technologies*, JST Presentation, January 2011 (courtesy of JST transformateurs SAS)
Figure 6.14:	Gábor Gurszky: *Comparison of two technologies*, JST Presentation, January 2011 (courtesy of JST transformateurs SAS)
Figure 6.15:	Gábor Gurszky: *Comparison of two technologies*, JST Presentation, January 2011 (courtesy of JST transformateurs SAS)
Figure 6.16:	Gábor Gurszky: *Comparison of two technologies*, JST Presentation, January 2011 (courtesy of JST transformateurs SAS)
Figure 6.17:	Gábor Gurszky: *Comparison of two technologies*, JST Presentation, January 2011 (courtesy of JST transformateurs SAS)
Figure 6.18:	Gábor Gurszky: *Comparison of two technologies*, JST Presentation, January 2011 (courtesy of JST transformateurs SAS)
Figure 6.19:	Gábor Gurszky: *Comparison of two technologies*, JST Presentation, January 2011 (courtesy of JST transformateurs SAS)
Figure 6.20:	Gábor Gurszky: *Comparison of two technologies*, JST Presentation, January 2011 (courtesy of JST transformateurs SAS)
Figure 6.21:	Gábor Gurszky: *Comparison of two technologies*, JST Presentation, January 2011 (courtesy of JST transformateurs SAS)
Figure 6.22:	Gábor Gurszky: *Comparison of two technologies*, JST Presentation, January 2011 (courtesy of JST transformateurs SAS)
Figure 6.23:	Gábor Gurszky: *Comparison of two technologies*, JST Presentation, January 2011 (courtesy of JST transformateurs SAS)
Figure 6.24:	Gábor Gurszky: *Comparison of two technologies*, JST Presentation, January 2011 (courtesy of JST transformateurs SAS)

(*Continued*)

Figure 6.25:	Gábor Gurszky: *Comparison of two technologies*, JST Presentation, January 2011 (courtesy of JST transformateurs SAS)
Figure 7.2:	Tamini archive
Figure 7.3:	Terna archive
Figure 7.4:	Terna archive
Figure 7.6:	Brochure 393: *Thermal performance of transformers*, Cigrè, 2009 (courtesy of Cigrè)
Figure 7.7:	Brochure 393: *Thermal performance of transformers*, Cigrè, 2009 (courtesy of Cigrè)
Figure 7.8:	Qiang Liu (University of Manchester): *Transformer thermal management for future power networks*, Presentation at CWIEME 2018, Berlin 20/06/2018 (courtesy of Qiang Liu)
Figure 7.9, left:	MESSKO® archive (courtesy of Maschinenfabrik Reinhausen GmbH)
Figure 7.9, right:	Neoptix presentation (courtesy of Neoptix)
Figure 7.10:	Brochure 393: *Thermal performance of transformers*, Cigrè, 2009 (courtesy of Cigrè)
Figure 7.12:	IEC 60076-14:2013
Figure 7.13:	Convett archive (courtesy of Convett S.r.l.)
Figure 7.14:	Brochure 659: *Transformer thermal modelling*, Cigrè, 2016 (courtesy of Cigrè)
Figure 7.15:	Brochure 659: *Transformer thermal modelling*, Cigrè, 2016 (courtesy of Cigrè)
Figure 7.16:	Brochure 659: *Transformer thermal modelling*, Cigrè, 2016 (courtesy of Cigrè)
Figure 7.17:	Brochure 659: *Transformer thermal modelling*, Cigrè, 2016 (courtesy of Cigrè)
Figure 7.18:	Brochure 659: *Transformer thermal modelling*, Cigrè, 2016 (courtesy of Cigrè)
Figure 7.19:	Brochure 659: *Transformer thermal modelling*, Cigrè, 2016 (courtesy of Cigrè)
Figure 7.20:	Qiang Liu (University of Manchester): *Transformer thermal management for future power networks*, Presentation at CWIEME 2018, Berlin 20/06/2018 (courtesy of Qiang Liu)
Figure 7.21:	Brochure 659: *Transformer thermal modelling*, Cigrè, 2016 (courtesy of Cigrè)
Figure 8.13:	F. Iliceto: *Considerations on the possibility of dispensing with tertiary winding in interconnecting autotransformers between networks*, L'elettrotecnica, October 1965 (courtesy of AEIT)
Figure 9.2:	MESSKO® archive (courtesy of Maschinenfabrik Reinhausen GmbH)
Figure 9.3:	G. Bertagnolli: *Short circuit duty of power transformers*, ABB, 3rd ed. 2006 (courtesy of Hitachi energy, formerly ABB)
Figure 9.4:	G. Bertagnolli: *Short circuit duty of power transformers*, ABB, 3rd ed. 2006 (courtesy of Hitachi energy, formerly ABB)
Figure 9.5:	G. Bertagnolli: *Short circuit duty of power transformers*, ABB, 3rd ed. 2006 (courtesy of Hitachi energy, formerly ABB)
Figure 9.6:	G. Bertagnolli: *Short circuit duty of power transformers*, ABB, 3rd ed. 2006 (courtesy of Hitachi energy, formerly ABB)

(*Continues*)

(*Continued*)

Figure 9.7:	G. Bertagnolli: *Short circuit duty of power transformers*, ABB, 3rd ed. 2006 (courtesy of Hitachi energy, formerly ABB)
Figure 9.8:	G. Bertagnolli: *Short circuit duty of power transformers*, ABB, 3rd ed. 2006 (courtesy of Hitachi energy, formerly ABB)
Figure 9.11:	Tamini archive
Figure 9.12:	Tamini archive
Figure 9.13:	Tamini archive
Figure 9.14:	Tamini archive
Figure 9.15:	Tamini archive
Figure 9.20:	Tamini archive
Figure 9.26:	EMTP simulation (courtesy of Francesco Palone, Terna)
Figure 9.28:	G. Bertagnolli: *Short circuit duty of power transformers*, ABB, 3rd ed. 2006 (courtesy of Hitachi energy, formerly ABB)
Figure 9.29:	B. Bošnjak, M. Stuhne: *Analysis and experimental verification of static tilting behavior of CTC in transformer windings*, 3rd International Colloquium Transformer Research and Asset Management, Split (Croatia), October 2014 (courtesy of Cigrè)
Figure 9.30:	B. Bošnjak, A. Hauck, H. Landes: *Coupled magneto-mechanical finite element analysis of a power transformer in short circuit conditions*, 3rd International Colloquium Transformer Research and Asset Management, Split (Croatia), October 2014 (courtesy of Cigrè)
Figure 10.5:	Kent Olsson: *First class shunt reactors*, Presentation ABB, Maggio 2007 (courtesy of Hitachi energy, formerly ABB)
Figure 11.4:	Nippon steel corporation: *Orientcore® grain-oriented electrical steel sheets*, 2015 (courtesy of Nippon Steel)
Figure 11.5:	EMTP simulation (courtesy of Francesco Palone, Terna)
Figure 12.1:	LTC archive (courtesy of Legnano Teknoelectric Company)
Figure 12.2:	LTC archive (courtesy of Legnano Teknoelectric Company)
Figure 12.3:	Tamini archive
Figure 12.12:	Tamini archive
Figure 13.8:	Tamini archive
Figure 13.11:	Tamini archive
Figure 13.13:	Tamini archive

Index

www.ingramcontent.com/pod-product-compliance
Lightning Source LLC
LaVergne TN
LVHW011838200926
848484LV00002B/373